MORPHOLOGY

MODERN LINGUISTICS SERIES
Series Editors

Professor Noël Burton-Roberts
University of Newcastle upon Tyne

Dr Andrew Spencer
University of Essex

Each textbook in the **Modern Linguistics** series is designed to provide a carefully graded introduction to a topic in contemporary linguistics and allied disciplines, presented in a manner that is accessible and attractive to readers with no previous experience of the topic, but leading them to some understanding of current issues. The texts are designed to engage the active participation of the reader, favouring a problem-solving approach and including liberal and varied exercise material.

Noël Burton-Roberts founded the **Modern Linguistics** series and acted as Series Editor for the first three volumes in the series. Andrew Spencer has since joined Noël Burton-Roberts as joint Series Editor.

Titles published in the series

Syntax and Argumentation Bas Aarts
Phonology Philip Carr
Linguistics and Second Language Acquisition Vivian Cook
Sociolinguistics: A Reader and Coursebook
Nikolas Coupland and Adam Jaworski
Morphology Francis Katamba
Semantics Kate Kearns
Contact Languages: Pidgins and Creoles Mark Sebba

Further titles are in preparation

Modern Linguistics Series
Series Standing Order
ISBN 0–333–71701–5 hardcover
ISBN 0–333–69344–2 paperback
(*outside North America only*)

You can receive future titles in this series as they are published by placing a standing order. Please contact your bookseller or, in the case of difficulty, write to us at the address below with your name and address, the title of the series and the ISBN quoted above.

Customer Services Department, Macmillan Distribution Ltd
Houndmills, Basingstoke, Hampshire RG21 6XS, England

Morphology

Francis Katamba

First published 1993 by
MACMILLAN PRESS LTD
Houndmills, Basingstoke, Hampshire RG21 6XS
and London
Companies and representatives
throughout the world

ISBN 0–333–54113–8
ISBN 0–333–54114–6

A catalogue record for this book is available
from the British Library.

10 9 8 7 6 5
05 04 03 02 01 00

Printed and bound in Great Britain by
Creative Print and Design (Wales), Ebbw Vale

The Scrabble tiles on the cover design are reproduced by kind permission of J.W. Spear and Son PLC, Enfield EN3 7TB, England.

To Janet,
Francis and Helen

Contents

Preface

This book is an introduction to morphology that presupposes little previous exposure to linguistics. It is meant to be useful both to students of English and to those of linguistics. Most of the first half of the book, as well as the final chapter, are devoted mainly to problems of English word-formation. The remaining chapters cover a range of morphological phenomena in other languages. But even the parts dealing with English raise issues of a general theoretical interest. The detail in which different parts are studied will vary, depending on the kind of student that uses the book.

I present morphology from the standpoint of current, mainstream generative grammar. My main concerns are the nature of word-formation processes and the ways in which word-formation interacts with phonology, syntax and the lexicon. I hope that the reader will come away not only with an understanding of the descriptive problems in morphology but also with a firm grasp of the theoretical issues and the analytical tools that are available within the model of generative grammar. On completing a course in morphology based on this book students should be equipped to tackle the growing morphological literature that has appeared in recent years.

There are many people whom I must thank for the help they have given me in writing this book. The book grew out of my morphology course at Lancaster University. I must thank the students who have taken this course over the last four years. Special thanks go to Elena Semino and Saleh al-Khateb, whose Italian and Syrian Arabic data I have used here.

I have benefited from discussions with a number of Berkeley linguists, especially Sharon Inkelas, Sam Mchombo and Karl Zimmer. Above all, I must thank in a special way Larry Hyman, with whom I have collaborated on Luganda morphology and phonology for the last ten years. I have learned much of what I know about phonology/morphology through our collaboration.

There are also many other linguists whose theoretical and descriptive studies I have drawn on. They have all contributed in an obvious way to my writing this book.

I also owe a special debt of gratitude to Noël Burton-Roberts, the editor of this series. His rigorous critical comments and positive suggestions have enabled me to avoid some of the pitfalls I would otherwise have encountered. There are two other people at Macmillan that I wish to thank for their technical support: they are Doreen Alig and Cathryn Tanner. I should also like to thank Valery Rose and David Watson, who both helped with the production of this book.

Finally, I thank my wife Janet for her support during the long months and years of writing this book.

Lancaster   FRANCIS KATAMBA

Acknowledgements

The author and publishers wish to thank the following who have kindly given permission for the use of copyright material:

Cambridge University Press for Figure 7.5 from P. Matthews, *Inflectional Morphology*, p. 132;

The International Phonetic Association for the International Phonetic Alphabet, revised in 1989, reproduced from *Journal of the International Phonetics Association*, vol. 19, no. 2.

Every effort has been made to trace all the copyright-holders but if any have been inadvertently overlooked, the publishers will be pleased to make the necessary arrangements at the first opportunity.

Finally, I would like to thank Joan Malig for helpful comments that have enabled me to improve the presentation of the material on recursion in the first chapter.

Abbreviations and Symbols

ADJ/Adj	adjective
AdjP	adjectival phrase
ADV/Adv	adverb
AdvP	adverbial phrase
BVS	Basic verbal suffix (in Bantu)
DET/Det	determiner
GF	grammatical function
GVS	Great Vowel Shift
Inf.	infinitive
N/n	noun
NP/Np	noun phrase
OBJ	object
OCP	Obligatory Contour Principle
OED	*Oxford English Dictionary*
P/Prep	preposition
PP	prepositional phrase
Pron	pronoun
RHR	Right-hand Head Rule
S	sentence
SPE	*The Sound Pattern of English*
SUBJ/Subj	subject
V	verb
VP	verb phrase
$V_{(intr)}$	verb$_{(intransitive)}$
$V_{(tr)}$	verb$_{(transitive)}$
VP	verb phrase
WFC	Well-formedness Condition
WP	Word and Paradigm (morphology)

The International Phonetic Alphabet

CONSONANTS

	Bilabial	Labio-dental	Dental	Alveolar	Post-alveolar	Retroflex	Palatal	Velar	Uvular	Phar-yngeal	Glottal
Plosive	p b			t d		ʈ ɖ	c ɟ	k ɡ	q ɢ		ʔ
Nasal	m	ɱ		n		ɳ	ɲ	ŋ	ɴ		
Trill	ʙ			r					ʀ		
Tap or Flap				ɾ		ɽ					
Fricative	ɸ β	f v	θ ð	s z	ʃ ʒ	ʂ ʐ	ç ʝ	x ɣ	χ ʁ	ħ ʕ	h ɦ
Lateral fricative				ɬ ɮ							
Approximant		ʋ		ɹ		ɻ	j	ɰ			
Lateral approximant				l		ɭ	ʎ	ʟ			
Ejective stop	p'			t'		ʈ'	c'	k'	q'		
Impulsive	ɓ ɓ			ɗ ɗ			ʄ ʄ	ɠ ɠ	ʛ ʛ		

Where symbols appear in pairs, the one to the right represents a voiced consonant. Shaded areas denote articulations judged impossible.

VOWELS

	Front	Central	Back
Close	i y — ɨ ʉ — ɯ u		
	ɪ ʏ ʊ		
Close-mid	e ø — ɘ ɵ — ɤ o		
Open-mid	ɛ œ — ɜ ɞ — ʌ ɔ		
	æ ɐ		
Open	a ɶ — ɑ ɒ		

Where symbols appear in pairs, the one to the right represents the rounded vowel.

OTHER SYMBOLS

ʍ Voiceless labial-velar fricative	⊙ Bilabial click
w Voiced labial-velar approximant	ǀ Dental click
ɥ Voiced labial-palatal approximant	ǃ (Post)alveolar click
ʜ Voiceless epiglottal fricative	ǂ Palatoalveolar click
ʡ Voiced epiglottal plosive	ǁ Alveolar lateral click
ʢ Voiced epiglottal fricative	ɺ Alveolar lateral flap
ɕ ʑ Alveolo-palatal fricatives	

k͡p t͡s

ɧ Simultaneous ʃ and X

ɜ Additional mid central vowel

Affricates and double articulations can be represented by two symbols joined by a tie bar if necessary.

DIACRITICS

◌̥ Voiceless	n̥ d̥	◌̹ More rounded	ɔ̹	ʷ Labialised	tʷdʷ	~ Nasalised	ẽ
◌̬ Voiced	s̬ t̬	◌̜ Less rounded	ɔ̜	ʲ Palatalized	tʲdʲ	ⁿ Nasal release	dⁿ
ʰ Voiced	tʰ dʰ	◌̟ Advanced	u̟	ˠ Velarized	tˠdˠ	ˡ Lateral release	dˡ
◌̤ Breathy voiced	b̤ a̤	◌̠ Retracted	i̠	ˤ Pharyngealized	tˤdˤ	̚ No audible release	d̚
◌̰ Creaky voiced	b̰ a̰	◌̈ Centralized	ë	~ Velarized or Pharyngealized ɫ			
◌̼ Linguolabial	t̼ d̼	◌̽ Mid centralized	e̽	◌̝ Raised	e̝ ɹ̝	(ɹ̝ = voiced alveolar fricative)	
◌̪ Dental	t̪ d̪	◌̘ Advanced Tongue root	e̘	◌̞ Lowered	e̞ β̞	(β̞ = voiced bilabial approximant)	
◌̺ Apical	t̺ d̺	◌̙ Retracted Tongue root	e̙	◌̩ Syllabic	ɹ̩	◌̯ Non-syllabic	e̯
◌̻ Laminal	t̻ d̻	◌˞ Rhoticity	ə˞				

SUPRASEGMENTALS

ˈ Primary stress	ˌfoʊnəˈtɪʃən
ˌ Secondary stress	
ː Long	eː
ˑ Half-long	eˑ
˘ Extra-short	ĕ
. Syllable break	ɹi.ækt
| Minor (foot) group	
‖ Major (intonation) group	
‿ Linking (absence of a break)	

TONES AND WORD ACCENTS

LEVEL TONES

˝ or ꜛ	Extra-high
´	High
¯	Mid
`	Low
ˎ	Extra-low
ꜜ	Downstep
ꜛ	Upstep

CONTOUR TONES

̌ or ꜛ	Rise
̂	Fall
᷄	High rise
᷅	Low rise
᷈	Rise fall
ꜛ	Global rise
ꜜ	Global fall

Reproduced courtesy of the International Phonetic Association

Part I
Background

1 Introduction

1.1 THE EMERGENCE OF MORPHOLOGY

Although students of language have always been aware of the importance of words, **morphology**, the study of the internal structure of words did not emerge as a distinct sub-branch of linguistics until the nineteenth century.

Early in the nineteenth century, morphology played a pivotal role in the reconstruction of Indo-European. In 1816, Franz Bopp published the results of a study supporting the claim, originally made by Sir William Jones in 1786, that Sanskrit, Latin, Persian and the Germanic languages were descended from a common ancestor. Bopp's evidence was based on a comparison of the grammatical endings of words in these languages.

Between 1819 and 1837, Bopp's contemporary Jacob Grimm published his classic work, *Deutsche Grammatik*. By making a thorough analytical comparison of sound systems and word-formation patterns, Grimm showed the evolution of the grammar of Germanic languages and the relationships of Germanic to other Indo-European languages.

Later, under the influence of the Darwinian theory of evolution, the philologist Max Müller contended, in his Oxford lectures of 1899, that the study of the evolution of words would illuminate the evolution of language just as in biology morphology, the study of the forms of organisms, had thrown light on the evolution of species. His specific claim was that the study of the 400–500 basic roots of the Indo-European ancestor of many of the languages of Europe and Asia was the key to understanding the origin of human language (cf. Müller, 1899; cited by Matthews, 1974).

Such evolutionary pretensions were abandoned very early on in the history of morphology. In this century morphology has been regarded as an essentially **synchronic** discipline, that is to say, a discipline focusing on the study of word-structure at one stage in the life of a language rather than on the evolution of words. But, in spite of the unanimous agreement among linguists on this point, morphology has had a chequered career in twentieth-century linguistics, as we shall see.

1.2 MORPHOLOGY IN AMERICAN STRUCTURAL LINGUISTICS

Adherents to **American structural linguistics**, one of the dominant schools of linguistics in the first part of this century, typically viewed linguistics not so much as a 'theory' of the nature of language but rather as a body of descriptive and analytical procedures. Ideally, linguistic analysis was expected to proceed by focusing selectively on one dimension of language structure at a time before tackling the next one. Each dimension was formally referred to as a **linguistic level**. The various levels are shown in [1.1].

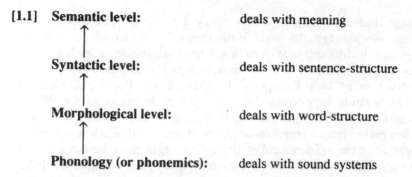

[1.1]   Semantic level: deals with meaning

         Syntactic level: deals with sentence-structure

         Morphological level: deals with word-structure

         Phonology (or phonemics):   deals with sound systems

The levels were assumed to be ordered in a hierarchy, with phonology at the bottom and semantics at the top. The task of the analyst producing a description of a language was seen as one of working out, in separate stages, first the pronunciation, then the word-structure, then the sentence-structure and finally the meaning of utterances. It was considered theoretically reprehensible to make use of information from a higher level, e.g. syntax, when analysing a lower level such as phonology. This was the doctrine of **separation of levels**.

In the early days, especially between 1920 and 1945, American structuralists grappled with the problem of how sounds are used to distinguish meaning in language. They developed and refined the theory of the **phoneme** (cf. Sapir, 1925; Swadesh, 1934; Twaddell, 1935; Harris, 1944).

As time went on, the focus gradually shifted to morphology. When structuralism was in its prime, especially between 1940 and 1960, the study of morphology occupied centre stage. Many major structuralists investigated issues in the theory of word-structure (cf. Bloomfield, 1933; Harris, 1942, 1946, 1951; Hockett, 1952, 1954, 1958). Nida's coursebook entitled *Morphology*, which was published in 1949, codified structuralist theory and practice. It introduced generations of linguists to the descriptive analysis of words.

The structuralists' methodological insistence on the separation of levels which we noted above was a mistake, as we shall see below in sections (1.3.2) and (1.3.3). But despite this flaw, there was much that was commendable in the structuralist approach to morphology. One of the structuralists' main contributions was the recognition of the fact that words may have intricate internal structures. Whereas traditionally linguistic analysis had treated the word as the basic unit of grammatical theory and lexicography, the American structuralists showed that words are analysable in terms of **morphemes**. These are the smallest units of meaning and grammatical function. Previously, word-structure had been treated together with sentence-structure under grammar. The structuralists introduced morphology as a separate sub-branch of linguistics. Its purpose was 'the study of morphemes and their arrangements in forming words' (Nida, 1949:1). The contribution of the structuralists informs much of the discussion in the first part of this book.

1.3 THE CONCEPT OF CHOMSKYAN GENERATIVE GRAMMAR

The bulk of this book, however, presents morphological theory within the linguistic model of **generative grammar** initiated by Chomsky. Before we begin considering how this theory works, I will sketch the background assumptions made by generative grammarians so that we can place the theory of morphology in the wider theoretical context of generative linguistics.

The central objective of generative linguistics is to understand the nature of **linguistic knowledge** and how it is acquired by infants. In the light of this objective, a fundamental question that a theory of word-structure must address is, 'what kinds of information must speakers have about the words of their language in order to use them in utterances?' Attempts to answer this question have led to the development of sub-theories of the **lexicon** (i.e. dictionary) and of morphology.

According to Chomsky (1980, 1981, 1986), the central goal of linguistic theory is to determine what it is people know if they *know* a particular language. Chomsky observes that knowing a language is not simply a matter of being able to manipulate a long list of sentences that have been memorised. Rather, knowing a language involves having the ability to produce and understand a vast (and indeed unlimited) number of utterances of that language that one may never have heard or produced before. In other words, **creativity** (also called productivity or open-endedness) is an aspect of linguistic knowledge that is of paramount importance.

Linguistic creativity is for the most part rule-governed. For instance, speakers of English know that it is possible to indicate that there is more than one entity referred to by a noun and that the standard way of doing this is to add -*s* at the end of a noun. Given the noun *book*, which we all have encountered before, we know that if there is more than one of these objects we refer to them as *books*. Likewise, given the nonsense word *smilts* as in the sentence *The smilts stink* which I have just made up, you know *smilts* would refer to more than one of these smelly things. Speakers

of English have tacit knowledge of the rule which says 'add -*s* for plural'
and they can use it to produce the plural form of virtually any noun. I have
emphasised the notion of rule, taking the existence of rules for granted.

I will now explain why **a generative grammar** is a system of explicit rules
which may apply recursively to **generate** an indefinite number of sentences
which can be as long as one wants them to be. **Recursiveness** has the
consequence that, in principle, there is no upper limit to the length of
sentences. A grammatical constituent like a noun phrase (NP) or prepositional phrase (PP) can directly or indirectly contain an indefinite number
of further constituents of that category as in the sentence *John saw a pile
of books on the table in the attic*. The recursion can be clearly seen in the
tree diagram in [1.2] which represents that sentence. As seen, on the one
hand NPs can have an optional determiner followed by N and PP, and on
the other hand PPs are expanded as P followed by NP. So, NPs can
indirectly contain other NPs since NPs can contain PPs which contain NPs:

[1.2]

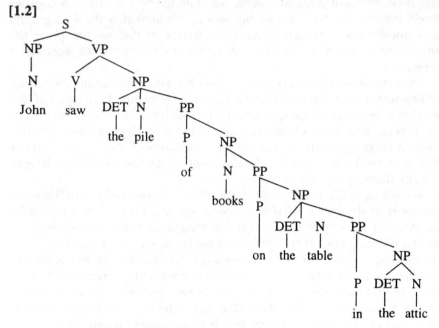

Notes: S – sentence; N – noun, NP – noun phrase; V – verb, VP – verb
phrase; P – preposition, PP – prepositional phrase; DET – determiner.

One of our concerns will be to determine whether morphology should be
recognised as a separate linguistic level (or **module**) that is independent of
syntax and phonology (see [1.1] above and [1.3] below). Do morphological
rules have certain properties which they do not share with rules in other
parts of the grammar? Are recursive rules of the kind found in syntax

needed in morphology? This book will address these issues in depth. Here I will only attempt to give you a flavour of one of the issues that I will be exploring.

There are morphological processes which are similar to syntactic processes. For instance, certain adjectives which describe periods in history, such as *industrial*, can have the prefix *post-* before them as in *post-industrial*. And, given the adjective *post-industrial*, we can place another *post-* before it to yield *post-post-industrial*. Clearly, the word-formation process we witness here is recursive. We have the rule attaching *post-* to a word reapplying to its own output. This raises an interesting question: if morphological rules that build words are similar to syntactic rules that build sentences, what reason is there for assuming that morphology is essentially different from syntax?

Before we go any further we need to clarify the terms **grammar** and **rule of grammar**. These terms are used by linguists in four distinct senses. Firstly, in generative linguistics 'grammar' can refer to the implicit, totally unarticulated knowledge of rules and principles of their language that people have in their heads. This tacit knowledge enables them to distinguish between well-formed and ill-formed words and utterances in their language. For example, many English speakers may not be able to explain in an articulate manner why it is 'correct' to say *a grain* but 'incorrect' to say *a oat*. Nevertheless their knowledge of English grammatical structure enables them to determine that the former is correct and the latter is not.

Secondly, whereas in traditional approaches 'grammar' only includes morphology and syntax, in generative linguistics the term grammar is employed in a much wider sense. It covers not only morphology and syntax but also semantics, the lexicon and phonology. Hence, there are rules of grammar in every linguistic module. Phonological rules, morphological rules, syntactic rules and semantic rules are all regarded as rules of grammar.

Thirdly, grammar and rules of grammar may refer to a book containing a statement of the rules and principles inferred by linguists to lie behind the linguistic behaviour of speakers of a particular language. These rules simply describe regular patterns observed in the linguistic data.

Lastly, some grammars are books containing prescriptive statements. Such grammars contain rules that *prescribe* certain kinds of usage. Outside linguistics this view of grammar is still prevalent. The reason for this is clear. In everyday life rules are normally mechanisms for regulating behaviour – the behaviour of pupils in a school, members of a club, inmates of a prison, etc. In many traditional pedagogical grammars rules serve the same purpose. They are statements like 'A sentences must not end with a preposition.' They prescribe what the 'officially or socially approved' usage is – in the opinion of the grammarian.

In much of modern linguistics, however, rules have a different function.

They are not prescriptions of behaviour which the grammarian imposes on speakers, but rather they are statements of principles responsible for the observed regularities in the speech or writing or users of a particular language. The characterisation of regularities in observed patterns of usage is what the American structuralists regarded as the primary objective of linguistic investigations. Their grammatical rules were descriptive statements like 'The article precedes the noun in the English noun phrase.' This statement reflects the fact that *the book*, as in *I read the book*, is allowed whereas **book the*, as in **I read book the* is disallowed. (An asterisk indicates a disallowed form.)

Chomsky has shifted the focus of linguistic theory from the study of observed behaviour to the investigation of the knowledge that underlies that behaviour. In generative linguistics rules are intended to go beyond accounting for patterns in the data to a characterisation of speakers' linguistic knowledge. The primary objective of generative grammar is to model a speaker's linguistic knowledge.

Chomsky characterises linguistic knowledge using the concepts of competence and performance. **Competence** is a person's implicit knowledge of the rules of a language that makes the production and understanding of an indefinitely large number of new utterances possible while **performance** is the actual use of language in real situations. Chomsky proposes that competence, rather than performance, is the primary object of linguistic inquiry. Put simply, knowledge of a language entails mastery of an elaborate system of rules that enables a person to encode and decode a limitless number of utterances in that language. One sub-set of this rule system is the rules of word-formation which this book introduces you to. In section (4.1.3) of Chapters 4 and section (12.3.3) of Chapter 12 it will be shown that speakers of a language do not just commit to memory all the words they know. Their competence includes the ability to manipulate rules in order to create new words and to unscramble the meanings of novel or unfamiliar words which they encounter.

If knowing a language essentially involves mastering a system of rules, how do humans accomplish this task? Chomsky contends that the linguistic capacity of humans is **innate**. The general character of linguistic knowledge is determined by the nature of the mind which is endowed with a specialised **language faculty**. This faculty is determined in turn by the biology of the brain. The human child is born with a blue-print of language which is called **Universal Grammar**.

According to Chomsky, Universal Grammar is the faculty of the mind which determines the nature of language acquisition in the infant and of linguistic competence. The properties that lie behind the competence of speakers of various languages are governed by restricted and unified elementary principles rooted in Universal Grammar. This explains the striking underlying similarity between languages in their essential structural properties. Admittedly, languages differ from each other, but the structural differences between them occur within the fairly narrow range sanctioned by Universal Grammar. As we shall see (especially in Chapters 3, 8, 9 and 12) with regard to word-formation, very similar word-building principles recur in language after language. The language faculty of the mind is essentially the same in all humans. Hence languages can only differ from each other within the limits predetermined by the neurology and physiology of the human brain, which determine the nature of Universal Grammar. And Universal Grammar in turn determines the kinds of grammars of particular languages that can be acquired by infants.

The differences between the grammars acquired by individual speakers of, say, English and Arabic can be attributed to experience. An individual's experience serves to specify a particular grammar for the particular language which that individual is exposed to – within the range permitted by Universal Grammar.

How is Universal Grammar structured? It is **modular** in structure: it consists of various sub-systems of principles. Many of its principles consist of **parameters** which are fixed by experience on the basis of simple evidence of the kind available to the child. Chomsky compares Universal Grammar to an intricate electrical system that is all wired up, but not switched on. The system contains a finite set of switches, each one of which has a restricted number of positions. Exposure to a specific language experience is required to turn on these switches and give them the appropriate setting.

The basic idea of parameters is meant to capture the fact that many rules are interdependent. If one choice is made, it may either preclude some other choices or set in motion other related choices. This makes the task of language acquisition simpler than it would be if each rule had to be worked out independently of all other rules. The parametric approach assumes that the infant acquiring a language makes very clever guesses or hypotheses about the rules of the grammar being acquired on the basis of rules already acquired after experience of a particular language.

For a concrete example of a parameter, we will consider **the Right-hand Head/Left-hand Head Rule** which will be discussed in Chapter 12. This parameter is concerned with the position of the **head** of a grammatical constituent. Some languages, like English, normally place the head on the right, i.e. it is the last element of a constituent. For example, in the noun phrase *these big books* the right-handmost word, the noun *books*, is the head. It must come last. (Alternatives like **books big these* and **these books big* are forbidden.)

As a rule, the head is the only obligatory element of a constituent like an NP. *Books* is a well-formed NP but neither *these* nor *big* is a permissible NP on its own. Furthermore, in terms of meaning, the head *books* is the key

word in this NP. The function of *these* and *big* is merely to specify further
the particular books referred to.

Likewise, at word level, in a compound like *farmhouse*, the head, *house*,
is the last element and it is the pivotal element from a semantic point of
view. (A *farmhouse* is a kind of *house*.) However, in some languages, such
as Japanese, the reverse is the case. The head of a grammatical constituent
is normally on the left. Once an infant has worked out the position of the
head for one construction this can be generalised with a considerable
degree of success to other constructions.

Universal Grammar consists of a number of modules which are inter-
related. This is shown in [1.3] (which you should compare with [1.1]
above):

[1.3]   (i)   Lexicon and Morphology
        (ii)   Syntax
        (iii) Phonetic Form (PF) (which deals with representation of
               utterances in speech)
        (iv)   Logical Form (LF) (which deals with meaning)

As seen, Universal Grammar includes the lexicon and morphology
module. Knowledge of word-structure is a central aspect of linguistic
competence. A case can be made for recognising morphology as a separate
module of Universal Grammar. Yet at the same time, morphology (and
the lexicon) are like a bridge that links the other modules of the grammar.
It is therefore necessary to examine morphology not in isolation, but in
relation to the other modules. Morphology interacts with both phonology
and syntax as well as semantics. So, it can only be studied by considering
the phonological, syntactic and semantic dimensions of words.

1.3.1 The Place of Morphology in Early Generative Grammar

Today the place of morphology in generative grammar is secure. But this is
a recent development. After being in the limelight when structuralism
peaked in the 1950s, morphology was at first eclipsed when generative
grammar came on the scene. Generative grammarians initially rejected the
validity of a separate morphological module.

From the point of view of advancing our understanding of word-
structure, this stance was unfortunate. Since generative grammar has been
the dominant school of linguistics in the second half of this century, it
meant that the study of word-structure was in the shadows for more than a
decade. Morphology did not re-emerge from oblivion until the mid-1970s.
Fortunately, the eclipse was not total. A few isolated (for the most part
non-generative) scholars such as Robins (1959) and Matthews (1972, 1974)

made important contributions to morphology during this time, as we shall see.

Part of the reason for the widespread neglect of morphology during the early years of generative grammar was the belief that word-formation could be adequately covered if it was partitioned between phonology and syntax. It was argued that no separate morphological level or component was needed in the grammar. Ways were found of describing the structure of words in a model of language that had a phonological component, a syntactic component and a semantic component but no morphological component. Those aspects of word-structure that relate to phonology (e.g. the alternation between *sane* [seɪn] and *sanity* [sænɪtɪ] would be dealt with using devices found in the phonological component. And those aspects of word-structure that are affected by syntax would be dealt with in the syntactic component.

The job of the syntactic component of the grammar was thought of as being to **generate** (i.e. to specify or enumerate explicitly) all the well-formed sentences of a language, without generating any ill-formed ones. Significantly, generating all the sentences of a language was seen as meaning generating all the permissible sequences of **morphemes** (not words), and showing which morpheme groupings formed syntactic constituents like noun phrases and verb phrases (also see p. 13 in this chapter). A specialised morphological component and a properly articulated lexicon were not part of the picture. Thus, Lees (1960), the first major descriptive study produced by a generative linguist, used syntactic rules to create derived words like the noun *appointment* from the verb *appoint*. As seen in [1.4a], Lees derived the sentence containing the noun *appointment* from a source sentence with the verb *appoint*. Likewise, he derived the abstract noun *priesthood* from a source sentence with the noun *priest*, as indicated in [1.4b].

[1.4]   a.   The committee appoints John.
             The committee's appointment of John.
                                        (Source sentence: Lees, 1960: 67)

        b.   John is a priest.
             John's priesthood.   (Source sentence: Lees, 1960: 110)

We will not examine the particulars of the syntactic rules which Lees uses. Our concern is that Lees saw this type of word-formation as taking place in the syntax and believed that he could dispense with morphology. We will revisit this issue in Chapter 12.

Let us now turn our attention to questions of phonological realisation. **Readjustment rules** (which were morphological rules in disguise) played a key role in this area. They operated on the final output of the syntactic component, making whatever modifications were necessary in order to

enable phonological rules to apply to the representation obtained after all
syntactic rules had applied.

Unfortunately, there seems to have been no constraint on the power of
readjustment rules. For instance, in *SPE* (*The Sound Pattern of English*)
which appeared in 1968 and was the pivotal work in the development of
generative phonological theory, Chomsky and Halle proposed (on p. 11)
that the syntax should generate both the regular past tense form *mended*
$[_v[_v mend]_v past]_v$ and the irregular past tense form *sang* $[_v[_v sing]_v past]_v$.
These bracketed strings, which were the output of the syntactic component,
would form the input to the readjustment rules. Next, the readjustment rules
would remove all the brackets associated with the past tense. In the case of
mend, a general readjustment rule would replace *past* by *d*, while in the case
of *sing* a special readjustment rule would delete the item *past*, together with
the associated bracket labels, giving $[_v sing]_v$. The same readjustment rule
would also attach the diacritic mark * to the vowel /ɪ/ indicating that
eventually a phonological rule would change it into /æ/. The readjustment
rules would give the forms $[_v[_v mend]_v d]_v$ and $[_v s*ng]_v$. These representations – and all other such representations yielded by readjustment rules –
were referred to as **phonological representations**. Finally, phonological
representations would be converted into the phonetic representations
[mendɪd] and [sæŋ] by rules in the phonology module.

With the benefit of hindsight, we can see that readjustment rules were a
mistake. They were rules with unbridled power. They could make whatever modifications were deemed necessary to enable phonological rules to
apply to strings of morphemes produced by the syntax. It is very undesir-
able to have a batch of rules that empower us linguists to do whatever we
like, whenever we like, so long as we come up with the answer we like. A
theory becomes vacuous if it has rules that can insert all manner of
elements, remove all manner of elements and make all manner of elements
exchange places whenever we choose to, with no principles restricting our
freedom. Effectively, this means that we are given *carte blanche* to start off
with any arbitrary input, apply the rules, and come up with the 'correct'
answer.

Furthermore, readjustment rules were a bad idea because they are
evidence of a lack of interest in words *qua* words and in morphology as a
linguistic level. Using rules of the syntax to specify permissible sequences
of morphemes, regardless of whether they occurred in words or sentences,
and using readjustment rules to turn strings generated by the syntax into
strings that the phonology could process and assign a pronunciation to was
merely skirting round the problem. Words are a central dimension of
language. They have certain unique properties that they do not share with
other elements of linguistic structure like sentences and speech sounds. A
theory of language must include a properly developed model of word-
formation that enables the linguist to describe words on their own terms –

without overlooking the ways in which word-formation rules interact with rules in other modules. As time went by, this became clear to generative linguists who, in increasing numbers, began to explore more satisfactory ways of dealing with word-structure.

1.3.2 The Morphology–Phonology Interaction

As regards the interaction with phonology, the selection of the form that manifests a given morpheme may be influenced by the sounds that realise neighbouring morphemes. Take the indefinite article in English. It has two manifestations. It is *a* before a word that begins with a consonant (e.g., *a pear*) and *an* before a word that begins with a vowel (e.g., *an orange*). We cannot describe the phonological shape of the indefinite article without referring to the sound at the beginning of the word that follows it.

1.3.3. The Morphology–Syntax Interaction

As regards the interaction with syntax, the form of a word may be affected by the syntactic construction in which the word is used. For instance, the verb *walk* has a number of forms including *walk*, *walks* and *walked*. The selection of a particular form of this verb on a given occasion is dependent on the syntactic construction in which it appears. Thus, in the present tense, the choice between the forms *walks* and *walk* depends on whether the subject of the verb is third person singular (in which case *walks* is selected as is *he/she/it walks*) or not (in which case *walk* is selected as in *I/you/we/they walk*). In the past tense, *walk* is realised as *walked*.

Chomsky (1957: 39) deals with all these facts as uncontroversial syntactic phenomena, using the phrase structure rule below:

[1.5]
$$C \rightarrow \begin{cases} S & \text{in the context } NP_{sing} \text{ —} \\ \emptyset & \text{in the context } NP_{pl} \text{ —} \\ past \end{cases}$$

Notes: (i) '→' stands for 'expand' or 'rewrite as'. (ii) C stands for the various verbal suffixes that may be realised as *-s* (as in *walks*), Ø (i.e. zero) as in *walk* and *-ed* as in *walked*.

Chomsky's analysis does not separate phrase structure rules (e.g. Sentence → NP + VP; VP → Verb + NP) which enumerate permissible combinations of words in phrases and sentences from rules of word-structure like the one in [1.5] that gives *walks* from *walk*. All these rules are banded together because they are concerned with enumerating permissible combinations of morphemes (see above).

Note, however, that this treatment of syntactically motivated alternation in the form of words is controversial. We have merely aired the problem for the present. We will postpone detailed discussion until Chapter 10.

Turning to semantics, the connection between morphology and the lexicon on the one hand with meaning on the other is obvious since a major role of the lexicon or dictionary is to list the meanings of words. This is because normally the relationship between a word and its meaning is arbitrary. There is no reason why a word has the particular meaning that it has. For instance, you just have to memorise the fact that the word *faille* refers to a kind of head-dress worn in the seventeenth century. There is no way that you could discover this fact from the sounds or the structure of the word. We will come back to this topic in section (12.3.2).

It is less immediately obvious that, in addition to indicating the meaning of words and morphemes, the lexicon must also store other kinds of information relevant to the application of syntactic and phonological rules. Syntax needs to have access to **morphosyntactic properties** (i.e. properties that are partly morphological and partly syntactic) such as whether a noun is countable like *spades* or uncountable like *equipment*. This affects its behaviour in phrases and sentences. We may say *this spade* or *these spades* but we can only say *this equipment* (not **these equipments*).

Furthermore, some phonological rules apply to words differently depending on their morphosyntactic properties. For example, some phonological rules are sensitive to the difference between nouns and verbs. Thus, in the word *permit*, the main stress (shown here by underlining) falls on the first syllable if the word functions as a noun ($permit_{[noun]}$). But if it functions as a verb ($permit_{[verb]}$), main stress falls on the second syllable. Obviously, for phonological rules that assign stress to apply correctly, access to such morphosyntactic information is essential. This information must form part of the entry of the word in the lexicon.

The study of morphology, therefore, cannot be self-contained. The structuralist doctrine of the rigid separation of linguistic levels sketched in (1.2) is untenable. True, there are some issues that are the internal concerns of morphology. But many morphological problems involve the interaction between morphology and other modules of the grammar. For this reason, much of the space in the chapters that follow is devoted to the interaction between the lexicon and morphology with the mother modules.

1.4 ORGANISATION OF THE BOOK

The book is organised as follows:

Part I (Chapters 1–4) introduces basic concepts and traditional notions which are fundamental to all morphological discussions.

Part II (Chapters 5–9) explores the relationship between morphology, phonology and the lexicon in current generative theory.

Part III (Chapters 10–12) deals with the relationship between morphology and syntax in current generative theory.

Over the years, there have been several morphological theories that have been proposed by linguists. One way of introducing you to morphology would be to present a historical and comparative survey. I could have examined various theories in turn, and perhaps compared them. Or, alternatively, I could have been polemical and proselytising. I could have tried to persuade you that my preferred theory is the best theory. That is not what I shall do in this book.

Instead, I present you, sympathetically but at the same time critically, with one theoretically coherent approach to morphology, namely the theory of morphology in current mainstream generative grammar. This decision is sensible not only because this is the dominant model in the field today, but also because I think it offers the most promising solutions to the perennial problems in morphological analysis.

Even so, the book is inevitably selective. I have not attempted to represent every shade of opinion within the generative school. Rather I have focused on ideas and practices that seem to me to form part of the emerging 'canon' in mainstream generative morphology. Obviously, to some extent this is a matter of subjective judgement. In some cases my judgement may not be the same as that of some other linguists.

Of course, morphological theory in current mainstream generative grammar does not enjoy a monopoly of insight. The debt owed to other approaches will be evident, especially in the early chapters and in the bibliography.

A major feature of the book is that you will be asked to be an active investigator, not a passive reader. I have endeavoured to engage you actively and practically in *doing* morphology rather than in merely learning about its history and watching from the stalls how it is done. As you read each chapter, you are asked to pause at places and answer in-text questions and exercises before proceeding (the questions and exercises are signalled by lines across the page). Each chapter (after this one) ends with further exercises dealing with points raised in the body of the text. This insistence on getting you to analyse data is due to my firm conviction that the best initiation for anyone who wishes to become a linguist is to *do* linguistic analysis right from the start rather than to read about it.

In the text new morphological terms appear in bold type and they are

using the term **lexeme**. The forms *pockling, pockle, pockles* and *pockled* are different **realisations** (or representations or manifestations) of the lexeme POCKLE (lexemes will be written in capital letters). They all share a core meaning although they are spelled and pronounced differently. Lexemes are the vocabulary items that are listed in the dictionary (cf. Di Sciullo and Williams, 1987).

Which ones of the words in [2.2] below belong to the same lexeme?

[2.2] see   catches   taller   boy   catching   sees
       sleeps   woman   catch   saw   tallest   sleeping
       boys   sleep   seen   tall   jumped   caught
       seeing   jump   women   slept   jumps   jumping

We should all agree that:

The physical word-forms	are realisations of	the lexeme
see, sees, seeing, saw, seen		SEE
sleeps, sleeping, slept		SLEEP
catch, catches, catching, caught		CATCH

The physical word-forms	are realisations of	the lexeme
jump, jumps, jumped, jumping		JUMP
tall, taller, tallest		TALL
boy, boys		BOY
woman, women		WOMAN

2.1.2 Word-form

As we have just seen above, sometimes, when we use the term 'word', it is not the abstract vocabulary item with a common core of meaning, the lexeme, that we want to refer to. Rather, we may use the term 'word' to refer to a particular physical realisation of that lexeme in speech or writing, i.e. a particular **word-form**. Thus, we can refer to *see, sees, seeing, saw* and *seen* as five different words. In this sense, three different occurrences of any one of these word-forms would count as three words. We can also say that the word-form *see* has three letters and the word-form *seeing* has six. And, if we were counting the number of words in a passage, we would gladly count *see, sees, seeing, saw* and *seen* as five different word-forms (belonging to the same lexeme).

2.1.3 The Grammatical Word

The 'word' can also be seen as a representation of a lexeme that is associated with certain **morpho-syntactic properties** (i.e. partly morphological and partly syntactic properties) such as noun, adjective, verb, tense, gender, number, etc. We shall use the term **grammatical word** to refer to the 'word' in this sense.

Show why *cut* should be regarded as representing two distinct grammatical words in the following:

[2.3] a. Usually I cut the bread on the table.
       b. Yesterday I cut the bread in the sink.

The same word-form *cut*, belonging to the verbal lexeme CUT, can represent two different grammatical words. In [2.3a], *cut* represents the grammatical word *cut[verb, present, non 3rd person]*, i.e. the present tense, non-third person form of the verb CUT. But in [2.3b] it represents the grammatical word *cut[verb, past]* which realises the past tense of CUT.

Besides the two grammatical words realised by the word-form *cut* which we have mentioned above, there is a third one which you can observe in *Jane has a cut on her finger*. This grammatical word is *cut[noun, singular]*. It belongs to a separate lexeme CUT, the noun. Obviously, CUT, the noun, is related in meaning to CUT, the verb. However, CUT, the noun, is a separate lexeme from CUT, the verb, because it belongs to a different word-class (see section 3.5 below).

The nature of the grammatical word is important in the discussion of the relationship between words and sentences and the boundary between morphology and syntax.

2.2 MORPHEMES: THE SMALLEST UNITS OF MEANING

Morphology is the study of word structure. The claim that words have structure might come as a surprise because normally speakers think of words as indivisible units of meaning. This is probably due to the fact that many words are morphologically simple. For example, *the, fierce, desk, eat, boot, at, fee, mosquito*, etc., cannot be segmented (i.e. divided up) into smaller units that are themselves meaningful. It is impossible to say what the *-quito* part of *mosquito* or the *-erce* part of *fierce* means.

But very many English words are morphologically complex. They can be broken down into smaller units that are meaningful. This is true of words like *desk-s* and *boot-s*, for instance, where *desk* refers to one piece of furniture and *boot* refers to one item of footwear, while in both cases the *-s* serves the grammatical function of indicating plurality.

The term **morpheme** is used to refer to the smallest, indivisible units of semantic content or grammatical function which words are made up of. By definition, a morpheme cannot be decomposed into smaller units which are either meaningful by themselves or mark a grammatical function like singular or plural number in the noun. If we divided up the word *fee* [fi:] (which contains just one morpheme) into, say, [f] and [i:], it would be impossible to say what each of the sounds [f] and [i:] means by itself since sounds in themselves do not have meaning.

How do we know when to recognise a single sound or a group of sounds as representing a morpheme? Whether a particular sound or string of sounds is to be regarded as a manifestation of a morpheme depends on the word in which it appears. So, while *un-* represents a negative morpheme and has a meaning that can roughly be glossed as 'not' in words such as *un-just* and *un-tidy*, it has no claim to morpheme status when it occurs in *uncle* or in *under*, since in these latter words it does not have any identifiable grammatical or semantic value, because *-cle* and *-der* on their own do not mean anything. (Morphemes will be separated with a hyphen in the examples.)

Lego provides a useful analogy. Morphemes can be compared to pieces of lego that can be used again and again as building blocks to form different words. Recurrent parts of words that have the same meaning are isolated and recognised as manifestations of the same morpheme. Thus, the negative morpheme *un-* occurs in an indefinitely large number of words, besides those listed above. We find it in *unwell*, *unsafe*, *unclean*, *unhappy*, *unfit*, *uneven*, etc.

However, recurrence in a large number of words is not an essential property of morphemes. Sometimes a morpheme may be restricted to relatively few words. This is true of the morpheme *-dom*, meaning 'condition, state, dignity', which is found in words like *martyrdom*, *kingdom*, *chiefdom*, etc. (My glosses, here and elsewhere in the book, are based on definitions in the *Oxford English Dictionary*.)

It has been argued that, in an extreme case, a morpheme may occur in a single word. Lightner (1975: 633) has claimed that the morpheme *-ric* meaning 'diocese' is only found in the word *bishopric*. But this claim is disputed by Bauer (1983: 93) who suggests instead that perhaps *-ric* is not a distinct morpheme and that *bishopric* should be listed in the dictionary as an unanalysable word. We will leave this controversy at that and instead see how morphemes are identified in less problematic cases.

List two other words which contain each morpheme represented below:

[2.4]   **a.**   -er   as in   play-er, call-er
                 -ness   as in   kind-ness, good-ness
                 -ette   as in   kitchen-ette, cigar-ette
        b.   ex-   as in   ex-wife, ex-minister
                 pre-   as in   pre-war, pre-school
                 mis-   as in   mis-kick, mis-judge

a.   Write down the meaning of each morpheme you identify. (If you are in
     doubt, consult a good etymological dictionary.)
b.   What is the syntactic category (noun, adjective, verb, etc.) of the form
     which this morpheme attaches to and what is the category of the
     resulting word?

I expect your answer to confirm that, in each example in [2.4], the elements
recognised as belonging to a given morpheme contribute an identifiable
meaning to the word of which they are a part. The form *-er* is attached to
verbs to derive nouns with the general meaning 'someone who does X'
(where X indicates whatever action the verb involves). When *-ness* is
added to an adjective, it produces a noun meaning 'having the state or
condition (e.g., of being *kind*)'. The addition of the diminutive morpheme
-ette to a noun derives a new noun which has the meaning 'smaller in size'
(e.g., a *kitchenette* is a small *kitchen* and a *cigarette* is smaller than a *cigar*).
Finally, the morphemes *ex-* and *pre-* derive nouns from nouns while *mis-*
derives verbs from verbs. We can gloss the morpheme *ex-* as 'former', *pre-*
as 'before' and *mis-* as 'badly'.

So far we have described words with just one or two morphemes. In fact,
it is possible to combine several morphemes together to form more complex words. This can be seen in long words like *unfaithfulness* and *reincar-
nation* which contain the morphemes *un-faith-ful-ness* and *re-in-carn-at-ion*
respectively. But on what grounds do we divide up these words in this
fashion? In the following sections we will examine the basis on which
morphemes are identified.

2.2.1 Analysing Words

Up to now, we have used the criterion of meaning to identify morphemes.
In many cases forms that share the same meaning may be safely assigned to
the same morpheme. Where the meaning of a morpheme has been somewhat obscure, you have been encouraged to consult a good etymological

dictionary. Unfortunately, in practice, appealing to meanings listed in etymological dictionaries has its problems.

Consider the following words:

[2.5] helicopter   pteropus   diptera
       bible   bibliography bibliophile

Historically *pter* was borrowed from Greek, where it meant 'feather or wing'. The form *bibl-* also came from Greek where it meant 'papyrus, scroll, book'. Do you think *pter-* and *bibl-* should be recognised as morphemes in modern English?

I do not know what you decided. But I think it is questionable whether *pter-* is a morpheme of modern English. A *helicopter* is a kind of non-fixed wing aircraft which most speakers of English know about; *pteropus* are tropical bats with membranous wings popularly known as 'flying foxes' and *diptera* are two-winged flies (which few of us who are not entomologists know about). Obviously, *pter-* does occurs in modern English words that have the meaning 'pertaining to wings'. What is doubtful is whether this fact is part of the tacit knowledge of speakers of English who are not versed in etymology. Most people probably go through life without seeing a semantic connection between 'wings' and 'helicopters'.

Similarly, as we have already noted, the words *bible, bibliography* and *bibliophile* have to do with books. Probably many English speakers can see the book connection in *bibliography* and *bibliophile*. But it is unlikely that anyone lacking a profound knowledge of English etymology (and a classical education) is aware that the word *bible* is not just the name of a scripture book and that it contains a morpheme which is found in a number of other words pertaining to books.

Clearly, we need to distinguish between etymological information, whose relevance is essentially historical, and synchronic information that is part of speakers' competence. Our primary task as morphologists is to investigate speakers' tacit knowledge of the rules of their language rather than to perform historical reconstruction. We shall discuss this further in Chapter 4. The point I am making is that over-reliance on meaning in isolating morphemes puts us in a quandary in cases where etymological meanings are shrouded in the mists of history and lose their synchronic relevance.

The common definition of the morpheme as the 'minimal meaningful unit' implies the claim that every morpheme has a readily identifiable meaning. But this is problematic. There are cases where we can justify recognising a recurrent word-building unit as a morpheme although we cannot assign it a consistent meaning.

This is true of *-fer* in words like *pre-fer, in-fer, de-fer, con-fer, trans-fer* and *re-fer*. An etymological dictionary will tell us that *-fer* comes from the Latin word meaning 'bear, bring, send'. However, we would be hard-pressed to identify a consistent meaning like 'bring' attributable to *-fer-* in every instance above. For this reason some linguists, such as Aronoff (1976: 8–10), have argued that it is the word in its entirety rather than the morpheme *per se* that must be meaningful. Whereas all words must be meaningful when they occur on their own, morphemes need not be. Some morphemes, like *ex-* 'former' as in *ex-wife* and *pre-* 'before' as in *pre-war*, have a transparent, unambiguous meaning while others like *-fer* do not. Their interpretation varies depending on the other morphemes that occur together with them in a word.

In view of the above remarks, while semantic considerations must play a role in the identification of morphemes, given the pitfalls of a *purely* semantic approach, linguists tend to give a higher priority to more formal factors.

2.2.2 Morphemes, Morphs and Allomorphs

At one time, establishing mechanical procedures for the identification of morphemes was considered a realistic goal by structural linguists (cf. Harris, 1951). But it did not take long before most linguists realised that it was impossible to develop a set of **discovery procedures** that would lead automatically to a correct morphological analysis. No scientific discipline purports to equip its practitioners with infallible procedures for arriving at correct theories. Creative genius is needed to enable the scientist to make that leap into uncharted waters that results in a scientific discovery. What is true of science in general is also true of linguistics (cf. Chomsky, 1957: 49–60). Writing a grammar of a language entails constructing a theory of how that language works by making generalisations about its structure that go beyond the data that are observed.

Nevertheless, although there are no effective mechanical procedures for discovering the grammatical structure of a language in general or, in our case, the structure of its words, there exist reasonably reliable and widely accepted techniques that have been evolved by linguists working on morphology. These techniques are outlined in this section.

The main principle used in the analysis of words is the principle of **contrast**. We contrast forms that differ (i) in phonological shape due to the sounds used and (ii) in meaning, broadly defined to cover both lexical meaning and grammatical function. Thus, the phonological difference between /bɔɪ/ and /gɜ:l/ correlates with a semantic difference. The difference in meaning between the two sentences *The girl plays* and *The boy*

Now compare the Luganda forms in [2.10] with those in [2.7] above.

[2.10]   twaalaba   kitabo   'we saw a book'
         twaagula   bitabo   'we bought books'
         twaatunda   kitabo   'we sold a book'

The first person plural is represented by the form *tu-* in [2.7] and by *tw-* in [2.10]. What determines the selection of *tu* vs *tw-*?

Observe that here again the difference in form is not associated with a difference in meaning. The morphs *tu-* and *tw-* both represent the first person plural in different contexts. *Tu-* is used if the next morpheme is realised by a form beginning with a consonant and *tw-* is selected if the next morpheme is realised by a form that begins with a vowel.

If different morphs represent the same morpheme, they are grouped together and they are called **allomorphs** of that morpheme. So, *tu-* and *tw-* are allomorphs of the 'first person plural' morpheme. (For simplicity's sake, for our present purposes, we are regarding 'first person plural' as a single unanalysable concept.) On the same grounds, /ɪd/, /d/ and /t/ are grouped together as allomorphs of the past tense morpheme in English.

The relationship between morphemes, allomorphs and morphs can be represented using a diagram in the following way:

[2.11]   **a.** *English*   **morpheme**
         e.g.   *'past tense'*

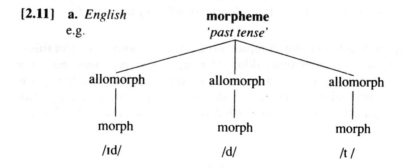

b. *Luganda*

e.g.

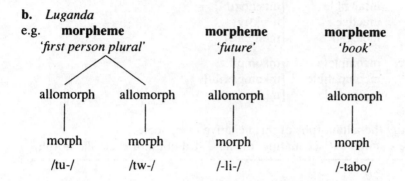

morpheme	morpheme	morpheme
'first person plural'	'future'	'book'

We can say that (i) /ɪd/, /d/ and /t/ are English morphs and (ii) we can group all these three morphs together as allomorphs of the past tense morpheme. Likewise, in Luganda we can say that (i) *tu-*, *tw-*, *-li-* and *-tabo* are morphs and furthermore (ii) *tu-* and *tw-* are allomorphs of the same morpheme since they represent the same superordinate concept, the morpheme 'first person plural'.

The central technique used in the identification of morphemes is based on the notion of **distribution**, i.e. the total set of contexts in which a particular linguistic form occurs. We classify a set of morphs as allomorphs of the same morpheme if they are in **complementary distribution**. Morphs are said to be in complementary distribution if (i) they represent the same meaning or serve the same grammatical function and (ii) they are never found in identical contexts. So, the three morphs /-ɪd/, /-d/ and /-t/ which represent the English regular past tense morpheme are in complementary distribution. Each morph is restricted to occurring in the contexts specified in [2.9]. Hence, they are allomorphs of the same morpheme. The same analysis applies also to Luganda *tu-* and *tw-*. Both morphs mean 'we' and they are in complementary distribution. *Tu-* occurs before consonants and *tw-* before vowels. They are therefore allomorphs of the first person plural morpheme. Morphemes realised by an invariant form (e.g., *future* and *book*) are said to have a single allomorph (cf. Matthews, 1974: 83).

Let us now examine some English words, focusing on the pronunciation of the underlined part of each word, which represents the negative morpheme *in-*. This morpheme can roughly be glossed as 'not':

[2.12]	**a.**	impossible	[ɪmpɒsɪbl]
		impatient	[ɪmpeɪʃnt]
		immovable	[ɪmuvəbl]
	b.	intolerable	[ɪntɒlərəbl]
		indecent	[ɪndiːsənt]

	intangible	[ɪntændʒɪbl]
	inactive	[ɪnæktɪv]
	inelegance	[ɪnelɪgəns]
c.	incomplete	[ɪŋkəmpliːt]
	incompatible	[ɪŋkəmpætɪbl]
	ingratitude	[ɪŋgætɪtjʊd]

a. Identify the allomorphs of this negative morpheme.
b. Write a statement accounting for the distribution of each allomorph.

I hope that you have isolated the following allomorphs of the morpheme *in-*: *im-* [ɪm-], *in-* [ɪn-] and *in-* [ɪŋ-].

The selection of the allomorph that is used in a particular context is not random. In [2.12] the *nasal* consonant in the various allomorphs of the morpheme *in-* is pronounced in a variety of ways, depending on the nature of the sound that immediately follows. To predict the allomorph that is selected in each case, a rule like [2.13] is required:

[2.13] **a.** select [ɪm] before a **labial** consonant (e.g., p, b, f, m) as in [ɪm]possible, [ɪm]patient, [ɪm]movable.

      b. select [ɪŋ] before the **velar** consonants [k] (here spelt with 'c') and [g] as in [ɪŋ]compliance, [ɪŋ]gratitude.

      c. select [ɪn] elsewhere, i.e. before an **alveolar** consonant like [t, d, s, z, n], as in [ɪn]**tolerable,** [ɪn]tangible and [ɪn]decent or before a vowel as in [ɪn]active, [ɪn]elegance.

The three allomorphs [ɪm], [ɪn] and [ɪŋ] of the morpheme *in-* are in complementary distribution. This means that selecting one precludes selecting the others. No two of them can occur in identical environments.

This example illustrates what is a very common state of affairs. If a morphene has several allomorphs, the choice of allomorph used in a given context is normally **phonologically conditioned**. This means that the allomorph selected to represent the morpheme in a particular context is one whose phonological properties are similar to those of sounds found in a neighbouring allomorph of some other morpheme.

The phonological resemblance between the nasal in the prefix and the first consonant representing the morpheme before which it is placed is due to **assimilation**. The pronunciation of the nasal in the prefix is adjusted to match the place of articulation of the first consonant representing the next morpheme. Thus, in [2.12] the labial consonant [m] occurs in [ɪm] before a labial consonant, the alveolar consonant [n] in [ɪn] occurs before alveolar consonants and the velar consonant [ŋ] in [ɪŋ] occurs before velar consonants. In each case the two consonants end up sharing the same place of articulation.

This example also illustrates another point, namely that spelling is a very poor guide to pronunciation in English (and many other languages). Where the point at issue would otherwise be obscured by the standard orthography, phonetic or phonemic transcription will be used as appropriate in this book. (See p. 14 for key.)

In the light of this discussion, let us return to the earlier example of the allomorphs of the English regular past tense morpheme in [2.9]. Clearly, the distribution of allomorphs is phonologically conditioned: /-ɪd/ is chosen after the alveolar stops /t/ and /d/ (with /ɪ/ being inserted to separate the alveolar stop of the suffix from the final alveolar stop of the verb to which it is attached); voiced /-d/ is chosen after voiced segments other than /d/ and voiceless /-t/ is chosen after voiceless consonants other than /t/.

So far, all the examples of morphs that we have seen have involved only vowels and consonants. But, as [2.14] shows, morphemes may also be signalled by **tone**, i.e. contrastive use of relative pitch (cf. Hyman, 1975; Katamba, 1989; Pike, 1948):

[2.14]	**a.**	ǹjálá	'hunger'	ǹjálá	'nails'
		mwèèzí	'sweeper'	mwéèzí	'moon'
		bùsá	'naked'	bùsâ	'dung'
		bùggyâ	'newness'	búggyà	'envy'
		àléètá	's/he brings'	àléétâ	'one who brings'
	b.	àsîká	's/he fries'	àsííkâ	'one who fries'
		àsómá	's/he reads'	àsòmâ	'one who reads'
		àgóbá	's/he chases'	àgòbâ	'one who chases'

Note: Interpret the tone diacritics as follows:
    ´ = High tone (H), ` = Low tone (L), and ^ = Falling tone (F)

In [2.14a], tonal differences are used to distinguish lexical items. The word-forms are identical in all respects except tone. In [2.14b], on the other hand, tone is used to signal grammatical distinctions. LHLH corresponds to LHHF in the first two verbs while in the last two, LHH corresponds to LLF. In each case, the first pattern represents a third person main clause present tense form of the verb and the second pattern represents the relative clause form.

2.2.3 Grammatical Conditioning, Lexical Conditioning and Suppletion

We have seen in the last section that the distribution of allomorphs is usually subject to phonological conditioning. However, sometimes phonological factors play no role in the selection of allomorphs. Instead, the choice of allomorph may be **grammatically conditioned**, i.e. it may be dependent on the presence of a particular grammatical element. A special allomorph may be required in a given grammatical context although there might not be any good phonological reason for its selection. For example, in [2.15a], which is typical, in English the presence of the past tense morpheme in the majority of cases has no effect on the selection of the allomorph that represents the verb itself. But, as [2.15b] and [12.5c] show, in certain verbs the presence of the past tense morpheme requires the selection of a special allomorph of the verb:

[2.15]		Present tense			Past tense	
	a.	walk	/wɔːk/		Walked	/wɔːkt/
		kiss	/kɪs/		kiss-ed	/kɪst/
		grasp	/grɑːsp/		grasp-ed	/grɑːspt/
	b.	weep	/wiːp/		wep-t	/wept/
		sweep	/swiːp/		swep-t	/swept/
	c.	shake	/ʃeɪk/		shook	/ʃʊk/
		take	/tʊk/		took	/tʊk/

In [2.15b], the choice of allomorph is grammatically conditioned. The presence of the past tense morpheme determines the choice of the /wep/ and /swep/ allomorphs in verbs that belong to this group. For the verbs in [12.15c] the past tense dictates the choice of the allomorphs *took* and *shook* of the verbs *take* and *shake* respectively.

In other cases, the choice of the allomorph may be **lexically conditioned**, i.e. use of a particular allomorph may be obligatory if a certain *word* is present. We can see this in the realisation of plural in English.

Normally the plural morpheme is realised by a phonologically conditioned allomorph whose distribution is stated in [2.16]:

[2.16] **a.** select allomorph /-ɪz/ if a noun ends in an alveolar or alveopalatal sibilant (i.e. a consonant with a sharp, hissing sound such as /s z ʃ ʒ tʃ dʒ/).
Examples: *asses   mazes   fishes   badges   beaches*
           /æsɪz/   /meɪzɪz/   /fɪʃɪz/   /bædʒɪz/   /biːtʃɪz/

   b. select allomorph /-s/ if a noun ends in a non-strident voiceless consonant (i.e. any one of the sounds /p t k f θ/).

Examples: *cups   leeks   carts   moths*
/kʌps/   /liːks/   /kɑːtsʃ   /mɒθs/

c. select allomorph /-z/ elsewhere (i.e. if the noun ends in a
voiced nonstrident segment; this includes all vowels and the
consonants /b d g d m n ŋ l r w j/).
Examples: *bards   mugs   rooms   keys   shoes*
/bɑːdz/   /mʌgz/   /ruːmz/   /kiːz/   /ʃuːz/

Can you explain why the rule in [2.16] fails to account for the realisation of
the plural morpheme in the word *oxen*?

I expect you to have failed to find a plausible explanation. There are cases
where for no apparent reason the regular rule in [2.16] inexplicably fails to
apply. The plural of *ox* is not **oxes* but *oxen*, although words that rhyme
with *ox* take the expected /ɪz/ plural allomorph (cf. /fɒksɪz/ *foxes* and
/bɒksɪz/ *boxes*). The choice of the allomorph *-en* is **lexically conditioned**. It
is dependent on the presence of the specific noun *ox*.

Finally, there exist a few morphemes whose allomorphs show no phonetic similarity. A classic example of this is provided by the forms *good/better*
which both represent the lexeme GOOD despite the fact that they do not
have even a single sound in common. Where allomorphs of a morpheme
are phonetically unrelated we speak of **suppletion**.

The pair *good* and *better* is not unique in English. Find one other example
of suppletion.

Other examples of suppletion in English include *bad* ~ *worse* (not **bad-
der*); *go* ~ *went* (not **goed*)

2.2.4 Underlying Representations

Above we have distinguished between, on the one hand, regular, rule-
governed **phonological alternation** (a situation where the choice between
alternative allomorphs is regulated in quite predictable ways by the phonological properties of the different morphs that occur near each other (see
section (2.2.2)) and cases of suppletion where there is phonologically
arbitrary alternation in the realisation of a morpheme (see section (2.2.3)).
This is standard in generative phonology (cf. Chomsky and Halle, 1968;
Kenstowicz and Kisseberth, 1979; Anderson, 1974: 51–61).

Merely listing allomorphs does not allow us to distinguish between eccentric alternations like *good ~ bett(-er)* and regular alternations like that shown by the negative prefix *in-* or by the regular *-s* plural suffix. The latter are general and will normally apply to any form with the relevant phonological properties, unless it is specifically exempted. Thus the regular plural rule in [2.16] above is used to attach /-s/, /-z/ or /-ɪz/ to virtually any noun that ends in the appropriate sound. By contrast, a rule of suppletion or lexical conditioning only applies if a form is expressly marked as being subject to it. Thus, for example, of all English adjectives, only *good* is subject to the suppletive rule that gives *bett-er* in the comparative; and only *ox* is subject to the lexically conditioned rule that suffixes *-en* to yield the plural *oxen*. Similarly, a grammatically conditioned rule will only be triggered if the appropriate grammatical conditioning factor is present. For example, the allomorph *slep-* of the morpheme *sleep* only co-occurs with the past tense (or past participle) morpheme. It cannot be selected to co-occur with the present tense: *sleep* + [past] yields *slept* /slept/ (not /*sli:pt/) while *sleep* + [present] gives *sleep* /sli:p/ (not /*slep/).

To bring out the distinction between regular phonological alternation, which is phonetically motivated, and other kinds of morphological alternation that lack a phonetic basis, linguists posit a single **underlying representation** or **base form** from which the various allomorphs (or **alternants** i.e. alternative phonetic realisations) of a morpheme are derived by applying one or more **phonological rules**. The stages which a form goes through when it is being converted from an underlying representation to a **phonetic representation** constitute a **derivation**.

For a concrete example, let us consider again the representation of the *in-* morpheme in [2.13], which is repeated below as [2.17] for convenience:

[2.17] **a.** select [ɪm] before a **labial** consonant (e.g. p, b, f, m)
              as in [ɪm]possible, [ɪm]patient, [ɪm]movable.
        b. select [ɪŋ] before the **velar** consonants [k] (here spelt with 'c')
              and [g] as in [ɪŋ]compliance, [ɪŋ]compatible, [ɪŋ]gratitude.
        c. select [ɪn] elsewhere, i.e. before an **alveolar** consonant like [t,
              d, s, z, n],
              as in [ɪn]tolerable, [ɪn]tangible and [ɪn]decent
              or before a vowel as in [ɪn]active, [ɪn]elegance.

The vital point to note is that the three parts of the rule in [2.17] are not independent of each other. By making three separate statements, we have missed a generalisation. A superior solution would be to restate [2.17] as [2.18]. The revised statement, in which we posit a single underlying representation from which the three allomorphs are derived, captures the fact that the alternation in the realisation of these allomorphs is due to a single factor, namely assimilation.

[2.18]   The nasal realising the morpheme *in-* /ɪn/ must appear in the phonetic representation as a nasal consonant that shares the place of articulation of the initial consonant of the form to which it is attached.

But how can we be certain that the base form is /ɪn/ rather than /ɪm/ or /ɪŋ/? We have seen that the nasal assimilates to the place of articulation of the consonant that follows it. The fact that when a vowel follows we still find [ɪn-] appearing as in [ɪn-ɔ:dɪbl] *inaudible*, and [ɪn-evɪtəbl] *inevitable* is very revealing. From a phonetic point of view, vowels do not have definite places of articulation, only consonants do. So, a consonant cannot assimilate to the place of articulation of a vowel. The occurrence of [ɪn-] before vowels is not due to place assimilation. Besides, the alveolar nasal is found regardless of whether the vowel that follows is made in the front of the mouth like [e], or in the back like [ɔ:]. So, the influence of the vowel cannot be responsible for the choice of [ɪn-].

A simple solution is to assume that [ɪn-] is the **default form**, i.e. the form selected unless there are explicit instructions to do otherwise. If we posit this form as the underlying representation, we do not need to change it before vowels or before alveolar consonants. We only need to change it before non-alveolar consonants. If, however we posited [ɪm-] or [ɪŋ-] as the underlying representation, we would need rules to modify them when they appeared not only before non-labial and non-velar consonants respectively but also before vowels. If two analyses can both account properly for the facts, the analysis that provides a simpler solution is preferred. Obviously, in this case it is the analysis in [2.18] (with /ɪn-/ as the base form) that wins.

Phonologically conditioned morphological alternations tend to be very general. Often allomorphs representing different morphemes will display the same phonological alternations if they occur in similar phonological environments. Thus, for example, the voice assimilation process displayed by the *-s* plural suffix in [2.16] is not unique to that morpheme. The *-s* third person singular present tense suffix in verbs shows exactly the same alternations, as you can see:

[2.19]   /-ɪz/   after sibilants e.g. /wɔ:ʃ/ *wash* ~ /wɔ:ʃɪz/ *washes*
  /-z/   after voiced segments other than sibilants e.g. /ri:d/ *read* ~ /ri:dz/ *reads*
  /-s/   after voiceless consonants other than sibilants e.g. /dʒʌmp/ *jump* ~ /dʒʌmps/ *jumps*

The same rule applies to genitives:

[2.20]   /-ɪz/   after sibilants e.g. /lɪz/ *Liz* ~ /lɪzɪz/ *Liz's*

/-z/   after voiced segments other than sibilants e.g. /dʒeɪn/ *Jane* ~
/dʒeɪnz/ *Jane's*
/-s/   after voiceless consonants other than sibilants e.g. /mɪk/
Mick ~ /mɪks/ *Mick's*

If we make three separate statements, one for the plural, another for the third
person singular and a third one for the genitive, we miss the generalisation
that a sibilant suffix agrees in voicing with the last segment of the form to
which it is attached, regardless of the morpheme the suffix represents.

However, this generalisation is captured if we posit just one underlying
representation (or base form) for any sibilant suffix, and if that underlying
representation is converted into different phonetic representations by the
(informal) phonological rules below:

[2.21]   a.   The underlying representation of any sibilant suffix is /z/.
         b.   It is realised as:
              (i)   /ɪz/ after alveolar and alveo-palatal sibilants (e.g. /s z ʃ ʒ tʃ
                    dʒ/)
              (ii) /z/ after voiced segments other than sibilants (e.g. vowels
                    and voiced consonants like /b m d v/)
              (iii) /s/ after voiceless consonants other than sibilants /p t k f θ/

The statement in [2.21] shows that the alternation in question is not the
idiosyncratic property of any one morpheme but rather a general phonological process in the language. The terms **morphophonemics** (in American
linguistics) and **morphophonology** (European linguistics) are used to refer
to rules of this kind that account for the realisation of phonologically
conditioned allomorphs of morphemes. The rule for the realisation /ɪn-/ in
[2.18] is another example of a morphophonemic rule.

2.3 THE NATURE OF MORPHEMES

Words can be divided into **segments of sound**. Thus, the word *book* can be
divided into the segments /b, ʊ, k/. Indeed, the division of words into
phonemes forms the basis of alphabetic writing systems like that of
English. But it is also possible, and natural to divide words into syllables.
For instance, Japanese uses fifty distinct symbols to represent the fifty
syllable types found in the language.
` So, it is important to avoid confusing morphemes with syllables.
Syllables are groupings of sounds for the purposes of articulation, while
morphemes are the smallest units of meaning or grammatical function. A
few examples should clarify the distinction. On the one hand, the words

sofa /səu fə/ and *balloon* /bə lu:n/ contain two syllables each while *camera* /kæ mə rə/ and *hooligan* /hu: lɪ gən/ contain three syllables each. (I have separated syllables with a space). But all these words have only one morpheme each. On the other hand, the word *books* /buks/ has one syllable, but two morphemes. They are the morpheme *book* /buk/ and the final *-s* /s/ which represents the plural morpheme although it is not a syllable in its own right.

When we divide a word into morphemes, we focus on strings of sound that are meaningful regardless of whether they constitute syllables at the phonological level. A question that lurks in the background concerns the precise nature of the relationship between strings of sounds and the meanings that they represent. This is the question to which we now turn. The discussion that follows draws on Matthews (1974).

At first, it might seem reasonable to assume that the relationship between morphemes and strings of phonemes, which are identified as morphs, is one of **composition**. In that case, we could say that the morpheme *book* /buk/ is *made up of* the phonemes /b/, /u/ and /k/.

As we will see in a moment, an approach which assumes that morphemes are made up of phonemes leads to a theoretical cul-de-sac. It is preferable to view morphemes as being **represented** or **realised** or **manifested** by morphs. It is unsound to assume that morphemes are actually composed of (sequences of) phonemes because this would suggest that the meaning of a morpheme is a function of its phonemic composition. It is not, since phonemes in themselves cannot have meanings. The same phoneme /ə/ (spelled *-er*) can represent either the comparative degree of adjectives, as in *kind-er* and the noun-forming suffix *-er* as in *worker*, which is formed from the verb *work*, or it can be a part of a word without a discernible meaning of its own, as in *water*. Clearly, it is the morphs rather than morphemes that are made up of (sequences of) phonemes. Possible relationships between morphemes and (sequences of) phonemes can then be summarised in this fashion:

1. There may be a one-to-one correlation between morphemes and morphs (which are made up of individual phonemes or strings of phonemes). For instance, in French, the word *eau* /o/ (water) has one morpheme which is realised by a morph that is composed of just one phoneme, namely /o/. This is the simplest situation.

2. The relationship between sound and meaning in language is for the most part **arbitrary**, that is to say that is no good reason why a particular sound or string of sounds has a particular meaning. Given this, several different pairings of sounds with meaning are possible:
(i)   As we saw above a moment ago with regard to /ə/, a single phonological form may be used to represent different morphemes. Now we will

examine this point in more detail. Consider the phonological form /saɪt/ which happens to have three orthographic representations, each one of which represents a different morpheme. Also think about the phonological form /raɪt/ which has four spellings which represent four separate morphemes:

[2.22]   **a.**   sight   site   cite
         b.   right   write   wright   rite

What we see here are **homophones**, i.e. forms which sound the same but differ in their meaning or grammatical function. From the point of view of the spoken language, there are only two morphs, namely the forms /saɪt/ and /raɪt/. The two morphs represent three and four morphemes respectively, but written English uses a different form to represent each morpheme in each case. Homophony is relatively common in language. Puns depend on it. And many a joke is due to the fact that morphs like *duck* and *bent* represent more than one morpheme.

(ii)   The converse is also common. A single morpheme may be represented by a variety of phonological representations. We have already seen this in the case of the plural morpheme, which has the three allomorphs [s], [z] and [ɪz] (see [2.16] above). The same applies to the negative prefix *in-* (which has the allomorphs [ɪm], [ɪn] and [ɪŋ] (see [2.17] above).

(iii) The same string of sounds may cumulatively represent several morphemes. The *-s* ending in English verbs (e.g. *walk-s*) signals three morphemes simultaneously, namely, third person, present tense and singular number. If morphemes consisted of morphs this would not be possible. A separate morph would be needed to represent each morpheme. This shows just how abstract morphemes are, as opposed to morphs. Morphemes themselves are not composed of sounds but they are represented by morphs which are made up of sounds.

The term **portmanteau morph** is used to refer to cases like the above where a single morph simultaneously represents a bundle of several different grammatical elements.

Morphemes are to morphs what lexemes are to word-forms. Morphemes and lexemes are the abstract entities found in the lexicon while morphs and word-forms are the physical entities found in speech or writing.

In addition to different morphemes being represented by the same morphs, we can also have a situation where different grammatical words are represented by the same word-forms. This is called **syncretism**. It is a result of **neutralisation**. The same form is used to represent distinct morphological concepts. Thus, in regular verbs, the same word-form represents two distinct grammatical words: e.g. *walk* + [past] *walked* (as in *I walked*) vs *walk* + [past participle] *walked* (as in *I have walked*). Irregular verbs like *see* and *take* exhibit no syncretism. They have distinct past tense and past participle forms: *see* + [past] (*saw*) and *take* + [past] (*took*); *see* + [past participle] (*seen*) and *take* + [past participle] (*taken*).

3. Finally, an approach that assumes a one-to-one correspondence between morphemes and morphs encounters difficulties when there is simply no match between morpheme and morph. There are two sets of circumstances in which this may happen:
(i)   The number of morphemes present may exceed the number of morphs available to represent them. This happens when a grammatical contrast which is marked overtly by a morph in some words is not overtly marked in others. Thus, for example, we know that in English, if the adverb *yesterday* or a phrase like *last week* is found in a sentence, the verb in that sentence must be in the past tense because that is the form of the verb that is required whenever a verb designates an event, action, state or process that happened prior to the moment of speaking or writing. As a rule, such a verb will end in *-ed*.

[2.23]   Last week the farmer sowed the corn.
        Yesterday Jane painted the roof.

In the light of the last remark, explain how the past tense is marked in the following.

[2.24]   Last week I cut the grass.
        I put those carnations in the vase yesterday.
        Yesterday they shut the factory down.
        The mob hit him last week.

We know that *cut*, *put*, *shut* and *hit* are every bit as past as *sowed* and *painted* in [2.23] because only verbs in the past tense can occur together with *yesterday* or *last week* in a sentence.

The past tense morpheme, which is represented by *-ed* in [2.23], is realised by a **zero allomorph** in [2.24]. In other words, we can infer from the structural patterns of the language that the verb is in the past tense although nothing about the shape of the word overtly shows this. If we allow ourselves to posit zero allomorphs, the assumption that morphemes 'consist' of phonemes must be rejected. Instead, we shall regard morphemes as abstract entities which are *represented* by morphs. In speech, the morphs are composed of phonemes but the morphemes themselves are not. (See also the discussion of conversion in section (3.5).)

(ii) The converse also occurs: the number of morphs that can be isolated may exceed the number of morphemes represented. In other words, there may be a surplus word-building element which does not realise any morpheme. Such an element is usually called an **empty morph**.

Describe in detail how the adjectives in [2.25a] and [2.25b] are derived from nouns.

[2.25]		Noun		Adjective	
	a.	medicine	/medɪsɪn/	medicin-al	/medɪsɪnəl/
		person	/pɜːsən/	person-al	/pɜːsənəl/
		tribe	/traɪb/	trib-al	/traɪbəl/
	b.	sense	/sens/	sens-u-al	/sensjʊəl/
		fact	/fækt/	fact-u-al	/fæktjʊəl/

In [2.25a] the adjectives are formed simply by adding the suffix *-al* to nouns. In [2.25b], however, there is an empty morph, *-u-* (/ju/) that does not represent any morpheme which is inserted immediately before *-al*.

'Empty morph' is an unfortunate choice of terminology – but we are stuck with it. If a morph is a morph by virtue of representing some morpheme, a surplus word-building element that does not represent any morpheme should not be regarded as a morph. Hence the appropriateness of the more neutral term **formative** for referring to any word-building element. Most formatives are morphs: they represent morphemes. And some are not. They are the so-called 'empty morphs'.

2.4 SUMMARY

The chapter opened with a discussion of the nature of the word (section (2.1)). We distinguished between lexemes, word-forms and grammatical words. Lexemes are abstract dictionary words like the verb SING. A lexeme is realised by one or more word-forms. Word-forms are concrete words that occur in speech and writing, e.g. *sing*, *sings*, *sang* and *sung*. We also saw that the word can be viewed as a lexeme associated with a set of morpho-syntactic properties, e.g. *sing*[verb, present, 3rd person, singular]. In this case we are looking at a grammatical word.

The next section introduced the segmentation of words into the smallest abstract units of meaning or grammatical function. These units are called morphemes. We saw that the analysis of words into morphemes begins with the contrasting of pairs of utterances which are partially different in sound and meaning. Word-forms are segmented into morphs, which are recurrent physical word-forming chunks. Any morphs that represent the same meaning are grouped together as allomorphs of that morpheme. Meaning plays a role in this, but the main principle used is that of distribution. Morphs are listed as allomorphs of the same morpheme if they are in complementary distribution, i.e. if they are realisations of the same morpheme in different contexts. (Sometimes a morpheme has a single allomorph.)

Normally, the distribution of allomorphs is *phonologically conditioned*. The relationship between allomorphs has a phonetic motivation. A single underlying (base) form is postulated and the phonetic representation of the various allomorphs is derived from it using phonological rules.

But sometimes allomorphs may be grammatically conditioned or even lexically conditioned, i.e. a particular allomorph is selected if either a particular grammatical element or a particular word is present.

Occasionally there is suppletion, which means that an allomorph bears no phonetic similarity to other allomorphs of the same morpheme.

The last section dealt with the relationship between morphological and phonological representations. It was established that the relationship between morphemes on the one hand and morphs on the other, is one of representation (or realisation) rather than composition.

EXERCISES

1. Define and give one fresh example of each of the following:
   (a) lexeme; grammatical word; word-form.
   (b) morpheme; morph; allomorph; portmanteau morph; suppletion; empty morph; zero morph.

2. (a) What is the allomorph of the plural morpheme that occurs in each group of words below?

(b) Explain whether the choice of allomorph is phonologically, grammatically or lexically conditioned:

(i) agenda data strata media desiderata
(ii) stimuli radii fungi alumni

3. Study the following data and answer the questions that follow:

*dis*like	*un*wind	*re*port	*dis*trust	*un*cover	*re*cover
*un*able	*re*write	*un*lock	land*less*	*dis*united	*re*draw
ex-monk	*dis*allow	penni*less*	*un*happy	*re*pel	*ex*-coach

(a) What is the meaning of the morphemes represented in writing by *ex-, in-, dis-, un-, re-* and *-less*?
(b) Are any of the meanings you recognise only of historical interest?
(c) Comment on cases of homophony where a single morph represents more than one morpheme.

4. (a) Distinguish between phonological conditioning and grammatical conditioning of allomorphs.
(b) Give one fresh example of each taken from any language that you know.

5. (a) Describe the tonal patterns found in the Rendille data below (from Ooomen-van Schendel, 1977).
(b) What are the functions of tone in these examples?

nyíràx	'young male camel'	nyīráx	'young female camel'
ħélèm	'ram'	ħēlém	'sheep'
kélèħ	'big male goat'	kēléħ	'big male goat'
dúfàn	'camel kept for slaughtering'	dūfán	'camels kept for slaughtering'
nâf	'a domestic animal'	náf	'domestic animals'

Note on tone marking: ´ marks High tone, ` marks Low tone, ⁻ marks mid tone, ^ marks (High–Low) Falling tone.

6. Illustrating your answer with examples of your own, construct an argument for setting up underlying representations or base forms. (Read section (2.2.4) again before attempting this question.)

3 Types of Morphemes

3.1 ROOTS, AFFIXES, STEMS AND BASES

In the last chapter we saw that words have internal structure. This chapter introduces you to a wide range of word-building elements used to create that structure. We will start by considering roots and affixes.

3.1.1 Roots

A root is the irreducible core of a word, with absolutely nothing else attached to it. It is the part that is always present, possibly with some modification, in the various manifestations of a lexeme. For example, *walk* is a root and it appears in the set of word-forms that instantiate the lexeme WALK such as *walk*, *walks*, *walking* and *walked*.

The only situation where this is not true is when suppletion takes place (see section (2.2.3)). In that case, word-forms that represent the same morpheme do not share a common root morpheme. Thus, although both the word-forms *good* and *better* realise the lexeme GOOD, only *good* is phonetically similar to GOOD.

Many words contain a root standing on its own. Roots which are capable of standing independently are called **free morphemes**, for example:

[3.1] Free morphemes

| man | book | tea | | sweet | cook |
| bet | very | aardvark | pain | walk |

Single words like those in [3.1] are the smallest free morphemes capable of occurring in isolation.

The free morphemes in [3.1] are examples of **lexical morphemes**. They are nouns, adjectives, verbs, prepositions or adverbs. Such morphemes carry most of the 'semantic content' of utterances – loosely defined to cover notions like referring to individuals (e.g. the nouns *John*, *mother*), attributing properties (e.g. the adjectives *kind*, *clever*), describing actions, process or states (e.g. the verbs *hit*, *write*, *rest*) etc., expressing relations (e.g. the prepositions *in*, *on*, *under*) and describing circumstances like manner (e.g. *kindly*).

Many other free morphemes are **function words**. These differ from lexical morphemes in that while the lexical morphemes carry most of the 'semantic content', the function words mainly (but not exclusively) signal

grammatical information or logical relations in a sentence. Typical function
words include the following:

[3.2]   Function words

articles: a the
demonstratives: this that these those
pronouns: I you we they them; my your his hers; who
                        whom which whose, etc.
conjunctions: and yet if but however or, etc.

Distinguishing between lexical and grammatical morphemes is normally
both useful and straightforward. However, there are cases where this
distinction is blurred. This is because there are free morphemes (i.e. simple
words) which do not fit neatly into either category. For example, a con-
junction like *though* signals a logical relationship and at the same time
appears to have considerably more 'descriptive semantic content' than,
say, the article *the*.

While only roots can be free morphemes, not all roots are free. Many
roots are incapable of occurring in isolation. They always occur with some
other word-building element attached to them. Such roots are called **bound
morphemes**. Examples of bound morphemes are given below:

[3.3]   a.   -mit   as in permit, remit, commit, admit
        b.   -ceive   as in perceive, receive, conceive
        c.   pred-   as in predator, predatory, predation, depredate
        d.   sed-   as in sedan, sedate, sedent, sedentary, sediment

The bound roots *-mit*, *-ceive*, *-pred* and *sed-* co-occur with forms like *de-*,
re-, *-ate*, *-ment* which recur in numerous other words as prefixes or suffixes.
None of these roots could occur as an independent word.

Roots tend to have a core meaning which is in some way modified by the
affix. But determining meaning is sometimes tricky. Perhaps you are able
to recognise the meaning 'prey' that runs through the root *pred-* in the
various words in [3.3c] and perhaps you are also able to identify the
meaning 'sit' in all the forms in [3.3d] which contain *sed-*.

These roots are **latinate**, i.e. they came into English from Latin (normally via French). I suspect that, unless you have studied Latin, you are
unable to say that *-mit* means 'send, do' and *-ceive* means 'take' without
looking up *-mit* and *-ceive* in an etymological dictionary. In present-day
English none of these meanings is recognisable. These formatives cannot
be assigned a clear, constant meaning on their own.

In the last chapter the morpheme was defined as the smallest unit of
meaning or grammatical function. In the light of the foregoing discussion,

the insistence on the requirement that every morpheme must have a clear, constant meaning (or grammatical function) seems too strong to some linguists. There are morphemes that lack a clear meaning. Instead, they suggest, it is the word rather than the morpheme that must always be independently meaningful whenever it is used. As we saw in section (2.2.1) above, the crucial thing about morphemes is not that they are independently meaningful, but that they are recognisable **distributional units** (Harris, 1951). As Aronoff (1976: 15) puts it, we can recognise a morpheme when we see a morph 'which can be connected to a linguistic entity outside that string. What is important is not its meaning, but its arbitrariness.'

The reason for treating those recurring portions of words that appear to lack a clear, constant meaning as morphs representing some morpheme is that they behave in a phonologically consistent way in the language which is different from the behaviour of morphologically unrelated but phonologically similar sequences. Take *-mit*, for example. Aronoff (1976: 12–13) points out that, notwithstanding the tenuous semantic link between instances of all the latinate root *-mit*, they nevertheless share a common feature which is not predictable from any properties of the phonetic sequence [mɪt]. All instances of latinate *-mit* have the allomorph [mɪʃ] or [mɪs] before the suffixes *-ion*, *-ory*, *-or*, *-ive*, and *-able / -ible*, as you can see:

[3.4] *latinate root* [mɪt]	[mɪʃ] before *-ion*	[mɪs] before *-ive, -ory*
permit	permission	permissive
submit	submission	submissive
admit	admission	admissive
remit	remission	remissory

By contrast, any other phonetic form [mɪt] (*-mit*) does not undergo the same phonological modification before such suffixes. Thus, although forms like *dormitory* and *vomitory* have a [mɪt] phonetic shape preceding the suffix *-ory*, they fail to undergo the rule that changes /t/ to [s]. If that rule applied, it would incorrectly deliver **dormissory* or **vomissory*, since the same phonetic sequence [mɪt] as that in [3.4a] precedes the suffix *-ory*. Clearly, the [mɪt] sequence in *vomitory* and *dormitory* is not a morph representing the latinate *-mit* morpheme. The rule that supplies the allomorph [mɪs] of verbs that contain [mɪt] is only activated where [mɪt] represents the latinate root *-mit*.

What this discussion shows is that even where the semantic basis for recognising a morpheme is shaky, there may well be distributional considerations that may save the day. Only latinate *-mit* has the allomorph

[-mis-]. Any word-form that displays the [mit] ~ [mis] alternation in the contexts in [3.4] contains the latinate root morpheme -*mit*.

3.1.2 Affixes

An **affix** is a morpheme which only occurs when attached to some other morpheme or morphemes such as a root or stem or base. (The latter two terms are explained in (3.1.3) below.) Obviously, by definition **affixes** are bound morphemes. No word may contain only an affix standing on its own, like **-s* or **-ed* or **-al* or even a number of affixes strung together like **-al-s*.

There are three types of affixes. We will consider them in turn.
(i) Prefixes
A **prefix** is an affix attached *before* a root or stem or base like *re-*, *un-* and *in-*:

[3.5]   re-make   un-kind   in-decent
        re-read   un-tidy   in-accurate

(ii) Suffixes
A **suffix** is an affix attached *after* a root (or stem or base) like *-ly*, *-er*, *-ist*, *-s*, *-ing* and *-ed*.

[3.6]   kind-ly   wait-er   book-s   walk-ed
        quick-ly   play-er   mat-s   jump-ed

(iii) Infixes
An **infix** is an affix inserted into the root itself. Infixes are very common in Semitic languages like Arabic and Hebrew as we will see in section (3.6) below and in more detail in Chapter 9. But infixing is somewhat rare in English. Sloat and Taylor (1978) suggest that the only infix that occurs in English morphology is /-n-/ which is inserted before the last consonant of the root in a few words of Latin origin, on what appears to be an arbitrary basis. This infix undergoes place of articulation assimilation. Thus, the root -*cub*- meaning 'lie in, on or upon' occurs without [m] before the [b] in some words containing that root, e.g. *incubate*, *incubus*, *concubine* and *succubus*. But [m] is infixed before that same root in some other words like *incumbent*, *succumb*, and *decumbent*. This infix is a frozen historical relic from Latin.

In fact, infixation of sorts still happens in contemporary English. Consider the examples in [3.7a] which are gleaned from Zwicky and Pullum (1987) and those in [3.7b] taken from Bauer (1983):

[3.7] **a.** Kalamazoo (place name) → Kalama-goddam-zoo

instantiate (verb) → in-fuckin-stantiate

b. kangaroo → kanga-bloody-roo

impossible → in-fuckin-possible

guarantee → guaran-friggin-tee

(Recall that the arrow → means 'becomes' or is 're-written as'.)

As you can see, in present-day English infixation, not of an affix morpheme but of an entire word (which may have more than one morpheme, e.g. *blood-y*, *fuck-ing*) is actively used to form words. Curiously, this infixation is virtually restricted to inserting expletives into words in expressive language that one would probably not use in polite company.

3.1.3 Roots, Stems and Bases

The stem is that part of a word that is in existence before any *inflectional* affixes (i.e. those affixes whose presence is required by the syntax such as markers of singular and plural number in nouns, tense in verbs etc.) have been added. Inflection is discussed in section (3.2). For the moment a few examples should suffice:

[3.8] Noun stem	Plural
cat	-s
worker	-s

In the word-form *cats*, the plural inflectional suffix *-s* is attached to the simple stem *cat*, which is a bare root, i.e. the irreducible core of the word. In *workers* the same inflectional *-s* suffix comes after a slightly more complex stem consisting of the root *work* plus the suffix *-er* which is used to form nouns from verbs (with the meaning 'someone who does the action designated by the verb (e.g. *worker*)'). Here *work* is the root, but *worker* is the stem to which *-s* is attached.

Finally, a **base** is any unit whatsoever to which affixes of any kind can be added. The affixes attached to a base may be **inflectional affixes** selected for syntactic reasons or **derivational affixes** which alter the meaning or grammatical category of the base (see sections (3.2) and (10.2)). An unadorned root like *boy* can be a base since it can have attached to it inflectional affixes like *-s* to form the plural *boys* or derivational affixes like *-ish* to turn the noun *boy* into the adjective *boyish*. In other words, all **roots** are **bases**. Bases are called **stems** only in the context of inflectional morphology.

Identify the inflectional affixes, derivational affixes, roots, bases, and stems in the following:

[3.9]   faiths   frogmarched
            faithfully   bookshops
            unfaithful   window-cleaners
            faithfulness   hardships

I hope your solution is like this:

[3.10]

Inflectional Affixes	Derivational Affixes	Roots	Stems	Bases
-ed	un-	faith	faith	faith
-s	-ful	frog	frogmarch	faithful
	-ly	march	bookshop	frogmarch
	-er	clean	windowcleaner	bookshop
	-ness	hard	hardship	window-clean
	-ship	window		window-cleaner
				hardship

It is clear from [3.10] that it is possible to form a complex word by adding affixes to a form containing more than one root. For instance, the independent words *frog* and *march* can be joined together to form the base (a stem, to be precise) *frog-march* to which the suffix *-ed* may be added to yield *[[frog]-[march]-ed]*. Similarly, *window* and *clean* can be joined to form the base *[[window]-[clean]]* to which the derivational suffix *-er* can be added to produce *[[[window]-[clean]]er]*. And *[[[Window]-[cleaner]]]* can serve as a stem to which the inflectional plural ending *-s* is attached to give *[[[[Window]-[cleaner]]]s]*. A word like this which contains more than one root is a called **compound word** (see section (3.4) below and Chapter 12).

3.1.4 Stem Extenders

In section (2.3) of the last chapter we saw that languages sometimes have word-building elements that are devoid of content. Such empty formatives are often referred to, somewhat inappropriately, as empty morphs.

In English, empty formatives are interposed between the root, base or stem and an affix. For instance, while the irregular plural allomorph *-en* is attached directly to the stem *ox* to form *ox-en*, in the formation of *child-r-nen* and *breth-r-en* it can only be added after the stem has been extended by attaching *-r-* to *child-* and *breth-*. Hence, the name **stem extender** for this type of formative.

The use of stem extenders may not be entirely arbitrary. There may be a good historical reason for the use of particular stem extenders before certain affixes. To some extent, current word-formation rules reflect the history of the language.

The history of stem extender *-r-* is instructive. A small number of nouns in Old English formed their plural by adding *-er*. The word 'child' was *cild* in the singular and *cilder* in the plural (a form that has survived in some conservative North of England dialects, and is spelled *childer*). But later, *-en* was added as an additional plural ending. Eventually *-er* lost its value as a marker of plural and it simply became a stem extender:

[3.11]　a.　Singular　　Plural　　b.　New singular　　Plural
　　　　　b.　cild 'child'　cild-er　　　cilder　　　　　cilder-en

3.2 INFLECTIONAL AND DERIVATIONAL MORPHEMES

As we have already hinted, affix morphemes can be divided into two major functional categories, namely **derivational morphemes** and **inflectional morphemes**. This reflects a recognition of two principal word building processes: **inflection** and **derivation**. While all morphologists accept this distinction in some form, it is nevertheless one of the most contentious issues in morphological theory. I will briefly introduce you here to the essentials of this distinction but postpone detailed discussion until Chapter 10.

Inflectional and derivational morphemes form words in different ways. Derivational morphemes form new words either:

(i)　by changing the meaning of the base to which they are attached, e.g. *kind* vs *un-kind* (both are adjectives but with opposite meanings); *obey* vs *dis-obey* (both are verbs but with opposite meanings). Or

(ii)　by changing the word-class that a base belongs to, e.g. the addition of *-ly* to the adjectives *kind* and *simple* produces the adverbs *kind-ly* and *simp-ly*. As a rule, it is possible to derive an adverb by adding the suffix *-ly* to an adjectival base.

Study the following data and answer the questions below:

[3.12] I <u>ducked</u>   He was <u>sheepish</u>
    two <u>ducks</u>   three <u>ducklings</u>
    He is <u>humourless</u> You are <u>ducking</u> the issue
    He <u>ducks</u>

a. Identify the suffixes in the underlined words. To what word-class do the words to which the suffixes are added belong, and what word-class results?
b. For each suffix determine whether it is inflectional or derivational. Briefly justify your decision.

I hope your answer is very close to the following:

	Suffix	Input	Output	Remarks
[3.13]				
a.	-ed	V	V	inflectional:- it marks past tense in *duck-ed*
	-s	N	N	inflectional:- it marks plural number (in *(two) duck-s* and *duckling-s*)
	-s	V	V	inflectional:- a portmanteau morph marking 3rd person, present tense and singular in *(he) ducks*
	-ing	V	V	inflectional:- it marks progressive aspect (i.e. incomplete action in *ducking*)
b.	-ling	N	N	derivational:- it changes meaning to 'small duck'
	-ish	N	Adj	derivational:- it changes word-class and meaning to 'like a sheep'
	-less	N	Adj	derivational:- it turns a noun into an adjective and adds the meaning 'lacking (e.g. humour)'

Sometimes the presence of a derivational affix causes a major grammatical change, involving moving the base from one word-class into another as in the case of *-less* which turns a noun into an adjective. In other cases, the change caused by a derivational suffix may be minor. It may merely shift a base to a different sub-class within the same broader word-class. That is what happens when the suffix *-ling* is attached to *duck* above.

 Further examples are given below. In [3.14a] the diminutive suffix *-let* is attached to nouns to form diminutive nouns (meaning a small something).

In [3.14b] the derivational suffix -*ship* is used to change a concrete noun base into an abstract noun (meaning 'state, condition'):

[3.14] **a.** pig   ~   pig-let   **b.** friend ~ friend-ship
        book   ~   book-let   leader ~ leader-ship

The tables in [3.15] and [3.16] list some common derivational prefixes and suffixes, the classes of the bases to which they can be attached and the words that are thereby formed. It will be obvious that in order to determine which morpheme a particular affix morph belongs to, it is often essential to know the base to which it attaches because the same phonological form may represent different morphemes depending on the base with which it co-occurs.

Note: These abbreviations are used in the tables below: N for noun, N (abs) for abstract noun, N (conc) for concrete noun. V for verb, Adj for adjective, and Adv for adverb.

[3.15]

Prefix	Word-class of input base	Meaning	Word-class of output word	Example
in-	Adj	'not'	Adj	in-accurate
un-	Adj	'not'	Adj	un-kind
un-	V	'reversive'	V	un-tie
dis-	V	'reversive'	V	dis-continue
dis-	N (abs)	'not'	N (abs)	dis-order
dis-	Adj	'not'	Adj	dis-honest
dis-	V	'not'	V	dis-approve
re-	V	'again'	V	re-write
ex-	N	'former'	N	ex-mayor
en-	N	'put in'	V	en-cage

[3.16]

Suffix	Word-class of input base	Meaning	Word-class of output word	Example
-hood	N	'status'	N (abs)	child-hood
-ship	N	'state or condition'	N (abs)	king-ship
-ness	Adj	'quality, state or condition'	N (abs)	kind-ness
-ity	Adj	'state or condition' etc.	N (abs)	sincer-ity
-ment	V	'result or product of doing the action indicated by the verb'	N	govern-ment

-less	N	'without'	Adj	power-less
-ful	N	'having'	Adj	power-ful
-ic	N	'pertaining to'	Adj	democrat-ic
-al	N	'pertaining to, of the kind'	Adj	medicin-al
-al	V	'pertaining to or act of'	N (abs)	refus-al
-er	V	'agent who does whatever the verb indicates'	N	read-er
-ly	Adj	'manner'	Adv	kind-ly

To sum up the discussion so far, we have observed that derivational affixes are used to create new lexemes by either: (i) modifying significantly the meaning of the base to which they are attached, without necessarily changing its grammatical category (see *kind* and *unkind* above); or (ii) they bring about a shift in the grammatical class of a base as well as a possible change in meaning (as in the case of *hard* (Adj) and *hardship* (N (abs)); or (iii) they may cause a shift in the grammatical sub-class of a word without moving it into a new word-class (as in the case of *friend* (N (conc)) and *friend-ship* (N (abs)).

With that in mind, study the data below which contain the derivational prefix *en-*.

(i)   State the word-classes (e.g. noun, adjective, verb, etc.) of the bases to which *en-* is prefixed.

(ii) What is the word-class of the new word resulting from the prefixation of *en-* in each case?

(iii) What is the meaning (or meanings) of *en-* in these words?

Consult a good dictionary, if you are not sure. Is there reason to regard *en-* as a homophonous morph?

[3.17]	Base	New word	Base	New word
	cage	en-cage	noble	en-noble
	large	en-large	rich	en-rich
	robe	en-robe	rage	en-rage
	danger	en-danger	able	en-able

You will have established that the new word resulting from the prefixation of *en-* in [3.17] is a verb. But there is a difference in the input bases.

Sometimes *en-* is attached to adjectives as seen in [3.18a], and sometimes to nouns, as in [3.18b]:

[3.18]	a.	Adj base	New word Verb	b.	Noun base	New word verb
		able	en-able		robe	en-robe
		large	en-large		danger	en-danger
		noble	en-noble		rage	en-rage
		rich	en-rich		cage	en-cage

Interestingly, this formal difference correlates with a semantic distinction. So, we conclude that there are two different prefixes here which happen be homophonus. The *en-* in [3.18a] has a causative meaning (similar to 'make'). To *enable* is to 'make able', to *enlarge* is to 'make large', etc., while in [3.18b] *en-* can be paraphrased as 'put in or into'. To *encage* is to 'put in a cage' and to *endanger* is to 'put in danger' etc.

Let us now turn to inflectional morphemes. Unlike derivational morphemes, inflectional morphemes do not change **referential** or **cognitive** meaning. We have already seen that a derivational affix like *un-* can change kind into *un-kind*. In this case, the derived word has a meaning which is opposite to that of the input. The addition of an inflectional affix will not do such a thing. Furthermore, while a derivational affix may move a base into a new word-class (e.g., *kind* (adjective) but *kind-ly* (adverb), an inflectional morpheme does not alter the word-class of the base to which it is attached. Inflectional morphemes are only able to modify the form of a word so that it can fit into a particular syntactic slot. Thus, *book* and *books* are both nouns referring to the same kind of entity. The *-s* ending merely carries information about the number of those entities. The grammar dictates that a form marked as plural (normally by suffixing *-s*) must be used when more than one entity is referred to. We must say *ten books*; **ten book* is ruled out, although the numeral ten makes it clear that more than one item is being referred to.

See the table in [3.19] for a sample of frequently used inflectional suffixes. English has no inflectional prefixes but some other languages do (see Luganda in [2.7]).

[3.19]	Suffix	Stem	Function	Example
	-s	N	plural	book-s
	-s	V	3rd person, singular, present tense	sleep-s
	-ed	V	past tense	walk-ed
	-ing	V	progressive (incomplete action)	walk-ing
	-er	Adj	comparative degree	tall-er
	-est	Adj	superlative degree	tall-est

Below I have presented an additional inflectional suffix. What is this suffix called and what is its function in each example?

[3.20] **a.** Janet's book
 b. The Winter's Tale
 c. in two days' time

The *-s* suffix in [3.20] is usually called the **genitive suffix**. Quirk and Greenbaum (1973) list these, among others, as the uses of the genitive suffix in English:
   (i)   marking the noun referring to the possessor of something (as in *Janet's book*),
   (ii) marking a noun that describes something (as in *The Winter's Tale*),
   (iii) marking a noun used as a measure (*in two days' time*).
We will return this and refine our analysis of genitive *-s* in (section 10.5).

3.3 MULTIPLE AFFIXATION

What we are now going to explore are some of the ways in which complex words are formed by creating bases which contain several derivational morphemes. Let us take the latinate root *-dict-* meaning 'speak, say' which is found in *diction, dictate, dictatorial, contradict, benediction*, etc. Starting with *-dict-*, we can form complex words such as *contradictory* and *contradictoriness* by attaching several affixes to the root, i.e. we can have **multiple affixation**. This process can take place in a number of rounds, with the output created by one round of affixation serving as the input to a later round:

[3.21]	Root -dict$_V$		Output
	base: -dict$_V$	(round one: prefixation: add *contra-*$_{Preposition}$)	→ *contra*dict$_V$
	base: contradict$_V$	(round two: first suffixation: add *-ory*$_{Adj}$)	→ contradict-*ory*$_{Adj}$
	base: contradictory$_{Adj}$	(round three: second suffixation: add *-ness*$_N$)	→ contradictori*ness*$_N$

Words may have multiple affixes either with different suffixes appearing in a sequence as in [3.21] or with the same prefix recurring as below in [3.22].

[3.22] **a.** the latest re-re-re-make of *Beau Geste*.
   b. the great-great-great-great grandson of the last Tsar of Russia.

What [3.22] shows is that, with a limited number of morphemes, morphological prefixation rules can apply recursively in English (see section 1.3). However, **performance** difficulties in working out what exactly *great-great-great-great grandson* or *re-re-re-make* means do severely restrict the chances of such words being used. But the point is that the grammar cannot exclude them as ill-formed. Recursive rules are one of the devices that make morphology open-ended. They make possible the creation of new words with the same morphemes being used again and again (cf. section 4.1).

Re-attaching the same morpheme again and again is permitted, but unusual. What is common is multiple affixation of different affixes. It is such affixation that we will concentrate on. We have already seen an example of it in *contradict-ori-ness* in [3.21].

Take the free morpheme *nation* and add to it as many prefixes and suffixes as you can. Attempt to go through at least four rounds of affixation.

I hope you have come up with something like this:

[3.23] nation
nation-al
national-ise   de-nationalise
denationalis-at-ion   (but there is no *denationalisate)
anti-denationalisation
pre-antidenationalisation

Observe that where several prefixes or suffixes occur in a word, their place in the sequence is normally rigidly fixed. Whereas there is usually some scope for rearranging words in different orders in sentences, as you can see:

[3.24] **a.** You can play badminton. **b.** What I need is a nice cup of tea.
   Can you play badminton?   A nice cup of tea is what I need.

there is virtually no possibility of re-arranging morphemes within a word.

So, for example, the morphemes in *de-nation-al-ise* must appear in that order. Rearranging the affixes produces ill-formed strings like **ise-nation-de-al-* or **al-ise-nation-de*. The main problem and interest, as we will see in section (6.2.1), is determining the order of derivational affixes where several of them occur in a word.

3.4 COMPOUNDING

As we briefly saw in (3.1.3), a **compound word** contains at least two bases which are both words, or at any rate, root morphemes.

Analyse the following compounds into their constituent elements: *teapot*, *week-end*, *hairdresser*, *kind-hearted*.

I expect you to have worked out an answer close to the following:

[3.25] **a.** $[tea]_N$ $[pot]_N$ → $[teapot]_N$

  $[week]_N$ $[end]_N$ → $[week-end]_N$

  b. $[hair]_N$ $[[dress]_V \text{ -er }]_N$ → $[hairdresser]_N$

  $[kind]_A$ $[[heart]_N \text{ -ed}]_A$ → $[open-ended)_A$

Compounding is a very important way of adding to the word stock of English as we will see. Sometimes it is bare roots that are combined in compounds as in [3.25a], and sometimes an input base contains an affixed form as in [3.25b]. We will discuss compounds again in a preliminary way in the next chapter and return to them in more detail in Chapter 12.

3.5 CONVERSION

We have seen that complex words may be formed either by compounding or by affixation, or by a combination of the two. We are going to see now that there is an alternative word-formation strategy which is commonly used in English. Words may be formed without modifying the form of the input word that serves as the base. Thus *head* can be a noun or verb. This is called **conversion**.

How do you know whether *head* is a noun or verb in the following?

[3.26] **a.** The head of the village school has arrived.
The heads of the village schools have arrived.
b. She will head the village school.
She headed that school.

It is partly the morphological structure, and partly the syntactic position that the word occupies that tells you whether it is a noun or a verb. From a syntactic point of view, we know that in [3.26a] *the head* is a noun phrase. The key word in a noun phrase must be a noun. As *head* occurs following *the* and is the key word in this construction, *head* must be a noun. But from a morphological point of view, we cannot tell whether *head*, is a noun or verb when it occurs with no affixes. However, in the case of *heads*, the presence of the *-s* morph which here realises the plural in nouns gives us a useful clue.

By contrast, in [3.26b] *head* must be a verb. It comes after the auxiliary verb *will* in a slot that is typically filled by verbs. In the second example, *head* has attached to it the *-ed* morph representing the past tense morpheme which is only found in verbs. Furthermore, from a syntactic point of view, we know that *she* is the subject and *that school* is the object. The sentence must also have a verb. The verb occurs between the subject and the object. (The order of sentence constituents in English is Subject Verb Object.) So, *headed* must be the verb, since it occurs between the subject and the object.

Conversion is also referred to as **zero derivation** in the literature (cf. Marchand, 1969; Adams, 1973) and is subsumed under affixation, by analogy to zero affixation in inflectional morphology (cf. section (2.3)). It is claimed that zero morphs (i.e. ones lacking any overt marking) are used as suffixes in derivational morphology as well. For instance, the verb *head* is derived by suffixing a zero morph to the noun *head*. This is done by analogy to the derivation of a verb like *victim-ise*$_V$ (from the noun *victim*$_N$ where the overt verb-forming suffix *-ise* is used).

The use of zero in derivational morphology is controversial. Since neither the original noun *head*, nor the derived verb *head*, has an overt suffix, if we assume that zero suffixation takes place here, we end up with a somewhat absurd situation where a zero suffix on the noun is said to contrast with a zero suffix on the derived verb. It is more prudent to recognise conversion as a distinct word-forming mechanism and to restrict zero morphs to inflectional morphology where it is supported by the evidence. See section (6.2.3) for further discussion.

3.6 MORPHOLOGICAL TYPOLOGY

We suggested in the opening chapter that although languages vary enormously in their structure they nonetheless show surprising similarities. The study of the significant shared structural properties which languages have in common is the domain of **language universals**. Many of the universals are abstract principles of **Universal Grammar** which determine the properties of rules that grammars of individual languages may have (e.g. the Strict Cycle Condition discussed in section (6.2.4)).

An integral part of the study of universals in language is the study of differences between languages. This might look odd to begin with. But it turns out that differences between the structural patterns found in different languages appear to occur within a fairly restricted range. There are **parameters** within which most differences between languages occur. Just as tram lines determine where trams can go in a city (while leaving them plenty of options), pre-set parameters determine the structural patterns from which different languages may select.

Structural patterns are not randomly distributed. There are a number of strongly preferred patterns which recur in language after language, while other patterns are rare, or non-existent (Greenberg, 1963; Comrie, 1981; and especially Chomsky, 1986). The study of the range of patterns within which languages may vary is the domain of **language typology**.

Our concern in this book is with both the similarities and differences between languages in the ways in which they form words. On the basis of typical patterns of word-formation linguists recognise five broad morphological types:

(i)   **analytic** (also called isolating) languages;
(ii)   **agglutinating** (also called agglutinative) languages;
(iii) **inflecting** (also called synthetic or fusional) languages;
(iv) **incorporating** (also called polysynthetic) languages)
(v)   **infixing** languages.

We will now consider the morphological types in turn, starting with examples of analytic morphology from Chinese:

[3.27] **a.** Tā bǎ   shū   mǎi le.
          he *OM* book buy *Asp*.
          'He bought the book.'
      b. Tā chǎo le   yíge cài   hěn xiāng.
          he cook *Asp*. a   dish very delicious.
          'He cooked a dish that was very delicious.'

<u>Note</u>: *Asp.* is short for 'perfective aspect'. It indicates that an action is

completed. *OM* is short for 'object marker', i.e. the morpheme that indicates the object of the verb.

(data from Li and Thompson, 1978)

As you can see from the morpheme by morpheme translation, in Chinese bound morphemes are infrequent. Usually the words are bare, unaffixed root morphemes.

Chinese is an example of an **analytic language**, i.e. a language where each morpheme tends to occur as a word in isolation. Words virtually never have inflectional affixes. Thus, the object marker *bǎ* is an independent word. By contrast, in other language types normally object markers are inflectional affixes that are part of a noun or pronoun. In English the subject pronoun *he* contrasts with the object pronoun *him* in *He saw Lauren* vs *Lauren saw him*. The change from he to him in the pronoun marks the change in grammatical function. Similarly, in English markers of aspect and tense are usually inflectional affixes of the verb such as *-ed*, as in *cook-ed* (vs *cook*). By contrast, in Chinese, in [3.27b] the aspectual morpheme is realised, not by an affix, but by the independent word *le*. (Note in passing that there are some Chinese words containing more than one morpheme. Usually they are compounds like *jue-she* (literally 'chew-tongue'), 'gossip' (noun) and *zhen-tou* ('pillow', literally 'rest-head').

Let us now turn to another language, Turkish.
a. Divide the following words into morphs and assign each morph to a morpheme.
b. How do the morphs match up with morphemes?

[3.28] el 'the hand' elimde 'in my hand'
elim 'my hand' ellerim 'my hands'
eller 'the hands' ellerimde 'in my hands'

Your answer to the first question should be: *el* 'hand', *-im* 'my' (genitive), *-ler* 'plural' and *-de* 'in'.

Turkish is a classic example of an **agglutinating** language. In this kind of language there tends to be a more or less one-to-one matching of morphemes with morphs:

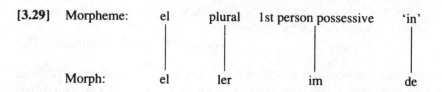

[3.29]   Morpheme: el   plural   1st person possessive   'in'

     Morph: el   ler   im   de

The Luganda words in the last chapter have already introduced you to another example of such a language. The word *tulilaba* 'we will see', for instance, is analysable as *tu-* 'we', *-li-* 'future', and *-laba* 'see'.

Now explain why any attempt to treat Latin as we have treated Turkish would fail. Show why it is impossible to isolate separate morphs representing the morphemes in the following Latin words:

[3.30]   Singular   Plural

    Nominative: mēnsa   mēnsæ
    Genitive: mēnsæ   mēnsārum
    Ablative   mēnsa   mēnsīs

Note: Nominative is used if the noun is the subject; genitive would mean 'of the table' and ablative 'from the table'.

It would be futile to try matching morphs with morphemes. How could one say, for instance, which part of the suffix *-æ* represents plural and which part represents nominative? How could one tell which part of the suffix *-is* represents plural and which part represents ablative ('from') in the word *mensis*? And so on. Here we see an unsegmentable morph representing simultaneously the plural and nominative morphemes, the plural and ablative morphemes, etc. See (section 3.7) for further discussion.

Latin is in a good example of an **inflecting language**. Words usually consist of several morphemes. But there is seldom a one-to-one matching of morphemes with morphs. Instead, a single morph is likely to represent several morphemes simultaneously. Other well-known representatives of this type of language include Sanskrit and Greek.

Now look at the following analysis of the Greenlandic Eskimo words *illuminiippuq* and *tuttusivuq* from Fortescue (1984):

[3.31]   **a.**   illu-   mi-   niip-   puq
                 house   his   be-in   3rd person-singular-indicative

'he is in his (own) house'
b. tuttu-   si-   vuq
caribou   come-across   3rd person-singular-indicative
'he saw (a) caribou'

Greenlandic Eskimo is a typical **incorporating language**. You can express
in Eskimo in one word (e.g. *tuttusivuq*), that may include a verb and its
object, what is said using a whole sentence containing several words in
English (and even more words in Chinese). Eskimo is a language with long
words (e.g. *illuminiippuq*) that tend to have very extensive agglutination
and inflection. Many Native American languages and aboriginal languages
of Australia are incorporating.

As we will see in section (11.7), recent studies have highlighted this
language type because it raises interesting questions about the relationship
between morphology and syntax. In incorporating languages the distinction between morphology, the study of word structure, and syntax, the
study of sentence structure, is blurred. Some processes which elsewhere
happen at the level of the sentence take place within the word.

We glanced at **infixation** in (3.1.2). Traditional typology neglected this
morphological processes typical of semitic languages like Arabic and
Hebrew. Much of semitic inflection involves infixing vowels in a root that
consists entirely of consonants.

Thus, in Egyptian Arabic the three-consonant root *ktb* means 'write'. It
provides the skeleton which is fleshed out with a variety of vowels in the
formation of word-forms which belong to the lexeme KTB, such as:

[3.32]   kitab   'book'
katab   'he wrote'
katib   'writer'

The description of this kind of morphological pattern has been the subject of
very fruitful investigations in the last few years as we will see in Chapter 9.

Use the opening sentence of *Moby Dick* to formulate a tentative hypothesis about English. Is it an isolating, inflecting, agglutinating or incorporating language?

[3.33]   Call me Ishmael. Some years ago – never mind how long precisely
– having little or no money in my purse, and nothing in particular
to interest me on shore, I thought I would sail about a little and see
the watery part of the world.

(Herman Melville, *Moby Dick*)

English is predominantly isolating. The vast majority of the 45 words in this sentence, which is typical of modern English, are simple. They contain just one morpheme. But English is not a thoroughbred isolating language. Five of the words, namely *year-s*, *precise-ly*, *hav-ing*, *no-thing*, and *water-y* contain two morphs representing two distinct morphemes. These words exemplify a degree of agglutination. In addition, there are also several words containing one morph which represents several morphemes concurrently, e.g. *me* (1st person, singular, accusative pronoun); *my* (1st person, singular, possessive pronoun), *I* (1st person, singular, nominative pronoun), *thought* (THINK, past) and *would* (WILL, past). In words like this, trying to designate a portion of the word as a morph representing one of the morphemes would be futile. Such words show that, to a certain extent, English is a synthetic language. Even infixation (which is not exemplified by [3.33]) is found occasionally in English, as in *incumbent*, *succumb*, and *decumbent*, where *-m-* is infixed in the root *-cub-* (see p. 44 above).

Greenberg (1954) made a proposal regarding *typology* that is widely accepted. He suggested that the number of morphemes in a representative sample of sentences should be divided by the number of words to work out the ratio of morphemes to words in a language. The result should form the basis of our typological classification.

(i)   If a language has between 1.00 and 1.99 morphemes per word it is **analytic** (isolating). With 1.68 morphemes per word in Greenberg's sample of sentences, English falls in the essentially isolating category. (It is similar to Chinese – see [3.27].)

(ii)   A language averaging between 2.00 and 2.99 morphemes per word, is **synthetic** (inflecting) if the realisation of the different morphemes tends to be simultaneous (as in Latin – see [3.30]).

(iii) A language averaging between 2.00 and 2.99 morphemes per word is **agglutinative** if each morpheme tends to be realised by a separate morph (as in Turkish in [3.28]).

(iv)   A language is incorporating if it averages 3.00 morphemes per word or more (e.g. Eskimo – see [3.31]).

It is important to realise that probably no language has an unalloyed analytic, agglutinating, inflecting or incorporating morphological system. All that the classification attempts to do is reflect the dominant tendencies found in a particular language.

3.7 WP AND THE CENTRALITY OF THE WORD

A central question which morphological theory needs to address is 'what is the key unit which morphological theory deals with?' In structuralist morphology the answer was unequivocally, 'the morpheme'. However, in recent years, various scholars have proposed that it is not the morpheme but rather the **word** that should be regarded as the central unit of morphological analysis. This debate has important repercussions for how we formulate our theory of morphology and the lexicon.

Word-and-paradigm morphology (WP) is one theory that puts the word at the centre. It was first mentioned in modern linguistics by Hockett (1954) who identified it as the approach assumed in traditional grammars based on Latin. This model was articulated in Robins (1959) and extensively revised by Matthews (1972). It has since been elaborated by S. R. Anderson (1977, 1982, 1984, 1988a). Unfortunately, in spite of its inherent merits, this approach has not been adopted by many linguists.

But although there are not many WP morphologists, the critique of morpheme-based approaches to morphology which this theory embodies has contributed to a healthy re-examination of the nature of morphological representations in recent years. WP is critical of the somewhat naïve view of the relationship between morphological representations and morphs found in some structuralist models of morphology. Matthews (1972) has shown that a theory of the morpheme that relies on the assumption that morphemes are always typified by a one-to-one pairing of morphemes with morphs is misguided. True, in straightforward cases of agglutination like the Turkish example in [3.29], a bit of the phonological representation may directly correspond to a bit of the morphological representation. But the phenomenon of **portmanteau morphs** that is found frequently in inflecting languages illustrates the difficulties that arise if morphemes are assumed to be always matched in a straightforward way with morphs.

Matthews (1972: 132) suggests that the Latin word /re:ksisti:/ 'you (sg.) ruled (or I have ruled)' could be analysed as in [3.34]:

[3.34]   Grammatical representation: REG– + Perfective + 2nd + Singular

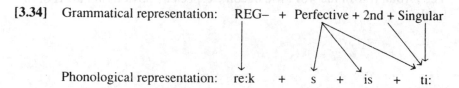

           Phonological representation: re:k   +   s   +   is   +   ti:

The morphemes second person and singular are both realised by the portmanteau morph *ti:* while the perfective is multiply signalled, partly in the selection of *rek-* (see below) and by the suffixes *-s-*, *-is-* and *-ti:*. The justification for this analysis will be clear if you compare parts of the perfect and imperfect forms of the verb *regere* 'rule':

[3.35]   Imperfect   Perfect

rege:bam	'I was ruling'	re:ksi:	'I have ruled'
rege:ba:s	'you (sing.) were ruling'	re:ksisti:	'you (sing.) have ruled'
rege:bat	'he was ruling'	re:ksit	'he has ruled'

If you examine the second person singular forms, for example, you observe that the root REG- has the phonological realisation /reg-/ in the imperfect but /rek-/ in the perfective. So, the distinction between perfective and imperfective is in part realised in the root itself. (See the diagram in [3.34].) The ending /-ti:/ marks second person singular if the grammatical representation also includes the perfective. If the verb is in the imperfective, the second person singular is marked instead by /a:s/. The crucial point is that these various morphs do not have a clear identifiable meaning on their own. They can only be interpreted in the wider context of the word as a whole of which they form a part. To know how second person singular is going to be realised we need to take into account the rest of the grammatical representation manifested in a particular word. A sensible solution, and one that WP morphology advocates, is one that recognises a combination of morphs as simultaneously signalling a particular meaning if they co-occur in a word that has a certain combination of grammatical properties.

I will not introduce you to the formalism of WP because that formalism is not important for the generative theory of morphology that I am outlining. If you wish to see WP rules, turn to Matthews (1972) and S. R. Anderson (1982). My aim has been to show that while morphemes are important theoretical entities, the word is the key unit of morphological representation. While still recognising the relevance of morphemes, present-day morphological theory in generative grammar is word-based. The pivotal role of the word will become especially obvious in Part II of the book.

EXERCISES

1. Examine carefully the following sentence:

Mr Nickleby shook his head, and motioning them all out of the room, embraced his wife and children, and having pressed them by turns to his languidly beating heart, sunk exhausted on his pillow.

(Charles Dickens, *Nicholas Nickleby*)

   (a)   List five free and three bound morphemes that occur in this sentence.

   (b)   List three functional morphemes in the sentence.

2. (a)   Identify the morphemes in the Swahili words below, distinguishing between roots and affixes.

   (b)   State the meaning of each morpheme.

   (c)   State whether the affix morphemes are: (i) prefixes or suffixes, and (ii) inflectional or derivational.

   (d)   On the basis of these data, would you classify Swahili as an isolating, agglutinating, synthetic or incorporating language?

nilipata	'I got'	niliwapiga	'I hit them'
walipata	'they got'	walitupiga	'they hit us'
nilipiga	'I hit'	walikipiga	'they hit it'
nilikipata	'I got it'	utatupiga	'you will hit us'
ulikipata	'you got it'	ulipata	'you got'
nitakipata	'I will get it'	watakupiga	'they will hit you'
ulipiga	'you hit'	ulitupiga	'you hit us'
watakipiga	'they will hit it'	nitakupata	'I will get you'

Note: Here the form 'hit' as in 'you hit' represents the past tense form of the verb *hit* and 'you' stands for 'second person singular'.

3. (a)   Make a morphological analysis of the following Latin data:

Present tense		Pluperfect	
regō	'I rule'	rēkseram	'I had ruled'
regis	'you (sing.) rule'	rēkserās	'you (sing.) had ruled'
regit	's/he rules'	rēkserat	's/he had ruled'
regimus	'we rule'	rēkserāmus	'we had ruled'
regitis	'you (pl.) rule'	rēkserātis	'you (pl.) had ruled'
regunt	'they rule'	rēkserant	'they had ruled'

Future simple	
regam	'I shall rule'
regēs	'you (sing.) shall rule'
reget	's/he will rule'
regēmus	'we will rule'
regētis	'you (pl.) will rule'
regent	'they will rule'

   (b)   Referring to your analysis, highlight the pitfalls of a theory of word-structure that assumes that there is always a one-to-one matching of morphs with morphemes.

4. (a) What is the morphological function of tone in the Lulubo words
        below?
   (b) State exactly how tone is used to perform this function.
   (c) Explain whether or not Lulubo fits in the morphological typology
        given in this chapter.

àzɔ́	'long'	àzɔ̀	'to become long'
ìnḍá	'good'	ìnḍà	'to become good'
ōsú	'good'	òsù	'to become good'
álí	'deep'	àlì	'to become deep'
àkēlí	'red'	àkèlì	'to become red'
áfɔ́rɔ̀	'yellow'	àfɔrɔ̀	'to become yellow'

(data from Andersen, 1987)

4 Productivity in Word-Formation

4.1 THE OPEN-ENDEDNESS OF THE LEXICON

One of the goals of morphological theorising is to account for the ways in which speakers both understand and form not only 'real' words that occur in their language, but also potential words which are not instantiated in use in utterances. While it is true that a large percentage of 'real' words listed in dictionaries (such as *pear* and *pair*) are memorised, it is equally true, and of great theoretical interest, that countless words used in conversation (and to a lesser extent in writing) are new, made up on the spur of the moment. So, morphology has to throw light not only on the structure of established words like *pair*, but also on that of freshly coined **neologisms** like *snail-mail* (meaning the postal service, as opposed to modern electronic mail).

The consensus appears to be that the words of a language are **listable** in a way in which sentences are not (see 12.3.2). The meanings of many words (e.g. *pear* and *pair*) must be listed in the **lexicon** because there is nothing about their sounds or morphological structure that would enable one to work out their meaning. In this respect morphology differs from syntax. Syntax cannot be restricted to cataloguing only those sentences that occur in some **corpus** (i.e. a body of texts), since language is vast and no list of sentences, no matter how long, could exhaust the set of possible well-formed sentences. Typically, speakers do not merely recycle sentences memorised from previous conversations. Rather, they tend to construct fresh sentences to suit the occasion.

However, by and large, people do not routinely make up new words each time they speak. Nonetheless, the lexicon cannot be seen as a *static* list. No dictionary, however large, (not even the complete Oxford English Dictionary, including all its supplements) can list every word in the English language. Why is this so?

Until recently, word-formation rules have tended to be seen as being largely passive in the sense that they are basically used to analyse existing words rather than to create new ones. It is significant in this connection that, whereas reasonably comprehensive dictionaries and wordlists for dozens of languages exist, there are no equivalent, all-encompassing sentence lists for any language. Lists of sentences such as those found in phrase-books for foreigners make no pretence of being exhaustive.

The verdict on whether or not morphology and the lexicon deal with what is effectively a closed list of words will hinge, to some extent, on our attitude to nonce-words (like *uncomplicatedness*), created by an individual, which do not catch on in the speech community. The more such words we recognise as part of the language the bigger and more open-ended will our lexicon be. But, perhaps, it will depend to an even greater extent on which forms we recognise as compound words and, hence, part of the province of morphology and the lexicon, and which forms we treat as phrases and, hence, the domain of syntax. Since both compounds and phrases are made up of words, determining which combinations of words are compounds and which combinations are syntactic phrases is not always straightforward.

For instance, consider *Lakeside Grammar School Former Pupils*. We have here the compound nouns *lakeside* and *grammar school*. Is the larger unit *Lakeside Grammar School* also a compound noun? It is even more unclear whether the whole string *Lakeside Grammar School Former Pupils* is also a compound noun. At first blush, we might decide that it is not a compound noun but a syntactic phrase. But we might change our minds if we discovered that *Lakeside Grammar School Former Pupils* is the name of a rugby team that plays in the local league. If we recognise strings like *Lakeside Grammar School Former Pupils* as compounds, we will have a very open-ended lexicon. (See Chapter 12 below for further discussion.)

Recall also that in the last chapter we mentioned another aspect of productivity in word-formation. Although no word is infinite in length, in principle there is no upper limit to the length of forms that may function as bases from which new words are formed. In many languages we cannot state categorically what the longest possible word is: words can be made longer, if the need arises, by the repetition of an affix or the addition of yet another affix as we saw in [3.22].

Finally, it is also possible to add to the lexicon of a language indefinitely by pillaging the vocabulary of other languages. This is called **borrowing**. English has borrowed **loanwords** from a great number of other languages with which it has been in contact. There are Latin loanwords like *port* (from *portus* 'harbour'); French loanwords like *omelette*; Italian loanwords like *fresco*, and so on. In this book we will not deal with the expansion of the vocabulary by borrowing. We will only be concerned with productive word-creation using the internal resources of a language (see section (4.2.1.2)).

The upshot of this discussion is that morphology is productive. There is no limit to the number of potential words in a language. Therefore a lexicon that merely attempted to list the words of a language in some corpus would be woefully inadequate.

4.1.1 What is Productivity?

But what exactly is productivity? We will provisionally view productivity simply in terms of **generality**. The more general a word-formation process is, the more productive it will be assumed to be. There are two key points requiring elucidation:

(i)   Productivity is a matter of degree. It is not a dichotomy, with some word-formation processes being productive and others being unproductive. Probably no process is so general that it affects, without exception, all the bases to which it could potentially apply. The reality is that some processes are relatively more general than others.

(ii) Productivity is subject to the dimension of time. A process which is very general during one historical period may become less so at a subsequent period. Conversely, a new process entering a language may initially affect a tiny fraction of eligible inputs before eventually applying more widely.

Study the following data:

[4.1]   chartist   morbid   worker
        communist   tepid   painter
        racist   timid   swimmer
        pianist   splendid   dancer
        anarchist   horrid   jogger

a.   Divide the above words into their constituent morphs.
b.   List all the suffixes.
c.   State the meaning of the morphemes represented in the data.
d.   Find five more words which are formed using each of the suffixes that you have identified.
e.   State the word-class of the base to which each suffix is added.
f.   To what class does the resulting new word belong?
g.   Was it equally easy to find more words which contain the different suffixes? If not, comment on any problems tht you encountered.

The words in [4.1] contain the roots listed in [4.2].

[4.2]	Noun	Adj/Verb	Verb
	chart	morb-	work
	commun(e)	tep-	paint
	rac(e)	tim-	swim
	pian(o)	splend-	danc(e)
	anarch(y)	horr-	jog

There are three suffixes present: *-ist*, *-id* and *-er*. All of them are derivational.

The latinate suffix *-ist* may be added to noun bases to form other nouns (typically they are noun bases which can also take the derivational suffix *-ism*, such as *chart-ism*, *commun-ism* etc.). It can also be added to noun bases to form adjectives (e.g. *race*$_N$ → *racist*$_{Adj}$). In view of this we can distinguish at least three morphemes which are realised by *-ist*:

(i)   N → N-*ist*: meaning 'advocate of' (as in *anarchist*, *communist*, etc.);

(ii) N → N-*ist*: meaning 'practitioner of' (as in *pianist*, *violinist*, etc.);

(iii) N → Adj-*ist*: meaning 'advocate of' (as in *racist*, *sexist*, etc.).

The suffix *-id* is also of Latin origin and got into English via French. It is added to bound adjectival bases with the meaning of 'having the quality specified by the verb'. In Latin *-id* was used to derive **attributive adjectives** from verbs, e.g. *timidus* from *timēre* 'to fear' gives us English *timid* (i.e. attributing to an individual the quality of being afraid) and *tepidus* from *tepere* 'to be warm' gives us English *tepid* (i.e. attributing to an individual the quality of being warm).

Finally, the native Germanic suffix *-er* is suffixed to verbs to create **agentive nouns** (with the meaning 'someone who does whatever is designated by the verb').

I expect that you have had some difficulty in finding five more adjectives which contain the derivational suffix *-id*. The morpheme *-id* is at the unproductive end of English morphology. It is frozen. It is no longer used actively (if it ever was) to produce new words. The words containing it could simply be listed in the lexicon.

At the other extreme, **agentive nouns** containing the suffix *-er*, are numerous and can be added to indefinitely. Most verbs can have a noun formed from them in this way. So, no reasonable case could be made for listing all agentive nouns ending in *-er* in the dictionary. Rather, what is needed is a rule stating that, by suffixing *-er*, an agentive noun can be derived from virtually any verb.

General derivational processes that apply more or less across the board (such as the formation of agentive nouns by using the *-er* suffix) and historical relics (like the formation of attributive adjectives using *-id*) excite little theoretical interest in discussions of productivity. What is fascinating is the grey area in between occupied by morphemes like *-ist*.

By suffixing *-ist* we can form a very large number of nouns with the meaning 'advocate of, follower of, supporter of or practitioner of whatever is designated by the input noun'. But we do not have a *carte blanche* to use it with absolutely any noun. There are unexplainable gaps. For example, a follower of the prophet *Mohammed* is not a **Mohammedist* though a follower of *Buddha* is a *Buddhist* and an adherent to *Calvin's* approach to Christianity is a *Calvinist*. And, note also that a *piano* is played by a *pianist*, a *guitar* by a *guitarist* but the *drums* are played by a *drummer*, not a **drummist*.

The innocent-looking question, 'How productive is this particular process?' turns out to be very troublesome (cf. Bauer, 1988). This is because, as we mentioned at the beginning, the term productivity suffers from an inherent ambiguity. On the one hand, a process is said to be productive if it is very general, i.e. affects a vast number of forms and creates very many words. In this sense, the agentive morpheme *-er* (as in *worker*, *writer*, etc.) is very productive, since an overwhelming majority of verbs can be turned into nouns by this suffix. It certainly is more productive than the semantically related suffix *-ent* found in *president*, *student*, *correspondent*, *proponent* etc. There are thousands of bases to which *-er* can be added as compared to the dozens which take *-ent*. If, on the other hand, we forget about the total number of words created using a given process and instead focus on the proportion of bases that are eligible to undergo a process which actually do undergo it, the results may be somewhat different. The chances of a particular affix appearing may crucially depend on characteristics of the base to which it is to be attached. Thus, while it is true that *-ness* (as in *goodness*) is more common than *-ity* (as in *gravity*, *banality*) in the English language as a whole, in the case of an adjective ending in *-ile* (e.g. *servile*, *docile*, *fertile*, *futile* etc.) *-ity* is the preferred suffix (Aronoff, 1976: 35–45). This is an instance of the phonological properties of the base influencing the likelihood of a morphological process taking place (see (4.2.1.1) below).

Study the following:

[4.3] appendicitis   bronchitis   vaginitis
       tympanitis   hepatitis   meningitis
       cephalitis   pneumonitis   tonsillitis
       arthritis   dermatitis   neuritis
       sclerotitis   pleuritis   bursitis

a. What is the meaning of *-itis*?
b. Is the *-itis* suffix below comparable to *-er*, *-ist* or *-id* in productivity?

The suffix *-itis* is borrowed from Greek where it formed the feminine of adjectives. Already in Greek it was used to form words referring to inflammatory diseases like arthritis. It is used in modern medical English to form names of diseases, especially inflammatory ones.

While it is true that the words formed by suffixing *-itis* are fewer than those formed by suffixing *-er*, nevertheless the suffix *-itis* attaches with an extremely high degree of regularity to most suitable bases. (And it can be generalised beyond the semantic niche of inflammatory diseases. In jocular parlance it is extended to even psychological ailments like *skiveritis* and *Monday-morningitis*.)

Furthermore, as we noted above, in the process of refining our understanding of productivity we must consider the time dimension. Let us assume, to begin with, that a word-formation process is productive if it is in current use. Frozen or atrophied processes like the suffixation of *-id* in *tepid*, *frigid*, etc., may be regarded, for practical purposes, as virtually unproductive in contemporary English. In contrast, the suffixation of the agentive *-er* suffix as in *worker*, which is attached freely and unfussily to most eligible forms, is said to be very productive. In between these two extremes there lies a vast grey zone in which there lurks a milliard of possible variations. This grey zone will be investigated in much of the rest of this chapter and in Chapter 5.

Let us begin by observing that productivity is affected by fashion. For a time, one method of forming words may be in vogue, but subsequently it may become less fashionble, or be abandoned completely. On first entering the language, an affix may affect only a fraction of bases to which it will eventually be attachable. For instance, the form *loadsa-* [ləʊdzə] (colloquial for *loads of*) was used in informal British English in the late 1980s in the word *loadsamoney*, a noun referring to the *nouveau riche* with conspicuously unrefined manners and tastes. In 1988, it began being used as a prefix in a few newly created words such as *loadsasermons*, *loadsaglasnost*, etc., that appeared in London newspapers (Spiegel, 1989: 230–1). Even if this prefix does survive, and even if it manages to spread to many other words, it remains true that by 1989 it had only affected a handful of bases to which it could be potentially attached.

Conversely, an affix that historically was used widely may atrophy or cease being applied to new forms altogether.

Study the forms of the verb *take* in Early Modern English prior to 1800 and determine which inflectional endings are no longer productive:

[4.4]	Singular	Plural
	I take	we take
	thou takest	you take
	he, she taketh	they take

The second and third person singular present tense form of the verb were realised by the regular suffixes *-est* and *-th* respectively. But these suffixes have dropped out of common use. They survive as relics in antiquated religious language and on stage when a pre-1800 play is performed.

4.1.2 Semi-productivity

Some linguists, like Matthews (1974: 52), recognise a special category which they call **semi-productivity** to cover idiosyncratic affixes which inexplicably fail to attach to apparently eligible forms. Furthermore, where such affixes are used, the meaning of the resulting word may be unpredictable.

Study the following data and show that the suffix *-ant* is capricious in these respects:

    (i)   in the selection of bases to which it attaches,
    (ii)   in the meaning of words which result from suffixing it.

[4.5] **a.**

communicant	defendant	applicant
assailant	servant	supplicant
entrant	contestant	participant
claimant	accountant	assistant
dependant	inhabitant	consultant

      b.   *writ(e)ant   *buildant   *shoutant

The suffix *-ant* turns a verbal base into an agentive nominal. (It is similar in meaning to *-er*). But it is very fussy. It accepts the bases in [4.5a] but not those in [4.5b]. The reasons for the particular restrictions on the bases to which *-ant* may be suffixed are, at least in part, historical. This suffix is descended from the Latin present participle ending *-antem/-entem*. Hence, it attaches to latinate bases only. Germanic bases like *write*, *build* and *shout* are ineligible.

But even then, attachment to latinate bases is unpredictable. For no apparent reason many bases of Latin (or French) descent fail to combine with *-ant*:

[4.6] destroy *destroyant (Old French *destruire*; Modern French *détruire*)
        adapt   *adaptant (Old French *adapter*)

Semantically *-ant* has unpredictable effects. The meaning of words created by suffixing *-ant* is inconsistent. For instance, a *defendant* has the narrow interpretation of a person sued in a law court not just any one who defends oneself; an *accountant* is not merely anyone who renders an account or calculation but a professional who makes up business accounts, and so on.

Unlike Matthews, we shall not give theoretical recognition to the concept of semi-productivity in this book because in practice it would be difficult to know which word-formation processes are properly classed as semi-productive. There is not a neat three-way opposition, between productive, semi-productive and unproductive processes. Rather, there are different shades of grey, with some processes being relatively more productive than others.

4.1.3 Productivity and Creativity

The term productivity has sometimes been used to refer to **creativity**, i.e. the capacity of all human languages to use finite means to produce an infinite number of words and utterances (see section (1.3)). In the domain of morphology, creativity manifests itself in two distinct ways: rule-governed creativity and rule-bending creativity.

For the most part words are formed following general rules and principles internalised by speakers in the process of language acquisition which most of this book is devoted to exploring. For instance, if the suffix *-ly* is added to an adjective (e.g. *quick*), an adverb (*quickly*) is produced; if the prefix *post-* is attached to a noun base (as in *post-war*), an adjective with meaning 'after' is formed, and so on.

However, speakers have the ability to extend the stock of words idiomatically by producing words without following meticulously the standard rules of word-formation. This can be seen in the way in which certain compounds are constructed:

[4.7]   **a.**   stool pigeon (police informer)
         b.   redlegs (poor whites in Tobago)
               deadline

No synchronic rules can be devised to account for the meaning of a semantically unpredictable compound like *stool pigeon*. But, in some cases, delving into history might show that some of these compounds originally had a literal meaning which got superseded by later metaphorical extensions. To take one example, during the American Civil War, a *deadline* was the line round the perimeter fence beyond which soldiers were not allowed to go. A soldier who wandered beyond that line risked being shot dead for desertion. (Thankfully, today, going beyond a *deadline* is unlikely to be fatal.) As for *redlegs*, it may be true that poor whites working in the hot sun as labourers on plantations in Tobago did literally have *legs* which were *red*; nevertheless the compound *redlegs* is semantically opaque. It is very unlikely that any one could work out the meaning of *redlegs* from the meaning of the words *red* and *leg*. Comparable examples in present-day English are not difficult to find. Consider words like *walkman* and *tallboy*. The former is not a kind of man but miniature personal stereo equipment and the latter is not a boy but a piece of furniture. (See section (12.5.4).)

Our primary concern in this book will be synchronic rule-governed word-formation. One of the perplexing problems we will deal with is the fact that rarely does a word-formation process apply consistently across the board to all the forms which, on the face of it, qualify for the application of a particular rule.

4.2 CONSTRAINTS ON PRODUCTIVITY

Although there is no limit to the number of words that can be produced in a language, not every conceivable word that in theory could be formed is allowed. In this section we are going to examine the factors that limit productivity.

4.2.1 Blocking

First, I outline in general terms some of the factors which frustrate the application of a word-formation process whose conditions of application appear to be met. The cover term **blocking** is used for these factors.

Blocking may be due to the prior existence of another word with the meaning that the putative word would have (Aronoff, 1976). Usually perfect synonyms are avoided. Thus, it may be because *thief* already exists that suffixing the otherwise very productive agentive suffix *-er* to the verb *steal* in order to form **stealer* is blocked. See section (6.2.4.2) for further discussion.

Interestingly, where there exist two semantically similar morphemes, one of which is more productive than the other, the more productive morpheme is less susceptible to blocking than its less productive counterpart. This can be seen in the behaviour of the suffixes *-ity* and *-ness*. Aronoff (1976) has shown that the suffixation of *-ness* is more productive than the suffixation of *-ity*. He goes on to point out (cf. Aronoff, 1976: 44) that where there is an existing noun derived from an adjective base ending in *-ous*, it is not possible to create a new noun by adding *-ity*. However, the existence of an established noun does not stop the derivation of a fresh noun using the more productive suffix *-ness*:

[4.8]	X + ous (Adjective)	Pre-existing Noun	Noun (-ty)	Noun (-ness)
	acrimonious	acrimony	*acimoniosity	acrimoniousness
	glorious	glory	*gloriosity	gloriousness
	fallacious	fallacy	*fallacity	fallaciousness
	spacious	space	*spaciosity	spaciousness
	furious	fury	*furiosity	furiousness

The concept of blocking can be further refined by highlighting a number of factors that play a role in it. These factors may be phonological, morphological or semantic.

4.2.1.1 Phonological factors

Blocking can be motivated by phonological considerations. Siegel (1971) and Halle (1973) have observed that verbs with an **inchoative meaning**, roughly interpretable as 'to begin to X', can usually be formed from adjectives by suffixing -en to an adjectival base provided it meets the following phonetic conditions:

(i)   the base must be monosyllabic;
(ii) the base must end in an **obstruent** (i.e. stop, fricative or affricate), which may be optionally preceded by a **sonorant** (e.g. a nasal consonant or an approximant like /l/ or /r/). These phonological constraints mean that the derived verbs in [4.9a] are allowed but those in [14.9b] are not.

[4.9]	a.	black-en	/blæk-ən/	whit-en	/waɪt- ən/
		damp-en	/dæmp-ən/	hard-en	/hɑːd-ən
		quiet-en	/kwaɪt-ən/	length-en	/lenθ-ən/
		tough-en	/tʌf-ən/	rough-en	/rʌf-ən/
		soft-en	/sɒf-ən/	fast-en	/fɑːs-n/
	b.	*dry-en	/draɪ-ən/	*dimm-en	/dɪm-ən/
		*green	/griːn-ən	*lax-en	/læks-ən/

Obviously, as Halle (1973: 13) remarks, given the existence of numerous well-formed words like /laɪən/ (*lion*) and /ʌnjən/ *onion*, the phonetic restriction on /-ən/ following sonorants is not general. It is peculiar to inchoative verbs formed from adjectives.

Next, we shall focus on -ly. We have already seen that this derivational suffix is attached in a highly predictable manner to adjectives to form adverbs such as:

[4.10]

Adjective	Adverb	Adjective	Adverb
kind	kindly	elegant	elegantly
fierce	fiercely	serious	seriously

Note, the adverbs in [4.11] are disallowed, or at best awkward, even though they might be listed in dictionaries. Suggest a phonological motivation for this restriction.

[4.11]

Adjective	Adverb	Adjective	Adverb
silly	*sillily	friendly	*friendlily
miserly	*miserlily	sisterly	*sisterlily

What [4.11] shows is that the segmental phonology of the base can determine whether a form can undergo *-ly* suffixation. The *-ly* suffix tends to be avoided where an adjective ends in *-ly* (/-lɪ/). Suffixing *-ly* would result in a dispreferred /-lɪlɪ/ sequence in the derived adverb. But *-ly* is used freely where the adjective does not end in *-ly* (Aronoff, 1976).

Our final example of phonological constraints on word-formation is taken from French where the diminutive suffix *-et* (masculine)/*-ette* (feminine) is used freely to form diminutive nouns like these:

[4.12]

fille	[fij] 'girl'	fillette	[fijet]	'little girl'
camion	[kamjɔ̃] 'truck'	camionette	[kamjɔ̃net]	'light truck'
livre	[livʁ] 'book'	livret	[livʁɛ]	'booklet'

Suggest a phonological explanation for the blocking of the suffixation of *-et*/*-ette* in the following:

[4.13]

contrefort	(masculine) *contrefortet	'little buttress'
bastide	(feminine) *bastidette	'little country-house'
ride	(feminine) *ridette	'little wrinkle'
cachet	(masculine *cachetet	'little stamp, seal'
carotte	(feminine) *carottette	'little carrot'

The suffixation of *-et*/*-ette* is frustrated if the last consonant in the base is an alveolar plosive, /t/ or /d/. This is reminiscent of the restriction in English on deriving adverbs from adjectives by suffixing *-ly* (cf. *friendlily* in [4.11]). Languages seem to have rules of 'euphony' which tend to bar certain jarring sound sequences in word-formation.

4.2.1.2 Morphological factors

The morphological properties of a base may prevent the application of morphological rules. Often **native morphemes** behave differently from foreign morphemes. Some affixes are typically added either to native bases or to bases of foreign origin. For example, as we saw above in [4.5] *-ant* (as in *defendant*) is suffixed to bases of French origin.

Similarly, the rule of **velar softening** which changes /k/ (usually spelled with the letter *c*) to [s] is essentially restricted to words of Latin and French origin:

[4.14]   Velar Softening

/k/   →   [s] before a suffix commencing with a nonlow vowel (e.g. *i*)

The effects of velar softening can be seen in [4.15]:

[4.15]   cynic, cynical   →   cynicism
             critic, critical   →   criticism, criticise
             fanatic   →   fanaticism
             ascetic   →   asceticism
             sceptic   →   scepticism

Velar softening only affects words with Romance roots. So, if a thinker called *Blake* developed a new philosophy, we might call it *Blakism* [bleɪkɪzm]. But we could not call it **Blacism* [bleɪsɪzm], since *Blake* is not a Romance root (cf. also p. 124).

---

a   What is the meaning of *-hood* in [4.16]?
b.   Show the relevance of the distinction between native and foreign bases in the selection of bases to which *-hood* is suffixed.

Hint: Consult a good etymological dictionary. This exercise requires some knowledge of the historical sources of English words.

[4.16]   **a.** boy-hood   brother-hood   man-hood   maiden-hood
                     girl-hood   sister-hood   woman-hood   maid-hood
                     child-hood   king-hood   priest-hood   knight-hood
             b. *judge-hood   *governor-hood *colonel-hood   *minister-hood
                     *director-hood *author-hood   *prisoner-hood *general-hood

---

I hope you have correctly observed that normally *-hood*, (which means 'rank, state, quality) co-occurs with native roots like those in [4.16a] and is disallowed after latinate roots like those in [4.16b].

Clearly the distinction between native and borrowed morphemes is very important. However, we should be careful not to press this too far. The roots in [4.17] below are borrowed from French, yet they can take the suffix *-hood*. With the passage of time, foreign morphemes can be fully assimilated and nativised so that they behave in the same way as indigenous morphemes.

[4.17]   parenthood   statehood   nationhood

Finally, the selection of affixes that co-occur with a particular base may depend on that base being a member of a particular **paradigm**, i.e. a purely morphological sub-class. Morphemes belonging to different paradigms take different affixes. This is very often the case in inflectional morphology. Consider the French examples below. The regular verbs belong to one of these morphological classes: (i) *-er* verbs ([4.18a]); (ii) *-ir* verbs ([4.18b]); and (iii) *-re* verbs ([4.18c]).

[4.18] a. -er verbs

donn-er	'to give' (inf.)	demand-er	'to ask' (inf.)
je donn-e	'I give'	je demand-e	'I ask for'
tu donn-es	'you (sing.) give'	tu demand-es	'you (sing.) ask for'
nous donn-ons	'we give'	nous demand-ons	'we ask for'

b. -ir verbs

fin-ir	'to finish (inf.)	gém-ir	'to moan' (inf.)
je fin-is	'I finish'	je gém-is	'I moan'
tu fin-is	'you (sing.) finish'	tu gém-is	'you (sing.) moan'
nous fin-issons	'we finish'	nous gém-issons	'we moan'

c. -re verbs

vend-re	'to sell' (inf.)	romp-re	'to break' (inf.)
je vend-s	'I sell'	je romp-s	'I break'
tu vend-s	'you (sing.) sell	tu romp-s	'you (sing.) break'
nous vend-ons	'we sell'	nous romp-ons	'we break'

As you have seen, depending on which paradigm a verb belongs to, it co-occurs with different allomorphs of inflectional suffix morphemes. This discussion will be resumed in sections (4.3) and in Chapter 10 below. The existence of paradigms is very important for understanding the nature of allomorphy.

4.2.1.3 Semantic factors

Semantic considerations too may impinge on the application of word-formation processes. This is seen in the way the otherwise general process of forming compounds from Adjective plus past participle (Ved) which is shown in [4.19] is blocked in [4.20]:

[4.19]   short-sleeved (shirt)   one-armed (bandit)
        short-sighted (man)   three-legged (stool)
        green-roofed (house)   red-nosed (reindeer)
        blue-eyed (boy)   red-haired (woman)

[4.20]   *two-carred (family   (for 'a family with two cars')
        *big-Alsatianed (woman)   (for 'a woman with a big Alsatian')

Compound adjectives derived from the past participle (Ved) form of the verb are most likely to be permitted where the root to which *-ed* is added is **inalienably possessed** (i.e. obligatorily possessed) by the head noun that it modifies. The compound words in [4.19] are permissible because someone's eyes are an integral part of their body. Similarly, the legs of a stool, the sleeves of a shirt and the roof of a building are an obligatorily possessed part of some piece of furniture, garment or building. But it certainly is not the case that an Alsatian dog or a car must necessarily be possessed by someone.

The use of the italicised words in the dialogue below is odd. What would be the natural and preferred word choices that one would probably use instead of *unill*, *unsad*, *unpessimistic* and *undirty*? Why?

[4.21]   SURGEON: How are you today, Leslie?
        PATIENT: I am feeling much better. It's just wonderful to be so *unill* again.
        SURGEON: Oh, I'm so *unsad* to see you making such good progress. I am very *unpessimistic* about your chances of making a full recovery. The main thing now is to make sure we keep the wound *undirty* to avoid infection.

This example illustrates how semantics may restrict the application of morphological rules. If there are two adjectives with opposite meanings, one of which has a more positive meaning than the other, normally the negative prefix *un-* attaches to the positive adjective (see [4.22a]). If *un-* is attached to the negative member of the pair as in [4.22b] the resulting word is usually ill-formed.

[4.22]	a.	unwell	b.	*unill
		unloved		*unhated
		unhappy		*unsad
		unwise		*unfoolish
		unclean		*unfilthy, *undirty
		unoptimistic		*unpessimistic

As seen, if there are words representing the two poles of the same semantic dimension, we tend to prefer treating the positive end as **unmarked** (i.e. as normal). We are happier to derive the **marked** (i.e. 'unusual'), less favourable meaning by prefixing the negative prefix to a positive base than doing the reverse. That is why a *happy* person is not said to be *unsad* (Zimmer, 1964). To make the dialogue in [4.21] normal, the marked words, which are italicised, must be replaced by their unmarked counterparts in [4.22a].

4.2.1.4 Aesthetic factors and the adoption of words
In some cases word-formation is inhibited by vague aesthetic factors. There are many examples of words that are in principle well-formed whose adoption has nevertheless been resisted.

In the 1970s, the word *stagflation* was coined to refer to the combination of economic *stagnation* and a high level of *inflation* that afflicted the world economy at that time. So far, this word seems to have failed to get a firm foothold in the language. Aesthetic considerations may have had something to do with it. Some commentators consider it 'ugly'.

Other 'ugly' words which raise hackles include *talkathon*, *swimathon*, *knitathon*, etc. These are made up by analogy to *marathon*. Erudite purists are outraged not only by at the sight of a combination of a Greek pseudo-suffix with native Anglo-Saxon roots, but also by the misanalysis of *-athon* as a 'suffix' meaning 'undertaking a strenuous prolonged activity (specified in the part of the word that precedes *-athon*) for the benefit of a good cause'. In Greek *-athon* was not a morpheme. But the average speaker of English who is unaware of such niceties will probably contentedly coin more *-athon* words, regardless.

4.3 DOES PRODUCTIVITY SEPARATE INFLECTION FROM DERIVATION?

Productivity is often taken as a criterion for distinguishing inflection from derivation. Derivational processes are by and large much more unpredictable than inflectional ones. While inflectional processes usually affect most of the eligible forms in a regular manner, derivational rules tend to be capricious. They tend to have as input a class whose membership is subject to various arbitrary exclusions and to affect it spasmodically as we have already seen (see [4.5] and [4.6]). Unlike derivation, inflection is normally productive in these senses:

(i)   It is general. The addition of particular inflectional affixes is not subject to various arbitrary restrictions. Stems that belong to a given class normally receive all the affixes that belong to that class (cf. section (10.2.1.2)).
(ii)   The words resulting from the addition of inflectional affixes have regular and predictable meanings.

Another way to put it is to say that typically inflectional morphology displays **lexemic paradigms** but derivational morphology does not. Paradigms are regular and predictable sets of word-forms belonging to the same set of lexemes. They share morphological characteristics (e.g. prefixes, suffixes or infixes). Such words belong to a particular word-class or sub-class. The selection of a specific word-form is determined by the syntax. See [4.18] above where three **inflectional paradigms** of French verbs were shown.

English verb forms also belong to inflectional paradigms:

[4.23]   walk ~ walks ~ walked ~ walking
             love ~ loves ~ loved ~ loving

Usually inflectional morphology exemplifies automatic productivity. Most English verbs have these forms. So, if we encounter a new verb like the made-up verb *pockle* (meaning perhaps 'to go (away in a huff)'), which was introduced in Chapter 2, we can predict that it will have the forms *to pockle*, *pockles*, *pockled* and *pockling* with the standard meanings.

By contrast, as a rule, paradigms cannot be set up for derivational morphology. If we try to produce a paradigm with the derivational suffixes *-ate*, *-ant* and *-ation*, we soon get thwarted. Our putative paradigm in [4.24] is riddled with gaps:

[4.24]

Verb (X)	Noun (one who does X)	Noun (act of X)
communicate	communicant	communication
donate	——	donation
navigate	——	navigation
rotate	——	rotation
militate	militant	——
applicate	applicant	application
(obsolete)	accountant	——
——	——	natation (art of swimming)

It is tempting to assume that wherever paradigms can be recognised, as in [4.23], one is dealing with inflection, but where no regular paradigms exist, as in [4.24], one is dealing with derivation.

However, although using the presence of paradigms involving a given affix as a rule of thumb for distinguishing between inflectional and derivational morphemes usually works, it will not unfailingly lead to safe results. This is because, on the one hand, there are exceptions to inflectional processes and, on the other, there are derivational processes which appear to fall into very general paradigm-like patterns. We will consider these two problems in turn.

First, while most inflectional processes tend to fall neatly into paradigms, there are often some forms that escape. For instance, although the vast majority of nouns are count nouns, which means that they have both singular and plural forms and meanings (e.g. *book* ~ *books*, *dog* ~ *dogs*, *ass* ~ *asses*, etc.), there is a minority which are not marked for number. They belong to different subclasses of **noncount nouns** referring to entities that are not individually counted:

(i) Some noncount nouns have a plural form but lack a plural meaning:

[4.25]   *alm   ~   alms
         *outskirt   ~   outskirts
         *oat   ~   oats
         *linguistic   ~   linguistics
         *new   ~   news

(ii) Other noncount nouns lack a plural form altogether (they do not refer to itemised individual entities which are counted in English):

[4.26]   milk   *milks
         health   *healths
         equipment   *equipments
         courage   *courages

Conversely, sometimes a derivational process such as the suffixation of -*ly* to adjectives to turn them into adverbs (e.g. *kindly*, *quietly*, etc.) is almost exceptionless. (But see [4.11], p. 75.) Similarly, derivation of nouns referring to containers (e.g. *cupful*, *basketful*, *canful*, *bagful*, etc.) by suffixing -*ful* to nouns is very regular and predictable. Hence, regularity cannot be used as a litmus test that distinguishes inflection from derivation.

4.4 THE NATURE OF THE LEXICON

We have seen in the foregoing sections that the lexicon is open-ended. Not all the words of a language can be listed in the lexicon. The question that arises is, On what basis, then, are words (and morphemes) selected for inclusion? And for the words selected, what kinds of information must the lexicon include? In brief, what is the nature and function of the lexicon?

4.4.1 Potential Words

As we have already observed, knowing a language involves, among other things, knowing the rules of word-formation. Speakers are able not only to identify the meaningful units which words contain, but also to create new words, and to understand the meanings of unfamiliar words that they have not encountered previously. If we came across 'words' like *grestifier* and *disperidate* we would recognise them as **potential English words**. But how do we know this? At least part of the answer lies in the nature of our mental lexicon.

The lexicon has a set of **phonotactic constraints** which function as a filter allowing entry only to phonologically well-formed words. Before any putative word can enter the lexicon, it must have a combination of sounds that is permissible in the language. So, 'words' like **ltarpment* and **mpandy* are immediately rejected because the consonant combinations /lt/ and /mp/ are disallowed at the beginning of a word in English.

Non-nativised foreign words with sound sequences that are not permitted in English may be allowed in, as a special case. Presumably such words are kept in a special sector reserved for them. We need to make some allowance for, say, foreign place names like *Tblisi* or foreign personal names like *Zgutsa* which begin with the consonant clusters [tbl] and [zg] which are unorthodox in English. Foreign words entering the language may be allowed to by-pass the phonotactic filter. In that case they keep their foreign pronunciation, virtually unchanged. But, more often than not, they tend to be significantly modified so that they fit in the general phonotactic patterns of the language. So, a schwa may be inserted between the first two consonants in *Zgusta* and *Tblisi* to produce [zəgustə] and [təblisi] respectively.

4.4.2 Knowledge of Language and the Role of the Lexicon

The lexicon is a mechanism for capturing broad regularities involving words in a language. For instance, using phonotactic constraints English speakers can distinguishing very generally between, on the one hand, possible words (which may not be instantiated in sentences and utterances, e.g. *grestifier*) and, on the other hand, impossible words (e.g. **ltarpment*).

In the foregoing I have proposed that some of the general properties of morphemes and lexical items should be shown in the lexicon. However, traditionally the lexicon was not regarded as a place where regularities were captured. Rather, it was viewed as the repository of exceptions, in the form of a list. We read in Bloomfield (1933: 274) that 'The lexicon is really an appendix to the grammar, a list of basic irregularities.'

This view is colourfully caricatured by di Sciullo and Williams (1987: 3) as being one where the lexicon is conceived of as a prison which 'contains only the lawless, and the only thing that its inmates have in common is lawlessness'. And the lawless are a disparate bunch including words (e.g. *work*), morphemes (e.g. *-ed*) and idioms (e.g. *'eat one's words'*).

Nonetheless, it is the case that the lexicon in a generative grammar must list various kinds of information about words (and morphemes and idioms) which have to be memorised. For example, speakers of English who know the word *aardvark* need to memorise at least this information:

(i)   Meaning: it refers to a Southern-African insectivorous quadruped mammal.
(ii) Phonological properties: its pronunciation /ɑːdvɑːk/.
(iii) Grammatical properties: e.g. it is a count noun (you can have one *aardvark*, two *aadvarks*).

Admittedly, what needs to be listed in speakers' mental lexicons may vary. While for most people *aardvark* needs to be memorised, some erudite speakers know that this word is a compound borrowed from Afrikaans and is composed of *aarde* 'forest' and *vark* 'pig'.

Today most generative linguists reject the view that the lexicon is merely a list of irregularities. If there is a need for lists in a grammar – and there clearly is, since we need to list basic morphemes – then the lists belong to the lexicon. But, this does not mean that the lexicon consists just of lists. There are many extensive and far-reaching lexical regularities resulting from the operation of general principles.

Normally, the relationship between the meaning and form of a morpheme or word is completely arbitrary and idiosyncratic (notable exceptions being cases of onomatopoeia, e.g. *cuckoo* and *miaou*), but many other properties are not. There are numerous pervasive regularities in the phonological and syntactic behaviour of words.

In the next part of the book we are going to explore the organisation of the lexicon, concentrating on the representation of word-formation regularities that relate to the phonology. In the last part of the book we will come back to the lexicon and consider regularities that relate to the syntactic and semantic properties of words.

EXERCISES

1. (a) Make a list of ten words containing the suffix -*ic* as in *magnetic* and
        allergic.
    (b) What word-class do the bases to which -*ic* is added belong?
    (c) What is the word-class of the resulting word?
    (d) Is -*ic* an inflectional or derivational suffix? What is your evidence?
    (e) Does -*ic* attach freely to all eligible bases? If not give two examples
        of bases which are ostensibly suitable for -*ic* suffixation but which
        fail to take this suffix.

2. Study the examples below and, with help of a good dictionary, answer
    the questions that follow:

monoism	unique
monologue	unilateral
monolingual	unitary
monolith	unify
monogamy	unipolar
monogenesis	unicellular
monochrome	unidirectional
monorail	unicycle

    (a) What is the meaning of the prefixes *mono-* and *uni-*?
    (b) What bases can these prefixes be attached to normally? List as
        many relevant factors that play a role in the selection of these
        prefixes as you can think of.
    (c) What is the word-class of the word formed by *mono-* and *uni-*
        prefixation?

3. This is an open-ended question.
    (a) In the data below list three examples of (i) conversion, (ii) suffixa-
        tion, (iii) compounding.
    (b) Comment on problem cases which do not seem to fit neatly into
        any one of these categories.
    Note: When you consider conversion, take into account the pronun-
    ciation of vowels and consonants as well as where stress falls in a word.

advice (noun)	advise (verb)
plan (noun, verb)	concrete (noun, verb)
reject (noun, verb)	table (noun, verb)
greyhound (noun)	milkman (noun)
blood (noun)	bleed (verb)
song (noun)	sing (verb)
teapot (noun)	breakfast (noun, verb)

bio-science (noun)   telegraph (noun)
petticoat (noun)   tragi-comedy
hardship   hardwood
Guinea-pig   microwave
troublesome   paratrooper

4. Compare the following pairs:

author ~ authoress   lion ~ lioness   actor ~ actress
manager ~ manageress   mayor ~ mayoress   waiter ~ waitress
editor ~ editress   poet ~ poetess   emperor ~ empress

(a) What is the word-class of the bases that the -*ess* suffix attaches to?
(b) What class do the resulting words belong to?
(c) What is the meaning of this suffix?
(d) Do you find these words equally acceptable? If you do not, explain why.

Part II
Morphology and its Relation to the Lexicon and Phonology

5 Introducing Lexical Morphology

5.1 THE LEXICAL PHONOLOGY AND MORPHOLOGY MODEL

In this chapter I present the model of **lexical phonology and morphology** in an introductory way and apply it to issues in English derivational and inflectional morphology of the kind raised in Part I of the book. The next chapter goes deeper into the theoretical issues that are thrown up by English, while remaining largely uncontroversial. To ensure clarity of presentation, I will initially take it for granted that the theory is totally adequate and postpone a critical evaluation until Chapter 7.

An important feature of this model is that it is the **word**, rather than the morpheme, that is regarded as the key unit of morphological analysis. In this respect it resembles Word-and-Paradigm morphology and more traditional approaches (see section 3.7).

A major claim made by proponents of this theory is that there is a symbiotic relationship between the rules that build the morphological structure of a word and the phonological rules responsible for the way a word is pronounced. All these rules are found in the lexicon where they are organised in blocks called **strata** (or **levels** or **layers**) which are arranged hierarchically, one below the other. Normally the model is referred to simply as either **lexical phonology** or **lexical morphology** for short.

5.2 LEXICAL STRATA

Central to lexical morphology is the principle that the **morphological component** of a grammar is organised in a series of **hierarchical strata** (cf. Allen, 1978; Siegel, 1974; Pesetsky, 1979; Kiparsky, 1982a, 1982b, 1983, 1985; Mohanan, 1982/86; Mohanan and Mohanan, 1984; Halle and Mohanan, 1985; Strauss, 1982a; and Pulleyblank, 1986).

English affixes (both prefixes and suffixes) can be grouped in two broad classes on the basis of their phonological behaviour. One type is **neutral** and the other type is **non-neutral**. Neutral affixes have no phonological effect on the base to which they are attached. But non-neutral ones affect in some way the consonant or vowel segments, or the location of stress in the base to which they are attached.

Let us consider the neutral affixes first. As you can see in [5.1], the presence of the neutral suffixes *-ness* and *-less* makes no difference. The same syllable of the base receives stress regardless; and the base is left unchanged:

[5.1] **a.** 'abstract   'abstract-ness   **b.** 'home   'home-less

           'serious   'serious-ness   'power   'power-less

           a'lert   a'lert-ness   'paper   'paper-less

Note: Here ' before a syllable indicates the syllable that receives **main stress**, i.e. the most prominent syllable in a word. I will not mark any other stressed syllables in a word because that is not essential for our purposes.

---

a. Say carefully the following pairs of words.
b. Place the stress mark ' before the syllable that receives main stress in each word. Consult a dictionary if you are unsure.
c. Are *-ic* and/or *-ee* neutral or non-neutral suffixes?

[5.2] **a.** strategy   strategic   **b.** detain   detainee

           morpheme   morphemic   absent   absentee

           photograph   photographic   pay   payee

           democrat   democratic   employ   employee

The answer is clear. The suffixes *-ic* and *-ee* are **non-neutral**. They affect the location of stress. The *-ic* in [5.2a] is a **pre-accenting** suffix. So, stress is attracted to the syllable immediately preceding it. On the other hand, *-ee* in [5.2b] is an **auto-stressed** suffix. This means that the suffix takes the stress from the base onto itself.

Besides affecting stress, non-neutral suffixes also tend to trigger changes in the shape of the vowels or consonants of the base to which they are added. You can see this if you compare *satan* with *satanic*. The presence of the non-neutral suffix *-ic* induces the replacement of the vowel [eɪ] of *satan* ['seɪtən] with [æ] in *satanic* (sə'tænɪk].

---

a. Transcribe phonetically the adjectives *wide*, *long* and *broad*.
b. Transcribe phonetically the nouns *width*, *length* and *breadth* which are derived from the above adjectives by suffixing *-th*.
c. Transcribe phonetically the adverbs *widely* and *broadly* which are derived by adding the suffix *-ly* to the corresponding adjectives.
d. Determine whether *-th* and *-ly* are neutral or non-neutral suffixes.

Like [5.3] below, your transcription should reveal the changes in the root vowel triggered by *-th*. These changes show that the suffix *-th* is non-neutral. In the examples in column [5.3b], the root vowel is either shortened or changed in quality or, if it is a diphthong, it is turned into a monophthong. By contrast, in column [5.3c], where the neutral *-ly* suffix is added, no change takes place in the location of stress or in the realisation of the consonants and vowels in the base.

[5.3]	a.	Adjective	b.	Noun	c.	Adverb
		wide [waɪd]		wid-th [wɪdθ]		wide-ly [waɪdlɪ]
				(*[waɪdθ])		(*[wɪdlɪ])
		long [lɒŋ]		leng-th [lenθ] (*[lɒnθ])		———
		broad [brɔːd]		bread-th [bredθ]		broad-ly [brɔːdlɪ]
				(*[brɔːdθ])		(*[bredlɪ])

In SPE (this is the standard way of referring to Chomsky and Halle's 1968 book, *The Sound Pattern of English*) the difference between the behaviour of neutral and non-neutral affixes was dealt with in terms of the **strength of boundaries**. Between the base and a neutral suffix like *-ness* or *-ly*, there was said to intervene a **weak boundary** (symbolised by '#'). In contrast, a **strong boundary** (symbolised by '+') was assumed to separate the base from a non-neutral suffix like *-ic*, *-ee* or *-th*.

The distinction between non-neutral affixes (associated with '+ boundary' in SPE) and neutral affixes (associated with '# boundary'), corresponds to the more traditional distinction between **primary** and **secondary affixes** (Whitney, 1889; Bloomfield, 1933: 240). The idea that phonological rules may be paired with morphological rules that introduce affixes is not new. About 2000 years ago, Panini, the ancient Indian grammarian, envisaged this kind of pairing of word-structure rules and phonological rules.

In addition to being neutral or non-neutral in their phonological effects, English primary and secondary affixes display contrasting phonotactic behaviour. Whereas secondary affixation can produce segment sequences that are disallowed within a single morpheme in lexical representations, primary affixation cannot give segment sequences that deviate from those allowed in single morphemes in the lexicon.

For instance, English does not allow **geminate consonants** within roots. The orthographic doubling of letters in, say, *addle* [ædɫ] or *miss* [mɪs] never corresponds to any gemination (i.e. 'doubling') of the consonants in pronunciation. Likewise, when we attach primary affixes like *ad* as in *adduce* [ədjuːs], again no gemination occurs. However, there is gemination when a secondary affix like *sub-*, *un-* or *-ness* is adjacent to an identical consonant in the base. Secondary affixation can yield geminates as in *unnamed* [ʌnneɪmd], *sub-base* [sʌbbeɪs] and *thinness* [θɪnnəs].

The classification of affixes as primary or secondary is not arbitrary. Typically, primary affixes are Germanic in origin while secondary affixes are mostly, though not exclusively, Greek or latinate, having entered the language with loanwords from Greek, Latin or French. Naturally, as we will see in the next section, many affixes from foreign sources will only combine with bases borrowed from the same foreign language.

Unlike SPE, which dealt with the difference in the behaviour affixes in terms of boundary strength, lexical phonology and morphology approach it in terms of **level ordering** (i.e. the **ordering of strata**). It is proposed that affixes are added at different strata in the lexicon. Each stratum of the lexicon has associated with it a set of morphological rules that do the word-building. These morphological rules are linked to a particular set of phonological rules that indicate how the structure built by the morphology is to be pronounced.

There are, of course, some **underived lexical items** which must simply be listed in the lexicon. They are not the product of any word-formation rules. They are simple words (such as *but*, *key*, *write*, *good* and *black*) containing just one morpheme. About such items morphology has nothing interesting to say, though of course their meaning, pronunciation and grammatical properties have to be listed in the lexicon.

More interesting are forms which involve some inflectional or derivational process, or both. Some of these processes will apply at stratum 1 and others at stratum 2, as we will see presently. In sections (5.2.1) and (5.2.2) we will see how derivational and inflectional processes are treated in lexical phonology.

5.2.1 Derivation in Lexical Morphology

We will begin by observing that normally the ordering of strata in the lexicon reflects the ordering of word-formation processes. Primary affixes (e.g. *-ic* in *phonemic*), which are phonologically non-neutral, are attached first at stratum 1. But the processes of compounding as well as the attachment of secondary affixes (e.g. *-ly* as in *widely*), which are phonologically neutral, happen at stratum 2.

The underived root is like the kernel of the word (as in [5.4a] below). Stratum 1 takes the root as the base to which non-neutral affixes are attached (see [5.4b]). Then stratum 2 takes the root-plus stratum 1 affixes as its input (see [5.4b]). A natural consequence of assuming that the strata in the lexicon are ordered in this way is that stratum 1 affixes are closer to the root of the word, and neutral affixes are added on the outside as an outer layer:

[5.4] **a.** [*root*]
      b. [stratum 1 affix – *root* – stratum 1 affix]
      c. [stratum 2 affix – *stratum 1 affix* – *root* – *stratum 1 affix* – stratum 2 affix]

Let us look at the following concrete examples containing the derivational suffixes -*(i)an* and -*ism*. I have used the abbreviations 'r', 's1' and 's2' for 'root', 'stratum 1 affix' and 'stratum 2 affix' respectively.

[5.5] **a.** [r]   **b.** [[r]s1]
          Mendel   Mendel-ian
          Mongol   Mongol-ian
          grammar   grammar-ian
          Skakespeare   Skakespeare-an
                                   (Shakespearian)

      c. [[[r]s1]s2]   **d.** [[[r]s2]s1]
          Mendel-ian-ism   *Mendel-ism-ian
          Mongol-ian-ism   *Mongol-ism-ian
          grammar-ian-ism   *grammar-ism-ian
          Skakespeare-an-ism   *Skakespeare-ism-(i)an
          (Shakespearianism)

                                   (based on Kiparsky, 1983: 3)

---

a. The suffix (-*ian*) is on stratum 1 because it is phonologically non-neutral. In what senses is it non-neutral? Use the data in column [5.5a] and column [5.5b] as your evidence.
b. Is -*ism* a neutral or non-neutral suffix?
c. What prediction does the theory of level-ordered morphology make about the ordering of these suffixes if they co-occur in a word?

You will have made a number of interesting observations about -*(i)an*:

(i)   When they appear without any suffix in [5.5a], these words are stressed on the first syllable.
(ii) The suffix -*ian* is non-neutral because:
      (a) it is pre-accenting: when it is present, stress moves to the syllable immediately before -*ian*,
      (b) it affects the segmental phonology of the root to which it is attached in all the examples except *Shakespeare* where only stress shifts. Contrast:

Mendel	['mendɫ]	*Men'delian*	[men'di:ljən]
Mongol	['mɒŋgɒl]	*Mon'golian*	[mɒŋ'gəʊlɪən]
grammar	['græmə]	*gra'mmarian*	[grə'meərɪən]
Shakespeare	['ʃeɪkspɪə]	*Shakes'perian*	[ʃeɪks'pɪərɪən]

(The fact that unlike the other words *Shakespeare* is a native word may have someting to do with its untypical behaviour.)

(iii) By contrast, *-ism* is a neutral suffix. Stress stays on the syllable it was on before *-ism* was added. Its presence does not affect the segmental phonology of the root to which it is attached.

The theory predicts that stratum 1 affixes come closer to the root than stratum 2 affixes. In other words, stratum 2 affixes appear on the outer layer and stratum 1 affixes are on the inner layer. This is the case in our data. In [5.5c] the stratum 1 non-neutral suffix *-(i)an* comes immediately after the root and the stratum 2 suffix *-ism* is attached on the outer layer. If stratum 2 *-ism* precedes stratum 1 *-ian* (as in 5.5d), the result is an inadmissible word. This is an important point. We will return to it in (6.2.1).

We have established the motivation for keeping morphemes on different strata in the lexicon. But what kinds of information should lexical entries of affixes contain? In the lexicon affixes are listed with their meaning, information about the bases to which they attach, the grammatical category of the word resulting from affixing them as well as the stratum at which they are found. This last point is important because there are many aspects of the behaviour of an affix morpheme that are not peculiar to it but are shared by other affixes found at the same stratum. By assigning affixes to strata we will be able to capture in a simple way a wide array of general facts about their behaviour.

For each of the affixes in the following explain why it is a stratum 1 affix.

[5.6]	Suffix	attach to		Output
	$-ion_{Adj}$	$[[erode_V] - ion_N]$	→	$[erosion]_N$
	$-ive_{Adj}$	$[[compete(t)_V] - ive_{Adj}]$	→	$[competitive]_{Adj}$
	$-al_{Adj}$	$[[Pope_N] - al_{Adj}]$	→	$[papal]_{Adj}$

As a rule, stratum 1 affixes modify in some way the base to which they are attached. In [5.6] we see the consonant [d] of *erode* [ɪ'rəʊd] change to [ʒ] before *-ion* in *erosion* [ɪ'rəʊʒn]. In *competitive*, the *-ive* suffix requires the introduction of the stem extender *-it* (cf. section (3.1.4) above). In addition, *-ive* triggers a vowel change in the root. The stressed vowel changes quality and is shortened from [i:] to [e] when *-ive* is attached. Compare *competitive* [kəm'petɪtɪv] with *compete* [kəm'pi:t]. Finally, the *-al* suffix also conditions a change of the root vowel from [əʊ] in *Pope* ['pəʊp] to [eɪ] in *papal* ['peɪpɫ].

The lexical morphology model is presented with reference to English, using the diagram in [5.7] which is based on Kiparsky (1982b: 5), but its essential properties are applicable to other languages.

[5.7]

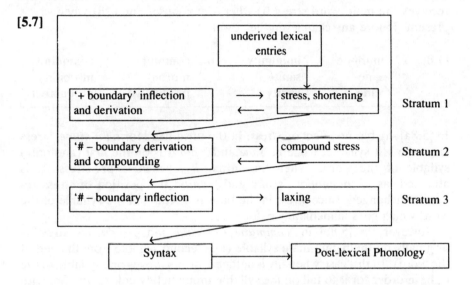

As you can see from the diagram in [5.7], the model is divided into two main parts, the **lexical** and the **post-lexical**. The point of the distinction is to reflect the generalisation that some phonological rules – the lexical rules – are closely tied to morphological rules that build word-structure in the lexicon. Normally, when a morphological rule applies, it activates some related phonological rule, whereas **post-lexical rules**, on the other hand, are not linked to word-formation rules. They are phonological rules that apply when fully-formed words are put in syntactic phrases. Since this book is concerned with word-formation, we will focus on the lexical rules. Post-lexical rules will only be dealt with very briefly in section (5.4) at the end of this chapter.

Some writers, such as Kiparsky (1982a) recognise three strata in their lexical morphology (see [5.7]). Kiparsky proposes that stratum 1 should deal with irregular inflection (e.g. *see ~ saw* (past tense)) and irregular derivation (e.g. *long* (adj) *~ length* (noun)). Stratum 2 deals with regular derivation (e.g. *kind* (adj) *~ kindly* (adv)) and compounding. Stratum 3 deals with regular inflection (e.g. *walk ~ walked* (past)). For reasons explained in section (7.2.3) I take a more restrictive stance and recognise only two lexical strata. I will assume that all irregular inflection and derivation takes place at stratum 1 and all regular derivation, inflection and compounding takes place at stratum 2.

In the next few paragraphs we are going to study in some detail the modifications in the base caused by the stratum 1 derivational suffix -*ity*. The discussion draws on Aronoff (1976). Start by working out the effect, if any, on stress of -*ity* in each of the following pairs. Mark the syllable that receives the main word stress (i) when -*ity* is absent and (ii) when -*ity* is present. Ignore any changes in the vowel.

[5.8]	a.	immune	~	immunity	b.	commune	~	community
		sane	~	sanity		morbid	~	morbidity
		vane	~	vanity		productive	~	productivity

In [5.8a] -*ity* has no effect on stress. In the unsuffixed form *im`mune*, stress is on the last syllable, while in `*sane* and `*vain* it is on the first (and only) syllable of the word. When the suffix -*ity*, which is pre-stressing, is attached, as in *im`munity*, `*sanity* and `*vanity*, the position of stress remains unchanged, since stress in the base is already on the syllable of the word which goes immediately before -*ity*.

However, in [5.8b] in `*commune*, `*morbid* and *pro`ductive* stress is originally on the penultimate syllable (i.e. second syllable from the end of the word). In this case when -*ity* is suffixed, stress moves one syllable to the right in order for it to fall on the syllable immediately before -*ity*. So, with -*ity* present, these words are pronounced *com`munity*, *mor`bidity* and *produc`tivity*.

As we have already seen in [5.3] and [5.6], besides affecting stress, stratum 1 affixes also tend to cause changes in the consonant and vowel segments of the stem to which they are attached.

Now study the following words:

[5.9] sanity, vanity, inanity
       extremity, obscenity, serenity
       morosity, verbosity, bellicosity
       divinity
       profundity

a. State the base to which -*ity* is attached in each case.
b. What are the changes in the segmental phonology of the base which are caused by -*ity*?

The presence of *-ity* brings about in the vowel of the base the changes that are indicated below:

[5.10] [eɪ] → [æ]   [i:] → [e]   [əu] → [ɒ]
         sane ~ sanity   extreme ~ extremity   morose ~ morosity
         vain ~ vanity   obscene ~ obscenity   verbose ~ verbosity
                             serene ~ serenity   bellicose ~ bellicosity

        [aɪ] → [ɪ]   [au] → [ʌ]
        divine ~ divinity   profound ~ profundity

The modification of the stem occasioned by the addition of *-ity* may go further still.

Aronoff provides numerous examples of drastic changes in the base which are sparked off by the suffixation of *-ity*. They include those illustrated in [5.11] below. First, recall from [3.16] that *-ity* attaches to adjective bases to form abstract nouns. The forms in [5.11] are adjectives. They represent a base plus the adjective-forming suffix '*-(i)ous*' (meaning 'having the quality X'). But this suffix is deleted from the base when *-ity* is suffixed in [5.11b] to form abstract nouns. The forms in [5.11c], where *-ous* is retained before *-ity*, are not allowed.

[5.11] a. audacious   b. audacity   c. (*audaciousity)
            sagacious   sagacity   (*sagacousity)
            rapacious   rapacity   (*rapaciousity)
            pugnacious   pugnacity   (*pugnaciousity)
            vivacious   vivacity   (*vivaciousity)

The effects of *-ity* are perhaps more drastic than those of many other derivational suffixes, but they are by no means extraordinary for a stratum 1 suffix.

Study the words in [5.12] and describe the phonological effects on the base of the stratum 1 suffixes *-ory* and *-acy*.

[5.12] a. Verb   Adjective   b. Adjective   Noun
            explain   ~   explanatory   intimate   ~   intimacy
            defame   ~   defamatory   legitimate ~   legitimacy
            inflame   ~   inflammatory   animate   ~   animacy

---

For our present purposes we will ignore what happens to consonants. We see that the suffixation of *-ory* or *-y* (both meaning 'having the quality of') causes the last vowel of the base to change from [eɪ] to [æ]. Comparable changes can be observed if you look at the effects of *-ity* in [5.11] (and perhaps less obviously in [5.10]). The point of grouping together affixes and regarding them as belonging to the same stratum in the lexicon should now be clear. It enables us to capture the generalisation that affixes belonging to the same stratum share certain properties. Thus the different stratum 1 suffixes which we have examined tend to affect stress and vowels in the base in broadly the same way.

The vowel alternations conditioned by various stratum 1 suffixes in present-day English are not phonologically conditioned. There is nothing about the articulation or acoustic properties of the sounds that can be blamed for the changes that occur in the base when any of these suffixes is present (see section 2.2.2).

Nevertheless, these phonological changes are very general and they are not merely capricious. Rather, they represent relics from an earlier era in the history of English when the **Great Vowel Shift** occurred. The Great Vowel Shift (often abbreviated to GVS was a drastic upheaval in the pronunciation of **long vowels**. It took place mostly between 1400 and 1600, during the Late Middle English period and was more or less completed in the Early Modern English period (see Jespersen, 1909; Dobson, 1957; SPE; Lass, 1976).

As a result of the GVS, the long vowels changed their quality. The non-high long vowels were raised and assumed the quality of the vowel immediately above them. The high vowels /i:/ and /u:/ became the diphthongs /eɪ/ and /ou/ in the sixteenth century and later changed to /aɪ/ and /əu/ respectively. The vowel changes are represented in [5.13]:

[5.13]   High

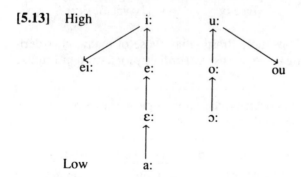

In [5.14] you can see an exemplification of the GVS:

[5.14]

Late Middle English	Early Modern English	Later English
/a:/ as in *spade*	→ /æ:/ > /ɛ:/	→ /e:/, /eɪ/
/e:/ as in *dēd*	/i:/	→ /i:/
(later spelt 'deed')		

/ɛ:/ as in *greet*   → /e:/   → /eɪ/
(later spelt 'great')
or
/ɛ:/ as in *lēaf*   → /e:/   → /i:/
(later spelt 'leaf')
/i:/ as in *wise*   → /əɪ/   → /aɪ/
/o:/ as in *boote* 'boot' → /u:/   → /u:/
/ɔ:/ as in *boot*   → /o:/   → /o:/, /əu/
(later spelt 'boat')
/u:/ as in *mouse*   → /əu/   → /au/

Note: A macron over a vowel (V̄) indicates a long vowel.

The GVS has left its mark on English phonology. The present-day vowel
alternations triggered by the stratum 1 suffixes which we have been considering are the result of **trisyllabic laxing**, a rule that has its roots in the
GVS. This rule changes a **tense vowel** (i.e. a long vowel or diphthong) in a
stem to a **lax vowel** (i.e. a short vowel). Essentially it applies when a
derived word of three syllables, or more, is created as a result of the
attachment of certain latinate suffixes, such as the noun-forming suffix
-acy, to a base:

[5.15]   **a.**   Tense [i:]   →   Lax [e]
                  su'prēme (adjective)   su'prĕmacy (noun]
                  [su'pri:m]   [su'prĕməsɪ]
                  (2 syllables)   (4 syllables)

         b.   Tense [eɪ]   →   Lax [ə]
                  advocāte (verb)   advocacy (noun)
                  ['ædvəkeɪt]   ['ædvəkəsɪ]
                  (3 syllables)   (4 syllables)

         c.   Tense [eɪ]   →   Lax [ə]
                  e'pɪscopāte (noun)   e'pɪscopacy (noun)
                  [ɪ'pɪskəupeɪt]   [ɪ'pɪskəpəsɪ]
                  (4 syllables)   (5 syllables)

Trisyllabic laxing primarily affects bases of latinate origin but is not restricted to them. It can affect native Germanic roots and affixes as well.
Observe the behaviour of the native root *wild*:

[5.16]   wild   wild-ness   wild-er-ness
         [waɪld]   [waɪld-nəs]   [wɪld ə-nəs]
         (one syllable) (two syllables)   (three syllables)

Let us now briefly turn to stratum 2 suffixes. Generally they all share the crucial property of being phonologically neutral. They do not cause drastic changes in the base to which they are attached. A sample of stratum 2 suffixes is given in [5.17]:

[5.17]	Suffix	attach to		Output
	-ness$_{nN}$	[[re'mote$_{Adj}$] ness]$_N$	→	re'moteness$_N$
	-less$_{Adj}$	['power$_N$] less]$_{Adj}$	→	'powerless$_{Adj}$
	-ful$_{Adj}$	['purpose$_N$] ful]$_{Adj}$	→	'purposeful$_{Adj}$
	-ly$_{Adv}$	[se'vere$_{Adj}$] ly]$_{Adv}$	→	se'verely$_{Adv}$
	-wise$_{Adv}$	['money$_N$] wise]$_{Adv}$	→	'money-wise$_{Adv}$
	-y$_{Adj}$	['velvet$_N$] y]$_{Adj}$	→	'velvety$_{Adj}$
	-er$_N$	[pre'tend$_V$] er]$_N$	→	pre'tender$_N$

Since the phonological effect of stratum 2 affixes is neutral, and their behaviour otherwise largely unproblematic, they will not detain us any longer in this chapter. But we will return to them again later, mainly in sections (6.2.1), (6.2.2) and (6.2.3) of the next chapter.

We have seen that derivational word-formation takes place at stratum 1 and stratum 2. Non-neutral processes are found at stratum 1 while neutral ones are found at stratum 2. In the next section I show that the same is true of inflection.

5.2.2 Inflection in Lexical Morphology

In the last section we saw that stratum 1 processes are more regular than their stratum 2 counterparts. Before we proceed, we need to clarify an important point. It is vital not to equate 'root-changing' or 'non-neutral' with 'irregular'. Some stratum 1 affixation processes are not root-changing but nevertheless they are irregular in the sense that they apply sporadically and are subject to ad hoc restrictions. An example of this is the plural allomorph -en in *ox-en*. Its presence causes no mutation in *ox*. But it is unpredicable. It attaches happily to *ox*, but for no apparent synchronic reason, it cannot be suffixed to phonologically similar nouns like *fox*, *box*, etc.

Most stratum 1 inflectional morphology consists of either erratic morphemes whose behaviour is largely unpredictable or inflectional processes that were once full of vitality but have long since atrophied, or borrowed affixes which came in with a few words borrowed from foreign languages and are largely restricted to those loanwords. We will consider examples of frozen historical relics first. We will see that much of the irregular inflection is a remnant of patterns that obtained (sometimes quite regularly) at an earlier historical period. We will illustrate this first with **ablaut** and then **umlaut**.

The term ablaut refers to the change in a root vowel which indicates a change in grammatical function. Thus, [aɪ] may alternate with [əu] to mark a change from present tense to past tense as in *ride* [raɪd] (present tense) ~ *rode* [rəud] (past tense). This pattern of vowel alternation goes all the way back to Indo-European, the reconstructed parent language from which English is ultimately descended.

Find at least two more examples of verbs which form their past tense by ablaut, involving substituting /aɪ/ for /əu/ to mark the past tense.

I hope you had no trouble finding some of these:

[5.18] Present tense   Past tense

drive	/draɪv/	drove	/drəuv/
write	/raɪt/	wrote	/rəut/
rise	/raɪz/	rose	/rəuz/
strive	/straɪv/	strove	/strəuv/
dive	/daɪv/	/dəuv/ (American); /daɪvd/ (British)	

This alternation constitutes a minor sub-regularity. Rather than simply say of each individual verb that it forms its past tense in this way, we shall establish a special class of verb and mark each member of this class. At stratum 1, any verb belonging to this class undergoes the vowel mutation whereby /aɪ/ → /əu/ in the past tense.

How does this irregular past tense formation interact with the regular rule that forms the past tense by suffixing *-ed*? If two ways of representing a morphological concept are available at both stratum 1 and stratum 2, the less general process applies at stratum 1 and it **blocks** (i.e. pre-empts) the more general process found at stratum 2 (cf. section (4.2.1)). Hence, once past tense forms like *rode* and *wrote* are created at stratum 1, they must be by-passed by the later, regular *-ed* past tense suffixation that takes place at stratum 2. If the latter applied as well, it would produce incorrect past tense forms like **roded* or **wroted*. This blocking effect follows naturally from the hierarchical ordering of strata in the lexicon (see 6.2.4.2).

Ablaut is a moribund phenomenon in English today. The chances of any new word entering the language being subject to it are extremely remote. If a new verb, say *to rine* (meaning perhaps 'to adore organic vegetables') entered the language, you would expect its past tense to be *rined*, which is formed by suffixing the regular stratum 2 *-ed* ending rather than *rone* which is formed using ablaut (on the same pattern as *ride ~ rode*). (But see section (12.2).)

Find some other examples where ablaut is used to form the past tense, with other changes of vowel.

Ablaut is not productive any more. Nonetheless, remnants of it affect a sizeable number of words in present-day English. Most of the words affected belong to the category of so-called **strong verbs**. In Old English (spoken circa 450–1100) many verbs formed their simple past tense form by a change in the root vowel. This can still be seen in their reflexes in modern English:

[5.19]

		Past tense	Past participle
Class I:	rise	rose	risen
Class II:	freeze	froze	frozen
Class III:	shrink	shrunk	shrank
Class IV:	bear	bore	born
Class V:	give	gave	given
Class VI:	know	knew	known
Class VII:	stand	stood	stood

Note: The past participle is the form of the verb that occurs after the auxiliary verb *have* as in *she has risen*, *she has known* etc.

If we know the class that a verb belongs to, we can predict how it is going to be affected by ablaut in the past tense and past participle. So, again, rather than deal with each verb individually, it is preferable to group together strong verbs belonging to the same class and to include in their lexical entries a marking that triggers the correct ablaut rule at stratum 1.

Ablaut may also be used in derivational morphology. Find two more examples of ablaut in English, but this time ablaut being used to derive nouns from adjectives.

If you have found this exercise difficult, now look back at the examples in [5.3] which illustrate this point.

Let us now turn to a different kind of stem-changing inflectional processes whose historical basis is **umlaut**. By umlaut we mean the fronting of a vowel if the next syllable contains a front vowel. Whereas ablaut was from the start a purely morphologically triggered root vowel alternation, umlaut was originally phonologically conditioned. Prior to 450 AD, in Pre-Old English (before the speakers of what was to become English migrated from Germany to England) there was a regular phonological process of umlaut that affected nouns in this way:

[5.20]

Nominative singular	Nominative plural
fōt 'foot'	fōtiz 'feet'
gōs 'goose'	gōsiz 'geese'
(where ō = long /o:/)	

First, /o:/ was fronted to /e:/ by umlaut because of the anticipation of the front vowel /i/ in the next syllable. This is the situation reflected by the spelling *fōt* (later *foot*) for /fo:t/ and *fēt* (later *feet*) for /fe:t/. (The double letters *e* and *o* represented long vowels.) Subsequently, the GVS raised the /e:/ to /i:/ and the pronunciation changed from /fe:t/ to /fi:t/ in the plural. In the case of the singular, the original /o:/ of /fo:t/ was raised to /u:/ by the GVS. (Later still, this /u:/ of /fu:t/ was shortened to [ʊ], resulting in /fʊt/; see [5.13] above.)

With the passage of time, the *-i* vowel of the plural suffix *-iz*, which had conditioned the umlaut in the original *fōtiz*, was lost. And with it went the phonological basis of umlaut. So, eventually umlauting became a purely morphological device for marking plurality in a small sub-set of nouns – just like ablaut is used to mark different forms of strong verbs. That is where the present-day *foot* /fʊt/ (singular) ~ *feet* /fi:t/ (plural) alternation comes from. A stratum 1 rule in the lexicon is required to deal with this method of plural-formation.

Umlauting is a fossilised process. There are very few nouns that use umlaut for plural formation. I list below some more examples to be dealt with at stratum 1 by the umlaut for plural rule:

[5.21]

Singular		Plural	
tooth	/tu:θ/	teeth	/ti:θ/
louse	/laʊs/	lice	/laɪs/
mouse	/maʊs/	mice	/maɪs/

The suffix *-en*, as I have already mentioned, is another irregular plural ending found at stratum 1. It too is a historical remnant. But in this case there is no historical ablaut or umlaut basis. It is found in:

[5.22]

	Singular	Plural
a.	ox	oxen
b.	brother	brethren (root vowel change /ɒ/ → /e/ and suffixation of -*en*)
	child	children (root vowel change /aɪ/ → /ɪ/ and suffixation of -*en*)

In *oxen* the plural is formed by simply suffixing *-en* to *ox*. But in the case of the *brother ~ brethren* and *child ~ children*, there is also a root vowel change, in addition to the suffixation of *-en*. Furthermore, as we saw in section (3.1.4), the stem extending formative *-r-* is interposed between the suffix and *-en* in these words.

Next we will consider examples of irregular plural inflection that have come in with loanwords borrowed mainly from Latin and Greek and have remained marginal to the system of English plural formation, being effectively confined to the original borrowed words with which they were introduced. Such plural suffixes are also placed at stratum 1.

[5.23]	Singular	Plural (Latin)
	addendum	addenda
	erratum	errata
	stratum	strata
	datum	data
	medium	media

As expected, irregular plural-formation at stratum 1 blocks the suffixation of the stratum 2 regular plural suffix *-s*. Hence we do not get *addendas* and *erratas* etc. (But see (6.2.4.2).)

Note in passing the interesting possibility of the **reanalysis** as a singular noun of what was originally a plural borrowed noun. This has happened to a few nouns borrowed from Latin which ended in *-um* in the singular and *-a* in the plural. Examples include *media* and *data*. Most English speakers normally treat *media* (in the sense of press, radio and television) and *data* as singular and noncount although they carry the latinate plural suffix *-a*.

Up to now we have assumed that the lexicon contains word-formation rules that put together bases and affixes to form words. But the precise form that these rules take has been left inexplicit. In the next section we are going to spell it out.

5.3 LEXICAL RULES

When morphological rules are formalised it is necessary to specify:

    (i)   the class of bases affected,
    (ii)   the affix that is attached,
   (iii) where exactly it is attached (is it a prefix or a suffix?),
   (iv) the class which the resulting word belongs to,

(v)   the stratum to which the affix belongs (and hence its general properties and the stratum at which it is attached).

Morphological rules that attach affixes to bases take this form:

[5.24]   At stratum n
Insert A in environment [Y_____ Z]$_X$
Output: *w*
(i.e. insert A in the environment of a preceding Y or following Z if a given morphological property or complex of properties symbolised as X is being represented; X, Y and Z are variables, and *w* is a mnemonic standing for word.)

Some nouns, e.g. *data*, *oxen* and *sheep*, form their plural at stratum 1. We need the rules in [5.25] to deal with them:

[5.25]   Stratum 1

Either	**a.**	Insert	/-ə/ in environment *-a*	[deɪt ⌐]$_{Noun\ +\ Plural}$ *dat-*
		Output:	/deɪtə/ *data*	
Or	**b.**	Insert	/-ən/ in environment *-en*	[ɒks⌐]$_{Noun\ +\ Plural}$ *ox*
		Output:	/ɒksən/ *oxen*	
Or	**c.**	Insert	⊘ in environment ⊘	[ʃi:p⌐]$_{Noun\ +\ Plural}$ *sheep*
		Output:	/ʃi:p/ *sheep*	
Or	**d.**	Insert	Replace /ʊ/ with /i:/ in nouns subject to umlaut (e.g. *foot*)	[fʊt ⌐]$_{Noun\ +\ Plural}$ ↓ /i:/
		Output:	/fi:t / *feet*	

In these lexical rules the various roots and affixes are in their phonological representation.

As we have already seen, if a noun receives its plural marking at stratum 1, this blocks plural assignment at stratum 2. But if no plural assignment takes place at stratum 1, the regular -*s* plural suffixation applies by default, as in the examples of *beds* and *pets* below.

[5.26] Stratum2
Insert -*s* in environment [Y_____]Noun + Plural
Output Y-*s*

Note: Y is a variable standing for any count noun whatsoever that is not assigned plural at stratum 1. So, we can replace Y with /bed/ or /pet/ etc., and the output will be /bed-z/, /pet-s/ etc., subject to the rules in [2.16].

In the lexical phonology and morphology model, lexical rules are **cyclic**. This means that phonological rules are coupled with morphological rules found at the same stratum in the lexicon. First, the morphological rules put together the morphs required to build a particular word. Then that word is cycled through (i.e. submitted to) the rules of the phonological module at that stratum which are triggered by that particular morphological rule (see [5.7]). So, in our example, when the words *bed-s* and *pet-s* are produced by the morphological rules, they are immediately submitted to the phonological rule of voice assimilation, which is standing by to apply. It applies, giving /bed-z/ and /pet-s/ respectively. In this way, the phonological rules indicate how the word created by the morphological rule is to be pronounced.

Each journey through all the morphological rules and all the phonological rules paired with them at a given stratum constitutes a **layer of derivation**. Thus, in our examples above a pass through the morphological and phonological rules in [5.25] constitutes part of the stratum 1 layer of derivation and a pass through the morphological and phonological rules in [5.26] constitutes stratum 2 layer of derivation.

Most of this chapter has focused on lexical rules that build word-structure at stratum 1 and stratum 2. We will end by briefly contrasting lexical rules with post-lexical rules which work on words once they are fully-formed and come out of the lexicon.

5.4 DIFFERENCES BETWEEN LEXICAL AND POST-LEXICAL RULES

As we saw in [5.7], the lexical morphology and phonology model has two compartments, the **lexical** and the **post-lexical**. So far, we have only considered rules in the lexical compartment. We have seen that the function of lexical rules is to build word-structure and we have emphasised the fact that they appear on two strata in the lexicon.

We are now going to examine the other compartment in the model which contains **post-lexical** rules. There are a number of key differences between lexical and post-lexical rules. I will highlight four.

(i) **Post-lexical rules** can apply across word boundaries
Whereas lexical rules only build words in the lexicon, post-lexical rules apply to words after they have been formed in the lexicon, and processed through the syntax. While the maximum domain of application of lexical rules is the word, post-lexical rules can apply, across word-boundaries, to words after they have been grouped together into phrases. Hence the post-lexical phonological rule system is also called **phrasal phonology**.

An example of a post-lexical rule applying across word-boundaries is the rule that optionally deletes a word-final alveolar stop in a consonant cluster if the next word begins with a consonant as in [5.27]:

[5.27]   last trip /lɑːst trɪp/ [/lɑːst trɪp] or [/lɑːs trɪp]
   lost property /lɒst prɒpətɪ/ [lɒst prɒpətɪ] or [lɒs prɒpətɪ]

(ii) Lexical rules are **cyclic**
As we have already seen, lexical rules are cyclic in the sense that at each stratum in the lexicon, it is necessary to make a pass through both the morphological and phonological rules which go with them. By contrast, post-lexical rules are not cyclic. There is no similar linkage of pairs to syntactic rules with phonological rules.

(iii) Lexical rules must be **structure-preserving**
This means that the output of each layer of derivation must be a word. A lexical rule may not produce a form that could not be a phonologically well-formed word in the language.

There are **canonical phonological patterns** (i.e. standard patterns) of segment structure, syllable structure and prosodic structure that severely restrict the kinds of morphemes and words that can appear in the lexicon of a language.

(a) Obviously, there are restrictions on segments that can appear in words of a language. Thus, the putative word /ɗasp/, which begins with an alveolar implosive /ɗ/, is ruled out because implosives are not part of the phoneme inventory of English.

(b) Lexical rules must not produce forms that violate phonotactic constraints on the canonical syllable structure patterns in a language (see 4.1.1). Thus **ltarpment* /ltɑːpmənt/ and **tsem* /tsem/ are not possible English words since [lt] and /ts/ are prohibited consonant sequences in syllable (and word) initial position (though, of course, both are allowed syllable finally as in *melt* /melt/ and *sets* /sets/).

(c) Every lexical item must have one main stress. So, no lexical rule may create a word like **'presta'pping* which has two main stresses.

By contrast, post-lexical rules are not subject to any structure-preserving constraint. They may have an output that is at variance with the canonical patterns of the language. For instance, though no word in the lexicon can begin with [ts], this sequence may occur phonetically when two words come together in utterances, in casual speech if vowels get elided, as in:

[5.28]   [tsnɒt]   'it's not'
        [tsbaʊt]   'it's about'
        [tsæm]   'it's Sam'

(iv) Post-lexical rules are **automatic** but lexical rules are not
Whereas lexical rules are exception-ridden, post-lexical rules are automatic and apply without exception to all forms with the requisite phonetic properties, regardless of any morphological properties. For our present purposes, by automatic I do not mean obligatory. Some post-lexical rules may be optional. Rather, by automatic I mean that if the necessary *phonetic* conditions are present, the rule can apply. There are no specific words or grammatical contexts in which its application is blocked.

There is hardly a lexical rule that applies to all the forms that it could affect in principle. We have already seen the unpredictability of word-formation processes in the lexicon, particularly at stratum 1. Thus, we cannot predict which bases will take the stratum 1 suffix *-th*, and which ones will not. We are allowed *leng-th*, *bread-th*, *dep-th* and *wid-th*, but not **tall-th*, **short-th*, **thick-th* or **narrow-th*.

Even 'regular' stratum 2 lexical rules are not entirely predictable. For instance, the meaning of the stratum 2 regular plural inflectional suffix *-s* (as in *books*) is not totally predictable. There are non-count nouns ending in the plural *-s* suffix that are always treated as syntactically plural – even when singular in meaning, e.g. *pliers*, *tights*, *trousers* and *shears*. But non-count nouns like *measles*, *rickets* and *mumps* with the same suffix are normally treated as syntactically singular, though some people treat them as plural. Here we see that the phonological behaviour of *-s* may be predictable, but the meaning realised by it may not be.

In contrast, post-lexical processes are automatic in the sense explained above, i.e. they are not subject to lexical exceptions nor can they be blocked in certain grammatical contexts. I will illustrate the exceptionless nature of post-lexical rules by considering the **glottalisation** rule that produces a glottal stop allophone [ʔ] of the phoneme /t/ in many varieties of British English, as seen below:

[5.29]   t → [ʔ]
      a.   in word final position
            /'kæt/ → ['kæʔ] 'cat' (noun)
            /ɪt/ → [ɪʔ] 'it' (pronoun)
            /'bʌt/ → ['bʌʔ] 'but' (conjunction)

b. before a consonant

/ˈketl] → [ˈkeʔl] 'kettle' (noun)

/ˈsetl/ → [ˈseʔl] 'settle' (verb)

/ˈkætfɪʃ/ → [ˈkæʔfɪʃ] 'catfish' (noun)

c. between vowels – if /t/ is initial in an unstressed syllable

/ˈwetə/ → [ˈweʔə] 'wetter' (adjective)

/ˈmiːtə/ → [ˈmiːʔə] 'meter' (noun'

/ˈbɒtəm/ → [ˈbɒʔəm] 'bottom' (noun)

/ˈputəˈbitəv ˈbʌtə/ → [ˈpuʔə ˈbɪʔə(v) ˈbʌʔə]

        'put   a bit   of butter'

        (verb)   (noun)   (noun)

The glottalisation rule applies blindly without taking into consideration the particular word that the relevant /t/ is in. This is typical of post-lexical rules. They apply anywhere so long as the phonetic conditions for their application are met.

Enough has been said to show that phonological rules applying in the lexicon, which are triggered by morphological rules, have different properties from phonological rules that are found in the post-lexical compartment. This is important from a theoretical point of view. It is one of the planks in the argument for recognising morphology and the lexicon as a separate component in our theory of language. But as our concern here is what happens in the lexicon, no more will be said about post-lexical rules.

EXERCISES

1.   Study the following data:

Suffix	attach to		Output
-al/(i)al/(u)al	[autumn]-al	→	autumnal
	[medicine]-al	→	medicinal
	[contract-(u)al]	→	contractual
	[resident]-(i)al]	→	residential
	[provinc]-ial]	→	provincial
	[sens]-ual]	→	sensual
-(ac)y	[democrat]-acy]	→	democracy
	[supreme]-acy]	→	supremacy
-er	[London]-er]	→	Londoner
	[village-er]	→	villager
-er	[hat]-er]	→	hatter
	[slate]-er]	→	slater

-er	[run]-er]s]	→	runners
	[sing]-er]s]	→	singers
-er	[quick]-er]	→	quicker
	[clean]-er]	→	cleaner

   (a) For each example determine the word-class of the bases that form
        the input to the suffixation process and the word-class to which
        the resulting word belongs.
   (b) At what stratum in the lexicon is each one of these suffixes found?
        Justify your answer.

2. (a) From which languages did English borrow the words in the two
        sets below?
   (b) Identify the number of suffixes in these words.
   (c) At what stratum in the lexicon is each plural suffix added? What is
        your evidence?

Set A		Set B	
Singular	Plural	Singular	Plural
stimulus	stimuli	phenomenon	phenomena
fungus	fungi	criterion	criteria
syllabus	syllabi	ganglion	ganglia
radius	radii	automaton	automata

3. At what stratum are the nouns *cook*, *guide* and *cheat* derived from the
   corresponding verbs? On what basis does one decide?

4. Write formal morphological rules using the notation introduced in this
   chapter to account for the formation of the past tense of the following
   verbs: *moved*, *baked*, *ran* and *hit*.

6 Insights from Lexical Morphology

6.1 INTRODUCTION

This chapter shows in detail how lexical morphology successfully deals with a number of analytical problems in English word-formation. The next chapter goes on to highlight recalcitrant problems that still defy this theory.

6.2 INSIGHTS FROM LEXICAL MORPHOLOGY

Lexical morphology provides us with the means of describing a number of morphological phenomena in an illuminating manner, with the word rather than the morpheme playing a pivotal role. This contrasts with the morphological models of the American structuralists in which the morpheme rather than the word enjoyed pride of place (see Bloomfield, 1933; Harris, 1942; Nida, 1949; Hockett, 1954, 1958). As mentioned (see section 5.1), in giving the word a key role, lexical morphology is more in tune with the word-based models of traditional, pre-structuralist approaches to morphology and modern word-and-paradigm morphology.

The centrality of words is enshrined in the stipulation that the output of each **layer of derivation** must be a possible word in the language. So, lexical rules must be **structure preserving**. The output of a layer of derivation cannot violate well-formedness constraints on *words*. (See the discussion in (5.4).)

The effect of this requirement can be seen, for example, in the constraints on segment sequences that apply to words *qua* words and not to morphemes or combinations of morphemes which will eventually be realised as words. In many Bantu languages, for instance, all words must end in a vowel, but morphemes need not do so. Most Bantu verb roots and verbal suffixes end in a consonant. So, because of the requirement that words must end in a vowel, there is a virtually meaningless vowel (usually it is *a*, but in a few tenses and in the subjunctive it is *e*) that is found at the end of every verb. This vowel is called by grammarians the **basic verbal suffix** (BVS). It can be seen in the following Luganda examples:

[6.1] **a.** ba-   lab-   a   'they see'
                they   see   BVS

      b. ba-   lab-   agan-   a   'they see each other'
                they   see   each other   BVS

    c. tu-   lab- is-   a   'we cause to see'
        we   see cause BVS

Whereas morphemes such as -*lab*-, -*agan*- and -*is*- which end in a consonant are permitted, words ending in a consonant are outlawed. The semantically empty formative -*a*, the basic verbal suffix, is attached to a verbal word to ensure that it meets the well-formedness requirement that words end in a vowel.

Recall the discussion in Chapter 3, where we pointed out that words rather than morphemes are the key elements in morphology. First, they are the minimal signalling units. All words must be independently meaningful but morphemes need not be. Further, we noted that when confronted with the problems of cumulative and overlapping representation of morphemes in fusional languages, a morpheme-based theory of word-structure runs into insurmountable problems. But these problems are avoided if the word is treated as the key morphological unit (cf. section 3.7).

The word is also a key unit for another reason: there are morphological processes whose input is normally a word and not just a morpheme. These processes include the following:

    (i)   **compounding**, where words like *school* and *teacher* are the input
          to the rule that produces *school teacher*;
    (ii)   **affixation** processes that have fully formed words as their input,
          such as the rules that prefix *re*- (meaning 'again') and the rule that
          suffixes -*ly*:

        [6.2] open ~ re-open quick ~ quickly
               write ~ re-write nice ~ nicely

    (iii) **conversion**, which changes the word-class of a pre-existing word
          without any overt change in the shape of the input (see section
          3.5):

        [6.3] staff$_{(N)}$ ~ staff$_{(V)}$ narrow$_{(A)}$ ~ narrow$_{(V)}$
               walk$_{(V)}$ ~ walk$_{(N)}$ cool$_{(A)}$ ~ cool$_{(V)}$

We will return to the treatment of conversion in more detail in (6.2.3) below.

6.2.1 Stratum Ordering Reflecting Morpheme Sequencing

We saw in the last chapter that the lexicon only lets in words that do not violate the canonical shape of morphemes in the language. However, having a permissible phonological representation is not sufficient to ensure that a string of sounds is a potential word. A further condition that has to be met is that the morphs representing morphemes in words must be arranged in a sequence that is allowed by the rules of word-formation in the language. Thus, the putative words *grestifier* and *dispregmentation*, are plausible potential words. This is because first, they contain sounds of the English phonological system arranged in ways that are phonologically permitted. Second, they contain morphs representing morphemes that are arranged in an order that is sanctioned by English grammar. The only thing odd about these 'words' is that while all the affix morphemes are found in the English lexicon, their root morphemes, *-grest-* and *-preg-*, are not.

We can segment the putative words *grestifier* and *dispregmentation* respectively as *grest-ifi(y)-er* and *dis-preg-ment-at-ion*. If, however, we juggled the affixes and produced *dis-preg-ion-ment-at(e)* or *grest-er-ify(i)*, such 'words', though still phonologically well-formed, would not pass for potential English words.

What is true of putative, nonsense words is true of established words. They have a fixed order of morphemes. You will be able to see this if you try to form words using the root and affix morphemes given below:

[6.4] Root morphemes: priv popul port
         Affix morphemes: de- -at(e) -ion

With the morphemes in [6.4] the words *deprivation, populate, population, depopulate, depopulation, port, portion, deport* and *deportation* can be formed. But morphology places restrictions on the order in which morphemes can be strung together in a word. For instance, it is imperative that *-ion* follows *-ate*. Putative words like **popul-ion-ate* or **deport-ion-ate* are strictly forbidden.

Evidently, our knowledge of word-structure includes knowledge of the sequence in which affixes are combined. Generally, the order of morphemes in a word is rigidly fixed. Rarely is there any scope for any departure from that pre-ordained order (but see section (11.6)). In this respect words differ drastically from sentences whose elements can be rearranged, within certain limits (e.g. *She came here often* can be turned into *She often came here* or *Often she came here*).

For the student of morphology, one of the challenges lies in providing an adequate account of the principles that determine the sequence in which affixes are added to roots. The theory of lexical morphology offers us insights into this problem.

Lexical morphology predicts that, when both stratum 1 and stratum 2 derivational affixes are present in a word, stratum 1 affixes are closer to the root than stratum 2 affixes. If you look back to the discussion in section (5.2.1), this follows automatically from the assumption that all affixes at stratum 1 are attached before any stratum 2 affixes, and is one of the major considerations in recognising the distinction between strata and layers of derivation in the lexicon.

Consider the data in [6.5]. Observe that in column C below, where one of the stratum 1 non-neutral pre-accenting suffixes (-*ic* or -*arian*) occurs together with the neutral, stratum 2 suffix -*ism*, the stratum 2 suffix -*ism* is at the outer edge of the word, following -*ic* or -*arian*.

[6.5]

A	B	C
'athlete	ath'letic	ath'let-ic-ism
'attitude	attitud(i)'(n)-arian	attitud(i)-'(n)arian-ism
an'tique	anti'qu-arian	anti'qu-arian-ism
'human	human-i't-arian	human-i't-arian-ism

Note: stem extenders are in brackets; -*ity* has the form -*it*- before the vowel of -*arian* (cf. hu'man-ity).

The theory correctly predicts that there are no words like *mongol-ism-ian, *athlet-ism-ic or *antiqu-ism-arian where stratum 2 suffixes are closer to the root than stratum 1 suffixes (cf. Kiparsky, 1982a, 1982b).

A further example should make this point even more clear. Consider the behaviour of the suffixes -*ise*/-*ize* and -*al*. First, using the data in [6.6] and [6.7], determine at what stratum each one of these suffixes is found:

[6.6]
computer	computerise
private	privatise
patron	patronise
real	realise

[6.7]
sentiment	sentimental
department	departmental
homicide	homicidal
medicine	medicinal

Hint: Mark the syllable that receives main stress in each pair of words.

I expect you to have decided that the suffix *-ize/-ise* is phonologically neutral and so it is a candidate for stratum 2. It has no effect on the placement of stress or on the segments in the base to which it is attached.

By contrast, the suffix *-al* must be at stratum 1 since it causes a (variable) relocation of stress. If the syllable immediately preceding the one with this suffix is **heavy** (i.e. if it contains a diphthong or long vowel (e.g. *homi'ci dal* [hɒmɪ'saɪdɫ]) or if it has a vowel followed by a consonant (e.g. *depart-'men tal*) it receives the stress. But if the immediately preceding syllable is **light** (i.e. contains just a short vowel nucleus), as in *me'dicinal* [me'dɪsɪnəl], stress falls two syllables before the one containing the suffix (see Katamba, 1989: 238–9).

Now, in what order do the suffixes *-ise/-ize* and *-al* occur if they are both attached to the following bases:

[6.8] industry   neuter   nation   verb   sentiment

As predicted by the theory, when both *-al* and *-ise/-ize* are present, they occur in the order *-al-* before *-ise/-ize*. Hence we get the words *industri-al-ise*, *neutr-al-ise*, *nation-al-ise*, *verb-al-ise* and *sentiment-al-ise*.

---

a. At what strata are the suffixes *-ity* and *-less* found?
b. Explain how the hierarchical ordering of strata rules out words like **homelessity*, **powerlessity* and **mercilessity*.

You will recall from the last chapter that *-ity* is a pre-accenting stratum 1 suffix while *-less* is a neutral stratum 2 suffix (5.2.1). Thus, words like *'home* and *'homeless* are stressed on the first syllable regardless of the presence or absence of *-less*. But, if the non-neutral stratum 1 *-ity* is attached, stress must fall on the syllable immediately before *-ity* (cf. *necessary* vs *ne'cessity*). Again, we see how constraints on morpheme sequencing are reflected in the hierarchical ordering of strata. Since *-ity* is attached earlier at stratum 1 and *-less* is suffixed later at stratum 2, it is obvious that adjectives derived by suffixing *-less* are unavailable to the rule that suffixes *-ity*. The stratum 2 suffix *-less* must be more peripheral in a word than the stratum 1 suffix *-ity*. Hence the ill-formedness of **homelessity*, **powerlessity* and **mercilessity*.

An important consequence of what we have seen is that the hierarchical ordering of strata means the ordering of morphological processes. The set of lexical processes taking place at stratum 1 precedes those taking place at stratum 2. Furthermore, all lexical processes precede post-lexical ones.

This ordering is not imposed on an ad hoc basis by a linguist but rather follows naturally from the banding together of broadly similar word-formation rules.

In the preceding examples we have been using the distinction between non-neutral (stratum 1) and neutral (stratum 2) in the lexicon to predict the sequence in which derivational affix morphemes appear in complex words. If this were the only way of predicting order, we might expect that if affixes of the same stratum co-occurred in a word, there would be no constraints on order. This is clearly not the case, as you will soon discover for yourself.

---

a   Which of the following suffixes are neutral: *full*, *-less*, and *-ness*?
b. Can any of the neutral suffixes occur together in a word?
c. If they can, are they ordered?
d. If they are ordered, account for the order.

All the suffixes *-full*, *-less*, and *-ness* are neutral (stratum 2) (see [5.17]). These neutral suffixes can occur together in a word but their ordering is subject to certain restrictions. The suffix *-ness* attaches to adjective bases to form abstract nouns while *-less* and *-ful* attach to nouns to form adjectives. These requirements dictate that *-less* or *-full* must be added first to a noun, turning it into an adjective, before *-ness* can be suffixed:

[6.9]   **a.**   'home$_N$-less$_A$-ness$_N$ (*'home$_N$-ness$_N$-less$_A$)
              'power$_N$-less$_A$-ness$_N$ (*'power$_N$-ness$_N$-less$_A$)
         b.   'care$_N$-ful$_A$-ness$_N$ (*'care$_N$-ness$_N$-ful$_A$)
              'cheer$_N$-ful$_A$-ness$_N$ (*'cheer$_N$-ness$_N$-ful$_A$)

As seen in [6.9], suffixes are entered in the lexicon with features like N, Adj, etc. When they are attached to words, the right-handmost suffix **percolates** (i.e. passes on) its category features to the entire word (cf. (12.4.1). Thus the right-handmost suffix functions as the grammatical **head** of the word: it determines the grammatical category of the entire word to which it is attached.

We saw earlier that the ordering of strata in the lexicon entails an ordering of processes. On the one hand, stratum 1 word-formation processes precede their stratum 2 counterparts. On the other hand, all lexical processes precede post-lexical ones. But within each stratum rules are *not* **extrinsically ordered**, i.e. the linguist does not stipulate the sequence in which the rules found at a given stratum will apply.

Instead, the rules are **intrinsically ordered**. There are universal principles which see to the sequencing of rules, where the order of application is an issue. For instance, if rule A **feeds** (i.e. creates the input to) rule B which is at the same stratum, then rule A must apply before rule B. Always the rule that does the feeding *will* apply first and create the forms that constitute the input required by the rule that is fed – otherwise the feed cannot apply.

This is what is going on in *homelessness* and the other examples in [6.9]. Until [*home*$_N$] has been turned into an adjective by the suffixation of *-less*$_A$ [*home*$_N$*-less*$_A$]$_A$) it is not possible to suffix *-ness*$_N$ which requires an adjective base. The suffixation of *-less* feeds the attachment of *-ness* ([*homeless*]$_A$*-ness*)$_N$. Hence, the order in which these suffixation rules apply must be *-less* before *-ness* (see below 6.2.4.1).

The way in which the rules interact also explains the non-occurrence of **power-ful-less* and **power-less-ful*. The suffixation of one of either *-less*$_A$ or *-ful*$_A$ **bleeds** (i.e. blocks) the suffixation of the other. The reason for this is simple. Both adjectival suffixes attach to noun bases, not to adjectives. Once you have added *-less*$_A$ or *-ful*$_A$ to a noun, the addition of another suffix that attaches to nouns (as in **[[[power]$_N$less$_A$]-ful]$_A$*) is blocked (cf. Koutsoudas et al., 1974; Ringen, 1972).

Presumably, the possibility of attaching both *-less* and *-ful* to the same noun is also ruled out for semantic reasons. Adjectives of the type **power-ful-less* or **power-less-ful* would have contradictory meanings.

Finally, let us consider the relative order of inflectional and derivational morphemes. Typically, where both derivational and inflectional morphemes are affixed at the same stratum as prefixes or suffixes, derivational morphemes occur nearer to the root than inflectional morphemes. From this we conclude that derivational processes apply before inflectional processes found at the same stratum. Thus, when both stratum 2 derivational suffixes like *-er* or *-ness* as well as the stratum 2 regular *-s* plural suffix are present, as in *work-er-s* (**work-s-er*) or *kind-ness-es* (**kind-s-ness*), the derivational suffixes are nearer to the root. This sequencing of inflectional and derivational morphemes was originally noted by Greenberg. It is his Language Universal 28:

> If both the derivation and inflection follow the root, or they both precede the root, the derivation is always between the root and the inflection.   (Greenberg, 1966: 93)

To summarise the discussion so far, the lexicon contains **morphological rules** that are paired with **phonological rules**. The **lexical rules** (i.e. morphological and phonological rules found in the lexicon) are organised in hierarchical strata. Rules belonging to the same stratum show morphological and phonological similarities. Furthermore, lexical rules are **cyclic**, i.e. the application of a morphological rule triggers the application of associated phonological rules. The word is built up from the root outwards. Consequently, stratum 1 affixes will be closer to the root than stratum 2 ones. Also, given the assumption of stratum ordering, affixes that are introduced at a later cycle presuppose the availability of information that only becomes available when rules of an earlier cycle have applied. Finally, normally derivational affixes are nearer the root than inflectional affixes.

6.2.2 Stratum Ordering and Productivity

As well as helping to predict morpheme order in a word, the hierarchical ordering of strata reflects the degree of generality of word-formation processes. Stratum 1 contains the more idiosyncratic word-formation processes while stratum 2 has the more general ones. In the last chapter we dealt with the phonological idiosyncrasies of the non-neutral stratum 1 affixes (see sections (5.2.1) and (5.2.2)) so we will not discuss them again.

Regarding meaning, stratum 1 affixes tend to be semantically less coherent (in the sense that they rarely have a regular predictable meaning) than stratum 2 ones. This will be evident from a comparison of the meanings of the typical stratum 1 and stratum 2 adjective-forming suffixes *-ous* and *-less*. The stratum 2 suffix *-less* has a regular predictable meaning, 'without', which it contributes to the word of which it is a part:

[6.10]   *X-less* means *'without X'*
pitiless   shameless   joyless   fatherless

By contrast, *-ous*, which is a typical stratum 1 suffix, does not have a meaning that can be so easily pinned down. The OED lists these among its meaning:

[6.11]   *-ous*
'abounding in, full of, characterized by, of the nature of'

As you can see in [6.12], the precise meaning of *-ous* is vague and very unpredictable. It seems to depend to some extent on the base with which it is combined. It is certainly not always clear which of the meanings listed by the OED *is* relevant in a particular word:

[6.12] | dangerous | curious | courageous | tremendous |
| pious | conspicuous | glamorous | rebellious |
| herbivorous | coniferous | ridiculous | odious |

The same point can be made about other stratum 1 affixes such as *-ist*, *-id* and *-ity* (see section (4.1.1) and *-ity*).

Aronoff (1976) contrasts the semantic unpredictability and vagueness of the stratum 1 (+-boundary) suffix *-ity* (which is added to nouns 'expressing a state or condition' (OED)) with the semantic predictability and unambiguousness of the comparable stratum 2 (#-boundary) suffix *-ness*. He considers data like those in [6.13] and [6.14]:

[6.13] The murder's *callousness* shocked the jury.

All nouns with the form *X-ous-ness*, which are formed by adding *-ness* to adjectives ending in *-ous* are predictable in meaning. Any such noun is paraphraseable as meaning either (i) 'the fact that something is X$_{Adj}$', or (ii) 'the extent to which something is X$_{Adj}$', or (iii) 'the quality of being X$_{Adj}$'.

However, semantic readings of the nouns formed with *-ous/-ity* are not so consistent. They may include one or more of the three meanings listed for *-ness*, as well as other meanings:

[6.14] a. The *variety* of the fruit in the market surprised me.
  b. *Variety* is seldom found in this desert.
  c. How many *varieties* of malt whisky do you stock?

Variety in [6.14a] has the meanings (i) and (ii) above; in [6.14b] it has meaning (iii) but in [6.14c] it has a somewhat different meaning.

According to Aronoff, the relative unpredictability of the meaning of an affix has implications for its productivity. Not wanting to lose face by misusing a word whose meaning they are unsure about, speakers tend to play it safe. Where there are two alternative affixes, one at stratum 1, with a regular predictable meaning and the other at stratum 2, with an unpredictable meaning, speakers will tend to opt for the predictable affix. So, when presented with the possibility of forming an unfamiliar word using either *-ity* or *-ness*, speakers tend to select *-ness*. For instance, many speakers prefer *perceptiveness* to *preceptivity* as the derived noun from the adjective *perceptive*.

Of course, the pressure on the speaker to play it safe is reinforced by the peculiar phonological effects induced by suffixing *-ity*. In addition to the possibility of getting the meaning wrong, there is the added hazard of getting the pronunciation wrong (by not making the necessary modifications in the base that *-ity* requires):

[6.15] | rapacious | rapacity | (not *rapaciousity) |
|---|---|---|
| pugnacious | pugnacity | (not *pugnaciousity) |
| credulous | credulity | (not *credulo(u)sity) |
| generous | generosity | (not *generity) |

Avoidance of potential embarrassment might well be a factor in the process of **language change** in so far as it contributes to the more regular stratum 2 word-formation processes becoming even more general and gradually squeezing out the less predictable stratum 1 processes.

I hope this has helped to persuade you that morphological theory is not an arcane science only of interest to those with the leisure to indulge their curiosity about words. It is a subject that has a direct bearing on the use of language by ordinary people.

6.2.3 Stratum Ordering and Conversion

Recall the discussion of **conversion** (also called **zero derivation** by some) in section (3.5). Conversion is a derivational process that involves no overt affixation. I claimed in [3.26] that the verb *head* is derived from the corresponding noun by conversion. You may have asked yourself, How do we know that the verb is derived from the noun and not the other way round? This is a valid question. The problem it raises is called the **directionality problem**. Is there a principled way of establishing the direction in which conversion goes?

Typically, the process of conversion adds an extra dimension of meaning. According to Marchand (1969), semantic considerations are paramount in determining the direction of conversion. The more basic member of the pair is the one whose semantic priority is implied by the other. It forms the basis of the semantic definition of the other member. So, following Marchand, in the case of *head*, we could say that the verb *head* is derived from the noun *head* since *to head* can be defined as 'to function as the head of'. The conversion process adds the semantic dimension 'to function as' to the basic meaning conveyed by the noun 'head'.

Unfortunately, while the semantic criterion looks plausible in the case of *head*, there are numerous other cases where it is questionable because neither member of the pair can convincingly be assigned semantic priority. For instance, it is unclear in the case of pairs like the noun *sleep* and the verb *to sleep* which is semantically more basic (though you might well have your own intuition).

Lexical phonology can throw some light on the phenomenon of conversion. Before we can see how, we need to digress and set up some necessary background. This concerns the differences in the characteristic stress patterns of disyllabic (two-syllable) nouns and verbs.

Mark the syllable that receives main stress in the following pairs of nouns and verbs derived by conversion:

[6.16]	Verb	Noun
	torment	torment
	protest	protest

> digest   digest
> progress   progress
> convict   convict

What generalisation can you make about differences in stress placement in nouns and verbs?

The pattern is simple. In English disyllabic nouns and verbs, a word stress rule that applies at stratum 1 in the lexicon places the main stress on the second syllable of a verb, but on the first syllable of a noun.

We know that: (i) all lexical words are required by a well-formedness condition to bear stress (there are no unstressed lexical words in English), and (ii) stratum 1 non-neutral suffixes like *-ee*, *-al*, *-ity*, *-ory* and *-ic* can cause stress to move around in a word. Since these affixes are attached at stratum 1, stress must be present at stratum 1 for them to move it. Lexical stress must be assigned and shifted at stratum 1.

We are now ready to tackle conversion. The analysis outlined here draws on Kiparsky (1982b: 12). Kiparsky suggests that the difference in stress placement displayed by nouns and verbs formed by conversion can be accounted for it we assume that: (i) some nouns are formed from verbs, and that (ii) some verbs are formed from nouns. The conversion of verbs into nouns takes place at stratum 1 while the conversion of nouns into verbs occurs at stratum 2. Evidence for these claims comes from stress placement.

When nouns are derived by conversion from verbs at stratum 1, this conversion is non-neutral. Morphological conversion feeds the derived nouns to the stratum 1 phonological rule that places stress on the first syllable in nouns.

[6.17] Verb   →   Noun
        sur'vey   'survey
        tor'ment   'torment
        pro'test   'protest

But when verbs are formed from nouns by conversion at stratum 2, this conversion is neutral. Stress does not shift from the first syllable to the second:

[6.18] Noun   →   Verb
        'pattern   'pattern (*pat'tern)
        'advocate   'advocate (*ad'vocate)
        'patent   'patent (*pa'tent)
        'lever   'lever (*le'ver)

Derived verbs escape the stratum 1 rule that places stress on the second syllable because they are formed, not at stratum 1, but at stratum 2.

The hypothesis that the lexicon contains strata that are arranged in a hierarchy not only captures the stress facts, but also matches the differences in productivity between **deverbal nouns** (i.e. nouns-from-verbs) as opposed to **demoninal verbs** (i.e. verbs-from-nouns) that are formed by conversion. The formation of nouns from verbs is less common than the formation of verbs from nouns. The majority of nouns can have a verb formed from them by conversion but the reverse is not true (cf. Clark and Clark, 1979).

This correlates with nouns being formed from verbs at stratum 1 while verbs are formed from nouns at stratum 2. Thus, as we have already established by comparing otherwise similar affixes like *-ity* (stratum 1) and *-ness* (stratum 2), lexical morphology correctly predicts that a morphological process which takes place at a lower stratum (e.g. stratum 2) is likely to be more productive than a morphological process that takes place at a higher stratum (e.g. stratum 1).

The assumption that verbs derived by conversion are created at stratum 2 receives further support form another source, namely the treatment of irregular verb inflection. For example, normally, verbs ending in *-ing* or *-ink* are **strong verbs** and form their past tense by ablaut. The stem vowel is changed from /ɪ / to /æ/, resulting in *-ang* [æŋ] or *-ank* [æŋk]:

[6.19]	**a.**	Present tense	Past tense	**b.**	Present tense	Past tense
		sink	sank		sing	sang
		stink	stank		spring	sprang
		shrink	shrank		ring	rang

Now explain why the verbs in [6.20], which are derived by conversion from nouns, fail to undergo ablaut in the past tense despite the fact that phonologically they are similar to the strong verbs in [6.19]. Why is the *-ed* past tense suffix used instead of the /ɪ / → /æ/ ablaut?

[6.20]	Noun	→	Verb	→	Past tense
	link		link		linked (*lank)
	ring		ring		ringed (*rang)

The answer is straightforward. Verbs are derived from nouns by conversion at stratum 2. Hence, although they have the *-ink* or *-ing* phonological shape, the denominal verbs in [6.20] elude the ablaut rule which only applies at stratum 1. Because denominal verbs are formed at stratum 2, they are unavailable at the stage when the ablaut rule applies. They can only receive the regular *-ed* inflection which is assigned at that stratum 2 (see section (5.2.2)).

We have shown that the theory of lexical morphology predicts that irregular processes operative at stratum 1 have no access to productive word-formation processes applying later at stratum 2. This means that historical relics like ablaut do not affect new forms entering the language as a result of the application of productive rules (like conversion) applying at a stratum lower down the hierarchy.

6.2.4 The Strict Cycle Condition

At the end of the last section we saw that the denominal verbs ending in *-ing* or *-ink* fail to undergo the ablaut rule. The past tense of *link* is not **lank*. This follows from a fundamental tenet of lexical phonology, namely the **Strict Cycle Condition** (also called **strict cyclicity**) (Mascaró, 1976; Kiparsky, 1982a: 154; Goldsmith, 1990: 249–73).

This principle means that, in the lexicon, rules are blocked off in such a way that each layer of derivation (see p. 106) is self-contained. Rules are only capable of affecting structures that are built by other rules belonging to the same stratum. Hence, in our example, the ablaut rule that replaces /ɪ/ with /æ/ (as in *sing* → *sang*) can only affect items that are available at stratum 1. Since *link* does not come into existence as a verb until stratum 2, there is no way that ablaut could apply to it.

Recall the discussion of the **trisyllabic laxing** rule in (section (5.2.1)). In derived words, this rule changes a **tense vowel** (i.e. a long vowel or diphthong) of a stem to a **lax vowel** (i.e. a short vowel) whenever suffixation creates a word of three or more syllables:

[6.21] di-vi-ne ~ di-vi-ni-ty   vain ~ va-ni-ty
         (/aɪ/ →/ɪ/)   (/eɪ/ → /ɪ/)

         con-fi-de ~ con-fi-dent   ath-lete ~ ath-le-tic
         (/aɪ/ →/ɪ/)   (/iː/ → /e/)

Bearing in mind the trisyllabic laxing rule, use the principle of Strict Cyclicity to explain why the tense stem vowel in words like *nightingale* /naɪtɪŋgeɪl/ and *ivory* /aɪvərɪ/, which appear to meet the requirements of the trisyllabic laxing, fail to undergo it. (If trisyllabic shortening applied, instead of [naɪtɪŋgeɪl] and [aɪvərɪ] we would say **[nɪtɪŋgeɪl] and **[ɪvərɪ].)

The explanation is that, although these words contain the minimum of three syllables required by the trisyllabic laxing rule, this does not arise through suffixation. Both words contain just a root morpheme on its own. They are entered in the lexicon as fully-formed **underived words**. The Strict Cycle Condition correctly predicts that since neither *nightingale* nor *ivory* is formed at the stratum where trisyllabic laxing applies, these words will not be affected by that rule.

The Strict Cycle Condition can be further illustrated with the example of **velar softening**, which was introduced on p. 76. The rule of velar softening turns /g/ into [dʒ] and /k/ into [s] when these consonants are followed by nonlow front vowels:

[6.22] **a.** /g/ → [dʒ]   before a front vowel (i.e. [ɪ], [iː], [e])
       b. /k/ → [s]   before a front vowel (i.e. [ɪ], [iː], [e])

(based on SPE: 219)

The effects of velar softening are shown in [6.23]. In the left hand column /g/ and /k/ are realised as [g] and [k] since they are not followed by nonlow front vowels. But in the right hand column they are realised as [dʒ] and [s] because they are followed by nonlow front vowels:

[6.23] **a.**

		g		*dʒ*
	analogue	[ænəlɒg]	analogy	[ənælɒdʒɪ]
	regal	[riːgəl]	regicide	[redʒɪsaɪd]
	rigour	[rɪgə]	rigid	[rɪdʒɪd]
b.		*k*		*s*
	critic	[krɪtɪk]	criticism	[krɪtɪsɪzm]
	medical	[medɪkəl]	medicine	[medɪsɪn]
	electrical	[elektrɪkəl]	electricity	[elektrɪsɪtɪ]

<u>Note</u>: the letter *g* often spells an underlying /dʒ/ as in *generation*, *allege*, *giant* etc. It is only the underlying velars that interest us.

Study these data:

[6.24] **a.** kilt   [kɪlt]   (*[kɪlt])
          bucket   [bʌkɪt]   (*[bʌsɪt])
      b. gelding   [geldɪŋ   (*[dʒeldɪŋ])
          geezer   [giːzə]   (*[dʒiːzə])

Using the principle of Strict Cyclicity, explain how the /k/s and /g/s in words like those above escape becoming softened to [s] and [dʒ] respectively.

Strict Cyclicity confines rules to affecting only those forms that are built on the particular stratum where a given rule is found. So, in [6.24], the nouns *kilt, bucket, gelding* and *geezer* escape velar softening because they are underived words. Velar softening only applies in derived words, when a suffix beginning with a nonlow front vowel is attached to a base that ends in a velar stop.

An ancillary principle, which restricts what rules can do and reinforces Strict Cyclicity, is the **Bracket Erasure Convention** (Mohanan, 1982; Pesetsky, 1979). At the end of each layer of derivation (recall section 5.3), information concerning bracketing and any morphological, phonological or other properties *internal* to the word is obliterated by the Bracket Erasure Convention. The output of each layer of derivation is a word – and it is only the fact that it is a word that matters. The existence of the Bracket Erasure Convention means that all words are treated in the same way when they enter the next stratum. Their internal structure is not taken into account.

So, in [6.25], a rule such as 'add -*s*' to a noun to make it plural will only look to see whether its potential input is a noun. All the information encoded in the brackets in column A is removed before the rule that adds -*s* applies. When suffixing -*s* to form the plural, the internal structure and derivational history of a noun are immaterial (see column B). Simple nouns with a stem consisting of just a root morpheme, stems that contain a root and an affix as well as compound nouns all get their plurals in the same way. And once the plural has been formed, as in column C, all plural nouns lose their internal brackets and are treated in the same way:

[6.25]   A   B   C

$[defend_V]ant]_N$, agent $\rightarrow$ $[defendant]_N$ -s$]_N$ $\rightarrow$ $[defendants]_N$

$[[book_N][shop_N]]_N$ $\rightarrow$ $[bookshop]_N$ -s$]_N$ $\rightarrow$ $[bookshops]_N$

$[pet]_N$ $\rightarrow$ $[pet]_N$ -s$]_N$ $\rightarrow$ $[pets]_N$

The effect of the Bracket Erasure Convention is to ban any surreptitious encoding of information which is relevant to phonological or morphological rules. Rules are not allowed access to information concerning the internal structure of a word that is provided at an earlier stratum. Thus the Bracket Erasure Convention puts Strict Cyclicity on a firmer footing. It makes it impossible to access information that belongs to a stratum other than the one where a given rule applies. The only information that phonological and morphological rules can access is that contained in their immediate input and specified on their stratum.

Furthermore, the Bracket Erasure Conventin ensures that all information that is in any way peculiar to a word is wiped out by the time the word reaches the **post-lexical** module. Hence, no post-lexical rule can refer to exceptional features or any other idiosyncratic properties of words. A rule like the one that optionally yields a glottal stop [ʔ] allophone of /t/ in *meter* and *water* is unable to see morphological information such as the fact that *meter* is a Greek loanword while *water* is a native word. Such information is irrelevant. Post-lexical rules apply exceptionlessly to any potential input that has the appropriate phonological characteristics.

We will now leave Strict Cyclicity and bracket erasure and turn to the **Elsewhere Condition**, a related principle that regulates the application of rules.

6.2.4.1 The elsewhere condition

Strict Cyclicity limits rules to applying only to those forms that are built on the particular stratum where a rule is found. This still leaves open the question of determining the order in which rules at the same stratum apply. The order in which the rules at each stratum apply is not determined by some stipulation imposed by the linguist. Rather they apply following the general principle that where several rules can potentially apply in a derivation, the interaction between them that maximises the chances of each rule applying is what the theory dictates. In other words, every opportunity to apply should be given to each rule at the stratum where its input requirements are met. We touched on this subject in section (6.2.1) when we saw that if rule A feeds rule B, rule A must apply first, creating the input to rule B, otherwise B cannot apply.

If rules are in a **bleeding** rather than a feeding interaction, a different approach is necessary. By rules being in a bleeding relationship I mean that two or more rules target the same type of inputs but they cannot all apply. Only one of them can apply – and in so doing it pre-empts the others. This is where the **Elsewhere Condition** comes in. For a long time, in linguistics, there has been a convention of stating rules starting with the most restricted, applying to narrowly specified sub-classes, and ending with the most general ones, which apply 'elsewhere'. The idea is that, unless you are told that a *special* rule is applicable, assume that 'elsewhere' the *general* rule applies.

The Elsewhere Condition guarantees the priority of the more restricted rule over the more general one. For example, consider the irregular stratum 1 plural-formation in English. The latinate form *larv-a* has the plural form *larv-æ*. This is a very restricted pattern, even by the standards of latinate roots. So, by the Elsewhere Condition, it must be assigned before other stratum 1 plural suffixes like *-a* in *strat-a* or *-i* in *cacti*, etc. We will discuss the implications of this fully in the next section.

6.2.4.2 Blocking

In section (4.2.1) we saw that one of the factors that limits the productivity of word-formation rules is **blocking**: the application of an earlier rule may thwart the application of a later one. Lexical morphology offers a coherent description of this phenomenon using the principles I have just outlined.

With the Elsewhere Condition in force, the presence of a less general rule may prevent the application of a related more general rule at the same stratum as we saw with the latinate -*ae* and -*a* plural suffixes at the end of the last section.

Blocking may also involve rules on different strata. As we have already seen, this is accounted for using the principles of stratum ordering discussed in sections (6.2.1), (6.2.2) and especially (6.2.3). We have seen in this chapter that stratum ordering mirrors productivity. More productive processes apply later than less productive ones. Thus, in the lexical module, stratum 1, which is at the top of the hierarchy, contains the least general lexical rules and stratum 2, lower down, contains the most general ones. The post-lexical module (which is lower than the lexical module) contains processes that may apply exceptionlessly whenever phonological conditions on their inputs are satisfied.

We will now examine the same phenomenon from a semantic angle. Both derivational and inflectional processes at higher strata tend to be more idiosyncratic in their meaning than their counterparts on lower strata. Normally, where both stratum 1 and stratum 2 affixes can be added to the same stem or base, semantic divergence occurs.

Consider the case of the word *brother*, which is typical. Suffixation of the stratum 1 plural suffix -*en* to *brother* blocks the suffixation of the regular stratum 2 suffix -*s*, with the same meaning. Consequently a semantic split has occurred. *Brother* has a regular plural *brothers*, formed at stratum 2, which means 'more than one brother' and an irregular plural *brethren*, which is formed at stratum 1. The latter has a distinct, very idiosyncratic meaning. It refers to members of a religious sect such as the *Plymouth Brethren*.

Turning from inflectional to derivational affixes, we observe a similar tendency. While the stratum 2 -*er* nominalising suffix has the standard meaning 'someone who does X', the equivalent stratum 1 -*ant* has that meaning too (as in *applicant*) as well as some idiosyncratic meanings (as in *defendant* and *communicant*).

Show how blocking can be used to account for the ill-formedness of the words in columns C and D below.

[6.26]	A	B	C	D
a.	profligate	profligacy	*profligaci(y)ness	*profligateness
	advocate	advocacy	*advocaci(y)ness	*advocateness

b.	decent	decency	*decenci(y)ness	*decentness
	complacent	complacency	*complacenci(y)ness	*complacentness
c.	aberrant	aberrancy	*aberranci(y)ness	*aberrantness
	constant	constancy	*constanci(y)ness	*constantness

Note: t → s before + y (/ɪ/).

We noted above the tendency to avoid creating synonyms of pre-existing words (see section (4.2.1); Aronoff, 1976: 55; and Kiparsky, 1983). Hence, the existence of nouns formed at stratum 1 by attaching the suffix -(c)y, to bases ending in -ate, -ant or -ent thwarts the suffixation of the more productive stratum 2 suffix -ness. Having received the suffix -(c)y at stratum 1 and become nouns with the meaning 'having the quality or state of', the bases in [6.26] are unavailable to the rule that attaches the equivalent -ness suffix which derives nouns from adjectives at stratum 2. (So, all the forms in [6.26C] are disallowed). The alternative of missing out the suffixation of -(c)y at stratum 1 and going for -ness suffixation, as in [6.26D], is also precluded.

In the following nouns, which are derived from verbs, why is attachment of the -er suffix (found in make$_V$ → maker$_N$, keep$_V$ → keeper$_N$, etc.) blocked?

[6.27]	a.	bore	*borer	b.	applicant	*applier
		guide	*guider		accountant	*accounter
		spy	*spier		participant	*participater (*participator)
		judge	*judger		intoxicant	*intoxicater (*intoxicator)

As shown above in section (6.2.3), nouns formed by **conversion** from verbs (as in 6.27a) are derived at stratum 1. We saw earlier that -er is a neutral suffix and so it is found at stratum 2. So, conversion done at stratum 1 will pre-empt the application of -er suffixation at stratum 2.

The same thing is also happening in [6.27b] where nouns are derived from verbs by adding -ant. The derived nouns have the meaning 'the agent (if it is a person) who does X' or 'the instrument with which one does X' (with X being whatever action or event the verb designates). The suffixation of -ant at stratum 1 blocks the subsequent formation of nouns with the same meaning by adding -er at stratum 2.

Does blocking unfailingly take place? No, is the answer. Notwithstanding its importance, blocking is only an important tendency. It is not a mandatory requirement. Sometimes, despite the existence of a noun formed at stratum 1 which contains the relevant affix, stratum 2 suffixation does take place, resulting in doublets. But, as a rule, when this happens there is semantic specialisation, as you can verify in [6.28]:

[6.28]	A Verb	B Noun	C Noun
	guide	guide	guider
	divide	divide	divider
	cook	cook	cooker
	drill	drill	driller
	defend	defendant	defender

(data from Kiparsky, 1982b)

a. What is the difference in meaning between the deverbal nouns in column B (which are the product of conversion) and the nouns in column C (which are formed by *-er* suffixation)?
b. How could we prevent blocking so that the creation of the nouns in column B by conversion at stratum 1 does not prevent the formation of the nouns in column C by *-er* suffixation at stratum 2?

I follow Kiparsky (1982b) in answering this question. The difference in meaning between the noun and verb pairs in columns B and C is as shown below:

[6.29]	a.	*guide*	person who guides
		guider	member of the Girl Guides
	b.	*divide*	thing (not a person) that divides e.g. a line, ridge
		divider	person or device that divides
	c.	*cook*	person that cooks
		cooker	appliance for cooking on or in
	d.	*drill*	instrument for drilling
		driller	person using a drilling instrument
	e.	*defendant*	person defending him or herself (especially in a law court)
		defender	any person defending (who is not a defendant)

Normally, where only the stratum 2 suffix *-er* can be selected, it has one of two predictable complementary meanings. Either it turns the verb into an agentive noun referring to a human (e.g. *write → writer*), or else it turns the verb into an instrumental noun referring to an inanimate entity (e.g. *rub → rubber*).

What is peculiar about the nouns in [6.29] is that the selection of a stratum 1 suffix does not block the suffixation of a related stratum 2 suffix. Either a stratum 1 or a stratum 2 suffix is allowed to occur with the same base, but normally with a different meaning. The meaning of a particular deverbal noun needs to be stated by a special stipulation. Although the usual instrumental and agentive meanings are found here, there is no way of knowing in any particular case whether a word will have the instrumental or the agentive meaning.

The only prediction of sorts which we can make is that usually the creation of words at stratum 2 which are synonymous with words created at stratum 1 is avoided. The selection of an agentive meaning at stratum 1 inhibits the selection of that same meaning at stratum 2. Thus, if an agentive meaning is assigned at stratum 1 (e.g. by conversion) as in *cook*, then the word derived by suffixing *-er* at stratum 2 will have the instrumental meaning (cf. *cooker*). Conversely, if the word formed by conversion at stratum 1 has an instrumental meaning (cf. *drill*), the deverbal noun derived at stratum 2 by suffixing *-er* will have an agentive meaning. The situation is similar to what we saw in inflectional morphology with respect to *brothers* ~ *brethren* (on p. 127). We have separate meaning associated with the words derived at the two different strata.

But the complementariness noted above is not due to a cast iron rule. A few words (e.g. *divide*$_{(N)}$ ~ *divider*$_{(N)}$) defy blocking. Conversion produces the noun *divide* meaning 'a thing (i.e. instrument) which divides' yet the instrument meaning assigned at stratum 1 in *divide*$_{(N)}$ fails to block the instrumental meaning assigned at stratum 2. So, *divider*$_{(N)}$ can be either a person or thing that divides.

How can blocking be used to explain the non-occurrence of the starred forms in [6.30].

[6.30]	Singular	Plural	
	ox	oxen	*oxes
	foot	feet	*foots

Finally, blocking throws some light on allomorphy in *inflectional morphology*. Irregular plural inflection that takes place at stratum 1 in words like *oxen* and *feet* prevents the stratum 2 regular plural inflection rule from creating words like *oxes* and *foots*. By the stage when stratum 2 is reached, the information that *oxen* is a plural noun, i.e. [oxen]$_{Noun + Plural}$, is encoded by the external brackets which are not removed by the Bracket Erasure Convention as we go from stratum 1 to stratum 2. So, this information is available to block regular plural formation at stratum 2. This example illustrates a general point: where there are several inflectional allomorphs available, any one stem will normally select just one of these allomorphs (see section (10.2.3) below).

Given blocking, how can we account for the existence of the inflectional doublets below?

[6.31]

Regular stratum 2 plural	Irregular stratum 1 plural
syllabuses	syllabi
cactuses	cacti
ganglions	ganglia
automatons	automata
formulas	formulae

Although blocking is the norm in inflectional morphology, in [6.31] we see that blocking is not always totally complete. Unlike the case of *brothers* and *brethren* which we discussed earlier, here the selection of different plural allomorphs does not necessarily correlate with any semantic difference. To deal the problem we need to recognise that some irregular (stratum 1) inflectional rules are **obligatory** and others **optional**. For example, the appropriate stratum 1 plural assigning rules obligatorily apply to forms like *foot* and *ox*, giving *feet* and *oxen*. However, some other stratum 1 inflectional processes like those in [6.31] are optional. If the stratum 1 rules that add *-i* or *-a* are chosen, plurals like *automata* and *syllabi* are produced. But if the stratum 1 plural formation rules are ignored and the regular stratum 2 *-s* plural ending is chosen instead, the output is *automatons* and *syllabuses*. (The same speaker may use either the irregular or the regular plural, with somewhat different meanings in different collocations – as in *mathematical formulae* vs *dental formulas*.)

To summarise, lexical morphology has thrown light on several aspects of word-formation, in particular: (i) constraints on the order of morphemes in a word, (ii) productivity, (iii) the direction of conversion, and (iv) blocking.

Key among the theoretical notions that enable the theory to perform these tasks are the following: Strict Cyclicity, the Bracket Erasure Convention and the Elsewhere Condition.

EXERCISES

1. Examine the following data:

Singular	Plural
sheep	sheep (*sheeps)
salmon	salmon (*salmons)
grouse	grouse (*grouses)

    (a)   On what stratum is the zero plural suffix found?
    (b)   How are plural forms *sheeps*, *salmons* and *grouses* blocked?

2. (a)   In the list of words below, separate those which undergo trisyl-
           labic laxing from those which do not.
    (b)   Write a rule to account for the application of trisyllabic laxing.

           provision   baloney   arena
           Oberon   sanity   tunic
           insane   angelic   stevedore
           inclination   application   chastity

3. Study the following data:

A	B	C
education	education-al	education-al-ly
theatric	theatric-al	theatric-al-ly
finance	financ-(i)-al	financ(i)-al-ly
universe	univers-al	univers-al-ly
artifice	artific-(i)-al	artific(i)-al-ly
adjective	adjectiv-al	adjectiv-al-ly

    (a)   Determine the stratum on which the suffixes -*al* and -*ly* are found.
    (b)   Explain the order of the suffixes -*al* and -*ly*.
    (c)   Write formal morphological rules using the notation introduced
           in the last chapter to derive *universal* and *universally*.

4. At what stratum should the prefix *un-* in these data be placed?

    unpalatable   unbearable   unmanageable   unloveable   unknown
    unreadable   unjust   uncooperative ungrateful   unnerve

5. Discuss the semantic specialisation observed where doublets are
    formed by adding to a base either the stratum 1 suffix -*ant* or the
    stratum 2 suffixes -*er* and -*or* as in the following:

A	B
servant	server
defendant	defender
protestant	protester
informant	informer
commandant	commander
stimulant	stimulator
refrigerant	refrigerator

7 Lexical Morphology: An Appraisal

7.1 INTRODUCTION: THE CLAIMS MADE BY LEXICAL PHONOLOGY

Before we consider the objections to lexical phonology, I will enumerate some of the key assumptions and claims of this model, as presented so far (see especially section 5.1).

- (i) Central to lexical morphology is the principle that the lexicon is hierarchically organised by **strata** which are defined on the basis of the properties of **affixes**.
- (ii) Affixes (unambiguously) belong to one stratum. Each stratum is uniquely associated with a particular set of affixes which share various phonological and morphological properties.
- (iii) All lexical morphologists ideally want a minimum number of lexical strata but they do not agree on what that minimum number should be.
- (iv) There is an intimate relationship between the phonological and morphological rules of each stratum in the lexicon. Every stratum is **cyclic**. The output of the morphological rules at each stratum is subjected to associated phonological rules at the same stratum.

    Furthermore, each morphological rule is restricted to one stratum. The Bracket Erasure Convention removes *internal* brackets which show the inner structure of the word. Rules applying at a later stratum cannot access information about the internal structure of a word that was created at an earlier cycle. The output of each stratum is a word.
- (v) Morpheme ordering reflects the hierarchical ordering of strata. Stratum 1 affixes are nearer to the root than stratum 2 affixes and derivational affixes are nearer the root than (regular) inflectional ones.

7.2 CRITICISMS OF LEXICAL PHONOLOGY

Each of the above claims of lexical morphology has been contested by sceptics (cf. Goldsmith, 1990; Spencer, 1991, among others). Below I present their objections in turn, together with a defence of lexical morphol-

ogy. As you will see, many of the criticisms have been answered satisfac-
torily, but a few have not.

7.2.1 Are Lexical Strata Determined by Affixes Rather than Roots?

While accepting the lexical morphologists' claim that the lexicon has
hierarchical strata, it has been proposed that it is *roots* rather affixes that
determine lexical strata. Thus, Goldsmith (1990: 260–2) argues that the
case for regarding affixes as the determinants of strata is at best not yet
proven. He suggests that instead of putting affixes on different strata as is
done in lexical phonology, or recognising differences in boundary strength
on the basis of the properties of affixes as is done in SPE where the +
boundary corresponds to stratum 1 and the # boundary corresponds to
stratum 2 (see section 5.2), one could distinguish between classes of **bases**
(more specifically **roots**). If this stance is adopted, the fact that *-ity*, *-ous*,
-ible and *-ic* do not attach to *tall* (**tallity*, **tallous*, **tallible* and **tallic*) is as
much a fact about the base *tall* as it is a fact about the suffixes.

Arguably, we could set up a class of roots that change when an affix is
present (see set A in [7.1]) and another class containing affixes that do not
change (see set B in [7.1]).

[7.1] a. Set A: Changing roots (Stratum 1)

wide [waɪd]: shortens to [wɪd] before *-th* in [wɪdθ] *width*;
broad [brɔːd] undergoes ablaut and changes to [bred] before *-th*
in *breadth*

b. Set B: Non-changing roots

warm [wɔːm] remains unchanged before *-th* as in [wɔːmθ]
warmth
grow [grəʊ] remains unchanged before *-th* as in [grəʊθ] *growth*

Defend the principle of stratum ordering in the lexicon based on affixes.
Use the data in [7.2] to construct a case against basing lexical strata on the
distinction outlined above between changing and non-changing roots.

[7.2]

A	B	C
various	variety	variously
[veɪrɪəs]	[vəraɪətɪ]	[veɪrɪəslɪ]
tenacious	tenacity	tenaciously
[tɪneɪʃəs]	[tɪnæsɪtɪ]	[tɪneɪʃʌslɪ]

A proponent of the principle of stratum ordering based on affixes could rightly point out that the proposal to base stratum ordering on roots is thwarted by the fact that the same root may show both the changing and non-changing patterns before different suffixes. Thus, before the suffix *-ity* in column B, all the roots are modified. But before suffixes *-ous* and *-ous-ly* in columns A and C they remain unchanged. From this evidence, we would be led to the confusing conclusion that the same root may simultaneously belong to stratum 1 and stratum 2.

To avoid this messy situation, in standard lexical morphology it is hypothesised that it is not the inherent characteristics of the root that determine whether it alternates before a given suffix. Rather, alternation is imposed on roots (and bases generally) from outside by affixes. Stratum 1 affixes are capable of triggering changes in the base to which they are attached while stratum 2 affixes are not. It is this difference in the behaviour of affixes that the hierarchical ordering of strata in the lexicon should endeavour to capture. Hence, stratum ordering should be justified on the basis of the properties of affixes rather than roots (or bases in general).

7.2.2 Do Affixes Uniquely Belong to One Stratum?

For the principle of recognising strata on the basis of affixes to work reliably, it should be the case that each affix belongs to just one stratum. If affixes belonged simultaneously to several strata, that principle would be subverted.

Goldsmith (1990: 262) argues that this is indeed the case. His analysis shows that, for instance, the suffix *-ize* is at stratum 1 in words like *'synchron[y]ize* and *Ca'tholicize*. It is non-neutral in these words since it conditions the deletion of the final [ɪ] of the base *'synchrony*; and in *Ca'tholicize* it shifts stress from the first to the second syllable in the base *'catholic*. But the trouble is that in *Ber'muda-ize* the same suffix occurs at stratum 2. There it is phonologically neutral. Besides, it even produces a vowel sequence where a non-high vowel ([ə]) is followed by another vowel ([bɜː'mjuːdə-aɪz]) – a sequence which is forbidden in monomorphemic roots and in forms derived at stratum 1. So, Goldsmith concludes, contrary to the claim of lexical morphology, the same affix can simultaneously belong to two strata. If this is the case, setting up strata on the basis of the properties of affixes becomes untenable. This, of course, is essentially the same kind of argument that we deployed (using the evidence in [7.2]) to oppose the claim that roots and bases are the determinants of lexical strata.

In reply to this criticism, I will suggest that the standard lexical morphology can be defended in most cases of this kind. I will argue that (normally) affixes belong to just one stratum. If the same affix appears to belong to two strata, we can attribute this to an ongoing language change. The affix morpheme may be splitting into two. So, what we have are

closely related homophonous forms, with different characteristics, occur-
ing at different strata.

The crucial data involves the behaviour of suffix *-able/-ible* meaning
'capable of X, worthy to or suitable for X'. Establish the stratum at which
the *-able* suffix in the words [7.3] and [7.4] is found.

[7.3]   de'cipher ~ de'cipherable
        re'pair ~ re'pairable
        de'bate ~ de'batable
        in'flate ~ in'flatable

[7.4]   'tolerate ~ 'tolerable (*'toleratable)
        ne'gotiate ~ ne'gotiable (*'negotiatable)
        'calculate ~ 'calculable (*'calculateable)

It appears that the *-able* suffix is phonologically neutral in [7.3] and hence
belongs to stratum 2. But in [7.4] the same suffix is non-neutral. It causes
mutation in the base. (We assume with Aronoff (1976: 124) that the
stratum 1 verb-forming suffix *-ate* in 'tolerate, 'negotiate and 'calculate is
first attached to 'toler-, 'negoti- and 'calcul- before *-able* is suffix.) From
this you might wish to conclude that, like *-ize*, the suffix *-able* simul-
taneously belongs to two strata and that hence the same argument we
deployed above (on the evidence of [7.2]) against basing the separation of
strata on the behaviour of roots could be used also against setting up strata
on the basis of the behaviour of affixes.

However, in reply to this objection, standard lexical morphology claims
that it is not just one suffix that appears on two strata but two distinct
suffixes, one of which is always spelled *-able* and the other having the forms
-able and *-ible*.

Support for this position comes from Aronoff (1976: 121–5). First,
Aronoff has shown that, there are two phonological arguments for recognising two distinct *-able* morphemes:

(i)   There is an *-able* suffix that is associated with stratum 2 (in his
        SPE-based '# boundary'). This is the commoner of the two. It
        does not affect stress or modify the base to which it is suffixed. It is
        the morpheme found in the words in [7.3].

(ii)   There is another *-able* suffix which is associated with stratum 1
        (i.e. '+ boundary'). This latter suffix is capable of modifying the
        base to which it is attached. It is the morpheme that causes the

truncation of *-at-* in [7.4]. Elsewhere, this suffix may cause a stress shift (contrast *hos'pitable* with *'hospital*).

Secondly, Aronoff points out that the phonological evidence is corroborated by morphological evidence. The stratum 1 *-able* suffix that conditions the mutations in the bases in [7.4] does also cause other mutations which are typical of latinate stratum 1 suffixes. When the *-ible* allomorph of the stratum 1 *-able* morpheme is suffixed to a base, a special allomorph of the base may have to be selected. So, many latinate bases have one allomorph when they occur before the stratum 1 *-ible* suffix and a different allomorph before the stratum 2 *-able*, as you can see:

[7.5]

	Base	Non-neutral stratum 1 -ible	Neutral stratum 2 -able
a.	circumscribe /sɜːkəmskraɪb/	circumscriptible /sɜːkəmskrɪptɪbl/	circumscribable /sɜːkəmskraɪbəbl/
b.	extend /ekstend/	extensible /ekstensɪbl/	extendable /ekstendəbl/
c.	defend /dɪfend/	defensible /dɪfensɪbl/	defendable /dɪfendəbl/
d.	perceive /pəsiːv/	perceptible /pəseptɪbl/	perceivable /pəsiːvəbl/
e.	divide /dɪvaɪd/	divisible /dɪvɪzɪbl/	dividable /dɪvaɪdəbl/
f.	deride /dəraɪd/	derisible /dərɪzɪbl/	deridable /dəraɪdəbl/

(based on Aronoff, 1976: 123)

Observe that the allomorph of the base found before *-ible* in [7.5] is normally the same as the one found before well-known stratum 1 latinate suffixes like *-ory*, *-ion* and *-ive* in [7.6]:

[7.6]

	Base	-ible	Non-neutral stratum 1 suffixes -ion, -ive, -ory
a.	circumscribe /sɜːkəmskraɪb/	circumscriptible /sɜːkəmskrɪptɪbl/	circumscription /sɜːkəmskrɪpʃən/
b.	extend /ekstend/	extensible /eksensɪbl/	extension, extensive /eksenʃən/, /eksensɪv/
c.	defend /dɪfend/	defensible /dɪfensɪbl/	defensive /dɪfensɪv/

d.	perceive	perceptible	perception, perceptive
	/pəsi:v/	/pəseptɪbl/	/pəsepʃənn/, /pəseptɪv/
e.	divide	divisible	division, divisive
	/dɪvaɪd/	/dɪvɪzɪbl/	/dɪvɪʒən/, /dɪvaɪzɪv/
f.	deride	derisible	derisory, derisive
	/dəraɪd/	/dərɪzɪbl/	/dəraɪzərɪ/, /dəraɪzɪv/

From this we conclude that *-able/-ible* in [7.4] and [7.6] is a phonologically non-neutral stratum 1 suffix which is distinct from the related homophonous stratum 2 *-able* suffix in [7.3].

Although the two *-able* (*-ible*) forms are closely related in meaning, nevertheless they can be distinguished from each other. They show a degree of semantic specialisation. Stratum 2 *-able* has the transparent meaning 'capable of X, worthy to or suitable for X' (as in *marriageable, readable, loveable*, etc.). But, stratum 1 *-able/-ible* is semantically opaque and unpredictable. For instance, it is not possible to give *-able* (in [7.7]) and *-ible* (in [7.8]) a consistent meaning. Depending on the base to which it attaches, the suffix can contribute a different meaning, or indeed several different meanings, to the word of which it is a part. It is reasonable to recognise stratum 1 *-able* and *-ible* as two separate morphemes.

[7.7]   Stratum 1
         -able   (i)   *They hired a reasonable number of buses* (= a number judged to be adequate for the circumstances)
         reasonable   (ii) *She is reasonable* (= has sound judgement)
                        (iii) *She paid a reasonable price for the car* (= moderate price)

[7.8]   Stratum 1
         -ible
   a. *sensible*   (*as in Janet is sensible*) (= able to reason, has sound judgement)
   b. *perceptible* (i)   *Is a feather falling down perceptible?* (= capable of being perceived)
                        (ii)   *There is a perceptible difference in price* (= 'significant, important')

If the same *-able/-ible* morpheme belonged to two strata, we would be in trouble. This would completely undermine the principle of justifying strata on the basis of the behaviour of affixes. However, as seen, careful analysis of the phonological, morphological and semantic evidence points to several distinct morphemes which are on two strata. The claim that affix morphemes typically belong to one stratum can be thus defended.

7.2.3 How Many Strata are Needed?

All advocates of lexical morphology agree that the lexicon is divided into strata. Yet one of the most contentious issues in lexical morphology is determining the exact number of strata that are required.

Even the same writer may not always recognise the same number of strata on different occasions. We shall assume in this book a model with only two lexical strata and a post-lexical stratum. This is a commonly held position today. For instance, Kiparsky (1983) suggests two strata are needed:

[7.9]   (i)   primary (regular) morphology
        (ii)   secondary (irregular) morphology (including compounding)

But in Kiparsky (1982b) he proposes three strata:

[7.10]   (i)   '+' boundary (i.e. irregular) inflection and derivation
        (ii)   '#' boundary (i.e. regular) derivation and compounding
        (iii) '#' boundary (i.e. regular) inflection

Halle and Mohanan (1985) are even less restrictive, recognising four strata:

[7.11]   (i)   Stratum 1: Class I (irregular) derivation and irregular inflection
        (ii)   Stratum 2: Class II (regular) derivation
        (iii) Stratum 3: Compounding
        (iv) Stratum 4: Regular inflection

Why is there this lack of unanimity? Everyone agrees that only the minimum number of strata required to account adequately for the facts should be set up. Permissiveness should be avoided because it would negate the objective of setting up ordered lexical strata, which is to reflect the fact that large bundles of morphological rules are linked with large bundles of phonological rules that share certain general properties. If, in the extreme case, instead of having a large number of morphological rules that are paired with corresponding phonological rules, we ended up with a single morphological rule linked to a single phonological rule, we would have lost the basic insight of lexical morphology. Nobody wants that absurd situation. Nevertheless it is still not at all obvious how restrictive our approach to recognising strata should be. As we have seen, for English, proposals have ranged from two to four strata (see [7.9]–[7.11] above).

The lack of agreement is largely due to a methodological indeterminacy. The theory allows two ways of ensuring that the correct output is obtained in cases where rules interact. One way is to place one rule on an earlier stratum than the other. Another way is to put the two rules on the same stratum and to let ordering principles determine which rule takes precedence. The former solution may result in more strata than the latter.

To illustrate this, let us consider briefly the interaction of plural formation with compounding. Everyone agrees that compounding must follow irregular plural formation done at stratum 1 and precede regular plural formation. That is why the plural of *walkman* is *walkmans* (not **walkmen*): compounding feeds regular inflection. This fact has been used by Kiparsky (1982b) to justify putting compounding on a stratum preceding that where regular inflection takes place. But we have argued here that this merely shows that compounding needs to be ordered before regular inflection at stratum 2. (See section (7.2.5.2) below for further discussion.)

7.2.4 Are Phonological Rules Restricted to One Stratum?

At first blush, the failure to agree on the number of strata required might be attributed to the fact that the theory allows a phonological rule to belong to more than one stratum at the same time. This seems to mean that we cannot be sure which stratum a given rule belongs too. Hence we may be unable to determine which stratum is which since strata are recognised on the basis of the properties of rules associated with affixes at each stratum. Fortunately, the fact that rules can belong to different strata does not make it impossible to tell in each individual case the stratum at which a rule applies.

The theory assumes that phonological and morphological rules behave differently in this respect. The morphological rules are allowed to operate only on one stratum. They are expressly forbidden to operate on two strata. So, for instance, the stratum 1 irregular strong verb ablaut rule that produces the past tense *rang* for the verb *ring* cannot affect verbs derived from nouns at stratum 2 by conversion. Hence, the past tense of the verb *ring* derived from the noun *ring* is *ringed* (as in the *The city is ringed by mountains* as opposed to *They rang the bell*) (see section (6.2.3)).

By contrast, phonological rules, are allowed to apply at several strata so long as those strata are **adjacent**. A given rule, let us call it Rule x, may apply everywhere (as in column A), at stratum 1 and stratum 2 in the lexicon (as in column B below), or on stratum 2 in the lexicon and post-lexically, but not on stratum 1 (as in column C). What is disallowed is a situation where a rule only applies on stratum 1 and post-lexically but misses out stratum 2 (as in column D):

[7.12]

A	B	C	D
Rule x stratum 1	Rule x stratum 1	___ stratum 1	Rule x stratum 1
Rule x stratum 2	Rule x stratum 2	Rule x stratum 2	___ stratum 2
Rule x Post-lexical	___ Rule x Post-lexical	Rule x Post-lexical	Rule x Post-lexical

If a phonological rule applies at more than one stratum in the lexicon, or when it applies both lexically and post-lexically, it displays different characteristics. So, having the same rule applying on different strata does not in itself make it impossible to keep strata apart. I will illustrate this with the behaviour of the by now familiar **voice assimilation** rule in English, drawing on Kiparsky (1985: 92–4).

In English, voicing is distinctive at the lexical level in **obstruents** (i.e. stops, fricatives and affricates). So the following pairs of consonants – /p b, t d, k g, f v, θ ð, s z, ʃ ʒ, tʃ dʒ/ – contrast. But voicing is not distinctive in **sonorants** in underlying lexical representations. All nasals (/m n ŋ/), liquids (/l r/) and glides (/w y/) are voiced. This means that while obstruents have to be specified as voiced or voiceless in the lexicon, sonorants need not be – we know they are all going to be voiced.

Explain the constraints on voice assimilation in the following:

[7.13] **a.**

adze	[ædz]	(*[æds])	act	[ækt]	([*ægt])
apse	[æps]	(*æpz])	soft	[sɒfr]	(*[sɒfd] *[sɒvt])
apt	[æpt]	([*æpd])	lift	[lɪft]	(*[lɪvt] *[lɪfd])

b.

mince	[mɪns]	(*[mɪnz])	wince	[wɪns]	(*wɪnz])
chance	[tʃɑːns]	(*[fʃɑːnz])	bacon	[beɪkn]	(*[beɪgn])
little	[lɪtɫ]	(*[lɪdɫ])	bottom	[bɒtm]	(*[bɒdm])

In all the above examples we are dealing with underived lexical items. We note that in [7.13a] adjacent obstruents within a syllable must agree in voicing but clusters of sonorants and obstruents need not do so (cf. [7.13b]).

How does the application of the voice assimilation rule in [7.13] differ from its application in [7.14]? (Also see section (2.2.4).)

[7.14] **a.** cat-s   dog-s   bak-ed
            [kæts]   [dɒgz]   [beɪkt]

     b. win-s   bell-s   farm-ed
            [wɪnz]   [belz]   [fɑːmd]

     c. swim   stray   plea
            [sw̥ɪm]   [str̥eɪ]   [pl̥iː]

     d. Jim's left.   Pat's coming   Gill's coming
            [dʒɪmz]   [pæts]   [dʒɪlz]

Clearly, voice assimilation is a rule that applies both lexically and post-lexically. The rule is more general when it applies post-lexically. Thus, in [7.13] where voice assimilation applies lexically within underived roots, only obstruents are affected.

All sonorants must be voiced in the lexicon. This being the case, the requirement that lexical rules must be **structure preserving** ensures that no lexical rule can change a sonorant from being voiced to being voiceless. So, in the lexicon, the assimilation of the voicelessness of the preceding consonant does affect obstruents (as in [7.13a]) but it cannot affect sonorants (see [7.13b]).

The contrasting effects of lexical and post-lexical voice assimilation can be seen again in [7.14]. When this rule applies in words derived at stratum 2 in the lexicon, it still only affects obstruents (see [7.14a] and [7.14b]). It causes the suffixed obstruent to agree in voicing with whatever segment precedes it. Consequently, as seen in [7.14b], at stratum 2 obstruents realising suffix morphemes agree with preceding sonorants in voicing (cf. *wins*). This is different from the situation where underived lexical items like *wince* [wɪns] in [7.13b] are involved. There, obstruents do not assimilate the voicing of the preceding sonorants.

In [7.14c] and [7.14d] we see voice agreement applying post-lexically. Interestingly, in this case the voice assimilation is not restricted to obstruents. In [7.14c], a sonorant following a voiceless obstruent in the same syllable is devoiced (as in *stray* [str̥eɪ] etc.) due to assimilation of voicelessness. Clearly, the post-lexical application of voice assimilation is not structure-preserving. It creates a class of voiceless sonorants (e.g. (w̥ r̥ l̥)) although no such sounds occur in lexical representations.

Lastly, in [7.14d] we see again voice assimilation applying post-lexically. When an auxiliary verb *is* or *has* is optionally reduced, it can be appended to the preceding word (cf. section 10.5). In that event, it agrees in voicing with the last segment of the word to which it is attached, regardless of whether it is an obstruent or a sonorant.

In the next section we consider arguments for the claim that, like phonological rules, morphological rules too are not always confined to a single lexical stratum. This has important consequences for the integrity of the model.

7.2.5 Are Morphological Rules Restricted to One Stratum?

Lexical morphology claims that **Morphological rules** can only access information found on the stratum where they operate. As we saw in section (6.2.4), peeping forward at later strata in a derivation or looking back to previous ones is prohibited by **Strict Cyclicity** and the **Bracket Erasure Convention**, which wipes out all bracketing at the end of each cycle in the lexicon. Consequently, on leaving each stratum all information encoded in internal brackets at that stratum is erased.

7.2.5.1 *Bracketing paradoxes*
Although the Bracket Erasure Convention has its merits, it also brings with it severe complications. Williams (1981a), Strauss (1982b), Pesetesky (1985) and Goldsmith (1990), among aothers, have highlighted a number of unresolved problems which derive from restricting morphological rules exclusively to a single stratum by the device of bracket erasure. These writers consider cases where morphological rules appear to need access to information found at more than one stratum. Just as phonological rules can apply at more than one stratum, it seems some morphological rules are not confined to a single stratum either. They need to be able to see information associated with brackets showing the structure of the word at a stratum other than the one at which they apply.

Some of the crucial evidence comes from words which contain the negative prefix *un-* and the noun-forming suffix *-ity*. We already have ample proof of the fact that *-ity* is a stratum 1 suffix (see section (5.2.1). What needs to be established now is the stratum at which *un-* is found.

Use the data in [7.15] to show that *in-* is a stratum 1 prefix and *un-* is a stratum 2 prefix:

[7.15]	a.	ungraceful	*ingraceful	b.	impossibility	*unpossibility
		unmerciful	*inmerciful		intangibility	*untangibility
		unchildish	*inchildish		incorrigibility	*uncorrigibility

As you see, the *in-* prefix goes with adjective bases containing stratum 1 suffixes while the *un-* prefix attaches to adjective bases containing stratum 2 suffixes or with underived adjectives (as in *unkind*). We know that *in-* is at stratum 1 because it is not prefixed to adjectives derived at stratum 2. Hence the ill-formedness of **ingraceful*, **inmerciful* and **inchildish* which contain the stratum 2 suffixes *-ful*, and *-ish*. At stratum 1 where *in-* is prefixed, *grace*, *mercy* and *child* are noun bases and so they are unavailable for *in-* prefixation since *in-* attaches to adjectives. By the stage when these words are turned into adjectives by the stratum 2 suffixes *-ful* and *-ish*, the stratum 1 process of prefixing *in-* has already passed them by. But *graceful*, *kindness* and *childish* can be made into the adjectives *ungraceful*, *unmerciful* and *unchildish* at stratum 2, since *un-* prefixation as well as *-ful* and *-ish* suffixation both take place at stratum 2. An adjective like *impossible*, with the stratum 1 *in-* prefix can be fed to the stratum 1 nominalising suffix *-ity* to form the noun *impossibility* as in [7.15b]. Finally, **unpossibility* which would be formed at stratum 2 is blocked by the prior existence of *impossibility* (cf. 4.2.1).

So far it seems all plain sailing. But there are some major complications. A number of theoreticians have identified some embarrassing problems posed by **bracketing paradoxes**. These problems arise when attachment of stratum 1 affixes seems to presuppose the presence of stratum 2 affixes, or when stratum 2 affixation needs to see the internal bracketing of a word that is only available at stratum 1. As we shall now see, difficulties of this nature confront us when we deal with affixes like *un-* and *-ity* which we have just described.

Paradox 1: Lack of required syntactic properties in the input

Pesetsky cites Siegel (1974) who considered the stratum 2 prefix *un-* which, as we have just seen, attaches to adjectives and has the meaning 'not' as in:

[7.16]   **a.**   unkind [$_A$ un [$_A$ kind]]   **b.**   *untree [$_A$ un [$_N$ tree]]
        unjust [$_A$ un [$_A$ just]]   *unsoon [$_A$ un [$_{Adv}$ soon]

Clearly then, this *un-* is only prefixed to adjectives, producing derived adjectives. (There is another *un-* prefix which attaches to verbs which has a reversive meaning (as in *unpeg*, *unwind* etc.). As this prefix is not germane to the present argument we will ignore it.)

Explain why, according to lexical morphology, these very well-formed words should in theory be disallowed.

[7.17]   unacceptability   uninterruptability
        unreliability   ungovernability

Recall that *-ity* attaches to adjectives to form nouns. *Governable* is an adjective so *-ity* can be suffixed to it. If we eradicate all internal brackets at the end of each stratum, by the time the derivation reaches stratum 2 in readiness for the prefixation of *un-*, the adjective bracket after *-able* will have been removed. Only the noun external bracket introduced by stratum 1 *-ity* will be visible in the noun *governability*, as seen in [7.18] (cf. also [7.27]):

[7.18]   stratum 1:   [[govern]$_{VERB}$able]$_{ADJ}$
                                            ↓
                              [governable]$_{ADJ}$ity$_{NOUN}$
                              (→ governability$_{NOUN}$)
            stratum 2:   [$_{ADJ}$[un [[governability]$_{NOUN}$]

We are put in a quandary by the fact that, on leaving stratum 1, the information that the form *governable* is an adjective is removed by the Bracket Erasure Convention. Since *governability*, the form that enters stratum 2 is a noun, it should not be able to combine with *un-* which requires an adjective base. But it does. This is the first paradox.

Paradox 2: Lack of required phonological properties in the input

We will now turn to the behaviour of **comparative degree** adjectives for our illustration of the second paradox. The discussion is based on Pesetsky (1985).

First, let us establish the relevant facts. Use the data below as the basis of a general and explicit verbal statement of the rule that predicts the selection of *-er* to represent the comparative degree of the adjectives below:

[7.19] **a.** tall   taller
                  big   bigger

        b. trendy   trendier
                  funny   funnier
                  happy   happier

        c. gentle   gentler
                  feeble   feebler

        d. shallow   shallower
                  narrow   narrower

        e. clever   cleverer

        f. foolish   *foolisher

intelligent	*intelligenter
beautiful	*beautifuler

I hope your answer has prepared the ground for understanding the second paradox. The comparative degree adjective suffix *-er* is typically attached to **monosyllabic adjectives** like those in [7.19a] which are repeated below:

[7.20] tall   [tɔːl] ~ taller [tɔːlə]   big [bɪg] ~ bigger [bɪgə]

or it is attached to a few **disyllabic adjectives** with a very light final syllable ending in [ɪ], [ɫ], [əu] or [ə] as in [7.19b–e], repeated below as [7.21a–d]:

[7.21]

a.	[ɪ]	trendy	[trendɪ]	~	trendier	[trendɪə]
		happy	(hæpɪ]	~	happy	[hæpɪə]
b.	[ɫ]	gentle	[dʒentɫ]	~	gentler	[dʒentlə]
		feeble	[fiːbɫ]	~	feebler	[fiːblə]
c.	[əu]	shallow	[ʃæləu]	~	shallower	ʃæləuə]
		narrow	[nærəu]	~	narrower	[nærəuə]
d.	[ə]	clever	[klevə]	~	cleverer	[klevərə]

The suffix *-er* does not readily attach to the longer adjectives in [7.19f] as the ill-formedness of **foolisher*, **intelligenter* and **beautifuler* testifies. Only a handful of longer adjectives such as *pleasanter* and *commoner* can optionally (and uncharacteristically) take *-er*. The forms *more pleasant* and *more common* are the ones that are usually preferred.

This is the bracketing paradox: notwithstanding the restriction of *-er* essentially to monosyllabic bases and a few disyllabic ones in the categories we have identified, *-er* is suffixed readily to certain **trisyllabic adjective** bases which have the prefix *un-*:

[7.22]   unhappier   unluckier   unfriendlier

In order to salvage the principle that *-er* is attachable to monosyllabic bases and a few disyllabic ones and is avoided in longer adjectives, we would need to argue that, first *-er* is suffixed to an adjective base and then, afterwards, *un-* is prefixed as in [7.23]:

[7.23]

a. happy
      ↓

b. [[happy$_A$] er$_A$]   (suffixation of *-er*)
      ↓

c. [$_A$ un[happy er$_A$]   (prefixation of *un-*)

Assuming that *-er* is suffixed while *happy* is still disyllabic solves the phonological problem. But it does not end all our troubles.

Explain why the solution outlined above is inadequate.
Hint: consider meaning and its relation to bracketing.

The bracketing in [7.23c] incorrectly points to the semantic analysis of *unhappier* as shown in [7.24].

[7.24] $[_A$un [happy er$_A$]
[not [more happy]]

The correct meaning of *unhappier* is 'more not happy'. For the purposes of working out the meaning the correct bracketing is:

[7.25] $[_A$ un happy] er$_A$]
[[not happy] more]

The paradox is that the bracketing in [7.25] which works well for the semantic interpretation contradicts the phonological generalisation that *-er* is not suffixed to bases of three or more syllables.

More research is needed to determine whether the Bracket Erasure Convention should be preserved. If in the end it becomes clear that the standard position that limits morphological rules to accessing information available on a single stratum is too restrictive, an important plank will have been removed from principles of Strict Cyclicity and the Bracket Erasure Convention which buttress the model of morphology with hierarchically ordered strata.

Paradox 3: Morpheme sequence out of kilter with morphological cycle

As we saw in (v) of section (7.1), in lexical morphology it is claimed that affixes are added in a cycle starting at stratum 1, going on to stratum 2 etc. Consequently, stratum 1 derivational affixes which are added first are closer to the root than stratum 2 affixes that are attached later. In other words, stratum 1 affixes are 'on the inside' and stratum 2 affixes are 'on the periphery' of words. This claim is vindicated by data like the following:

[7.26] office   (Underived root)
offic-(i)al   (Underived root – stratum 1 affix)
offic-(i)al-ly   (Underived root affix – stratum 1 – stratum 2 affix)

Unfortunately, this neat pattern is not what we find in all cases. Aronoff (1976: 84) and Kiparsky (1983: 21–2) point out that there are cases where stratum 2 affixes are nearer to the root than stratum 1 affixes.

We will illustrate the problem with *-able/-ible*. As we have already seen in (7.2.1), there are two *-able/-ible* suffixes. One is non-neutral and is found at stratum 1 (cf. *negotiate* ~ *negotiable*, *divide* ~ *divisible*). The other is neutral and occurs at stratum 2 (cf. *detect* ~ *detectable*). It is this latter stratum 2 *-able* that we will be considering here.

How is *-able* ordered with respect to stratum 2 suffixes like *-ise/-ize* (of Bermudaize (cf. section (7.2.2)?

The *-able* suffix follows the stratum 2 suffix *-ise/-ize* in words like in *adv-is(e)-able*, *bapt-ize(e)-able*. This is not surprising. The theory allows an affix at a given stratum to follow other affixes at the same stratum.

Now make up six words in which the stratum 1 suffix *-ity* and the stratum 2 suffix *-able/-abil* combine. In what order do these suffixes appear?

You will have discovered that, contrary to what the theory predicts, the stratum 2 suffix *-able precedes* the undisputed stratum 1 suffix *-ity*:

[7.27]   read-abil-ity   reli-abil-ity   sell-abil-ity
         depend-abil-ity   spread-abil-ity   elect-abil-ity

Although normally that stratum ordering reflects the order of affixes, with stratum 1 affixes, which are attached at an earlier cycle, closer to the root than stratum 2 affixes, which are added later, the counter evidence in [7.27] cannot be ignored.

7.2.5.2 *Bracketing paradoxes: should looping be permitted?*

The problems identified in the last section stem from the fact that rules are restricted by the Bracket Erasure Convention and Strict Cyclicity to apply-ing to one stratum only. Mohanan (1982) and Halle and Mohanan (1985) have argued that the remedy lies in relaxing the constraint on lexical rules that prohibits returning to a stratum through which a derivation has passed. Working in a lexical phonology model with *four lexical strata* they propose the incorporation of a **loop** linking stratum 2 with stratum 3. They state the role of the loop thus:

The loop is a device that allows a stratum distinction for the purposes of the phonology, without imposing a corresponding distinction in morphological distribution.   (Halle and Mohanan, 1985: 64)

With the loop in place, the pattern of rule application which is outlawed in the standard model of lexical phonology which has been presented so far in this book would be permitted.

[7.28]   Stratum 1:   Stratum 1 derivation; irregular inflection

Stratum 2:   Stratum 2 derivation

Stratum 3:   Compounding

Stratum 4:   Regular inflection

Since this section is based on Mohanan and Halle's word, we will momentarily assume, as they do, a lexicon with four rather than just two strata. The presence of the loop means that sometimes the difference between strata is maintained with respect to phonological rules, but suspended with respect to morphological rules. Morphological rules may have access to morphological information at a stratum other than the one where they operate. There is only one proviso: the strata in question must be **adjacent**.

   For example, in [7.28], for the purposes of the morphology, both **stratum 2 affixation** (attaching neutral affixes like the prefixes *re-* and *ex-*, or suffixes like *-ness* and *-ful*) and **compounding** belong to the same stratum. But for purposes of the phonology they are put on distinct strata and they are linked by a loop.

Given the evidence in [7.29] and [7.30], why is it not possible to establish unequivocally whether stratum 2 affixation rules precede or follow compounding? Show how this evidence can be used to motivate the loop.

[7.29]   boy-friend   ex-boy-friend
         house wife   ex-housewife
         carbon-date   re-carbon-date
         over-throw   re-overthrow

[7.30]   neighbour-hood   neighbourhood watch
         teach-er   headteacher
         fight-er, bomb-er   fighter bomber

In the above examples we see the formation of words using neutral stratum 2 affixes in some words precedes and in other words follows compounding. So, it is impossible to establish one ordering relation that is always observed (Selkirk, 1982). Thus, in [7.29] compounding (which is at stratum 3 in the model in [7.28]) feeds the neutral stratum 2 affixation rules which attach *ex-* and *re-* to compound bases like *house-wife* and *carbon-date* to form *ex-housewife* and *re-carbon-date*. But the converse is true in [7.30] where we see neutral stratum 2 affixation feeding compounding. Words like *neighbour-hood* or *fighter* and *bomber*, which have undergone neutral stratum 2 derivational suffixation, can serve as bases used to build compounds like *neighbourhood watch* or *fighter bomber*.

Halle and Mohanan argue that since stratum 2 and stratum 3 morphological rules can create inputs for each other, a **loop** is needed to enable stratum 2 affixation rules and stratum 3 compounding rules to have access to each other's outputs. However, phonologically neutral stratum 2 affixes are on a separate stratum from that where compounding occurs. Compounding is non-neutral – it may affect stress. Compare `head and `teacher with `head ˌteacher (where ˌ marks a secondary stress). But, unfortunately, what seems to make good sense phonologically has the undesirable morphological consequence of placing secondary affixation and compounding on different strata, although these processes mutually feed each other.

To solve this problem Halle and Mohanan propose the use of the loop. This allows them to keep phonologically neutral secondary affixation (of prefixes like *re-* and suffixes like *-ness*) on separate strata for phonological purposes while at the same time allowing them access to each other's output via the loop, for morphological purposes.

But the introduction of the loop is not the only possible solution. And in my view it is not the best solution. The alternative assumption made by Kiparsky (1982a), that both stratum 2 affixation and compounding belong to the same stratum, is preferable. Recall that the theory requires morphological rules of the same stratum to be unordered. They apply automatically in whatever sequence ensures that all those (obligatory) rules that have their input present at a given cycle are given the opportunity of applying. It is expected that all rules whose input requirements are met at a given morphological stratum are all applied before moving on to the stratum. Furthermore, having moved to a later stratum, there is absolutely no going back to an earlier stratum.

The above conditions are enough to ensure that if compounding and neutral (stratum 2) affixation both have their inputs present at the same stratum, they must both apply at that stratum in the order that ensures that neither rule is blocked. What this entails is compound-formation feeding neutral stratum 2 affixation in [7.29] and neutral stratum 2 affixation feeding compounding in [7.22]. No looping is needed.

From a methodological point of view, an economical theory that uses a minimum of apparatus is always to be preferred. If we can account for the facts with 2 strata rather than 4 strata plus a loop, that is what we should do.

We will return to the thorny problem of morpheme sequencing from a different perspective in section (11.6) for further discussion and refinement.

7.3 CONCLUSION

We have seen in the last two chapters how lexical morphology offers many fruitful insights into the lexicon and morphology. In this chapter we have seen that there remain some problems that have not yet been fully resolved. The problems all revolve round the important question of determining the criteria for recognising strata in the lexicon.

Lexical morphology is not 'The Perfect Theory'. Nevertheless, in spite of its blemishes, this model is intuitively satisfying and offers us a useful way of dealing with many of the major dimensions of word-formation in English and other languages (see Mohanan, 1982, 1986; Pulleyblank, 1986). The obstinate problems that have been identified may be resolvable by future research.

But some linguists, such as Rubach (1984), are not so optimistic. While espousing the notion of lexical morphology, they reject the principle of hierarchically ordered strata which brings with it so many complications.

Lexical morphology is a cog in the bigger wheel of the theory of morphology and the lexicon that this book presents. In the next two chapters we are going to investigate other dimensions of the theory that deal with the interface between phonology, morphology and the lexicon.

EXERCISES

1. Explain why the word-formation rules can generate the words in (a) but not those in (b):

   (a) implausible   incorrect   indelible   insignificant
   (b) *imboyish   *inchildish   *instylish   *immanish

2. Study the following data and answer the questions below.

   stuck-up-ness   pickup-ful (e.g. a *pickup-ful* of tomatoes)
   grown-up-ness   dugout-ful (e.g. a *dugout-ful* of soldiers)
   standby-less   (e.g. a *standby-less* flight)

kickback-less	(e.g. a *kickback-less* deal)
stand-off-ish	(e.g. a *stand-off-ish* remark)
send-off-less	(e.g. a *send-off-less* departure)

(based on Selkirk, 1982: 109)

(a) Identify the stratum 2 affixes in each example.

(b) Provide a derivation that shows the rules that apply in the formation of each compound.

(c) In what order do the rules apply?

(d) To what word-class does each compound belong? What determines the word-class of the entire compound in each case? Make your statement as general as possible.

3. Using the words below as evidence, describe the interaction between compounding and regular inflection. What implication has this interaction got for the ordering of these processes?

a.	booksellers	*bookssellers
	bus drivers	*busses driver
	tax-collectors	*taxes-collector
	shopkeepers	*shopskeepers
b.	menservant	*manservants
	grants-in-aid	*grant-in-aids
	passers-by	*passer-bys
	commanders-in-chief	*commander-in-chiefs
	Governors-general	*Governor-generals

4. (a) Show in detail how compounding interacts with tense marking and plural formation in the two tables below.

   (b) Should this interaction be used to justify the loop?

A	present	past	
	undergo	underwent	*undergoed
	underfeed	underfed	*underfeeded
	withhold	withheld	*withholded
	withstand	withstood	*withstanded
	withdraw	withdrew	*withdrawed
	overtake	overtook	*overtaked
	self-teach	self-taught	*self-teached

B singular   plural
   field-mouse   field-mice   *field-mouses
   head-louse   head-lice   head-lice
   pen-knife   pen-knives   *pen-knifes
   policeman   policemen   *policemans
   grandchild   grandchildren   *grandchilds
   milk tooth   milk teeth   *milk tooths

5. (a) In your view, what is the worst shortcoming in the theory of lexical phonology?
   (b) How could the theory be amended to overcome this shortcoming?

8 Prosodic Morphology

8.1 INTRODUCTION

In the last three chapters we have explored the theory of lexical phonology and morphology, focusing on prefixation and suffixation processes in English. In this chapter we are going to branch out and investigate other morphological systems where, in addition to prefixation and suffixation, there is extensive **infixing** (see section 3.1.2).

We will treat infixing using the theory of **prosodic morphology** or **template morphology** which draws heavily on the theoretical apparatus and formalisms of the generative phonology model known as **autosegmental phonology**. I assume that you are not a phonologist of this school and so I briefly outline autosegmental phonology in section (8.2) below before introducing prosodic morphology. If you are well versed in autosegmental phonology, move directly to section (8.3).

8.2 PHONOLOGICAL PRELUDE: AUTOSEGMENTAL PHONOLOGY

My intention is not to turn you into an accomplished autosegmental phonologist in these short paragraphs, but rather to provide you with a basic grasp of the issues in the model of 'phonology with tiers' that is a pre-requisite to understanding prosodic morphology.

Autosegmental phonology has several precursors, as John Goldsmith, the originator of the model, points out (1990: 1–4). It is a direct descendant of the theory of generative phonology, given its fullest expression in *The Sound Pattern of English* (SPE) by Chomsky and Halle in 1968. Autosegmental phonology continues to regard as central the goals of phonological investigations set out in SPE, although it has rejected many of the original assumptions, formalisms and principles of SPE.

Furthermore, the work of the London School on **prosodies** (see Firth, 1948; Palmer, 1970) and that of some leading American structuralists such as Hockett (1947, 1955) and Harris (1944) on **long components** foreshadows the central tenet of autosegmental phonology to some extent. Like linguists of these earlier schools, autosegmental phonologists hypothesise that phonological representations consist of several independent, parallel **tiers**, i.e. **levels** of representation. Autosegmental phonology articulates this insight more clearly than its precursors. It also provides the formal means of representing these autonomous tiers and of showing how they link up with each other.

In our discussion of English stress, we have already implied a multi-tiered analysis. We have assumed that stress is a **prosody**, i.e. it is a phonological element that is not an inherent feature of consonants or vowels but rather a property of the entire word. Hence, it can hop from syllable to syllable when certain stratum 1 suffixes are added as in [8.1]:

[8.1]   'democrat   ~   demo'cratic
       'pompous   ~   pom'posity
       'adjective   ~   adjec'tival

If stress were an integral part of the segments in a syllable it would not enjoy such mobility and independence.

8.2.1 Autosegmental Phonology: Mapping Principles

The theory of autosegmental phonology was initially used to describe tone. Although it has been extended to cover other phenomena, its essential properties are still best illustrated using tone. So, before seeing how the theory was used later to describe morphological data we will come to grips with its main characteristics through a brief account of its application to phonological phenomena, in particular tone.

    As we saw above, stress is independent of consonants and vowels. Similarly, in **tone languages**, tone can be shown to be an independent prosody, and not an integral part of vowel and consonant segments. Tones are represented on the **tonal tier** and vowels and consonants on the **segmental tier**. Processes affecting elements on one tier may in some cases have no impact on elements on a different tier. For instance, when vowels are deleted, the tones associated with them are not necessarily lost. Thus, as seen below, in Luganda, when two vowels are adjacent, the first is deleted unless it is high (in which case it becomes a glide [y] or [w]) but the tone associated with it is not necessarily affected:

[8.2]   Segmental tier: kusa   ebyo   →   kuse:byo   'to grind those'

Tonal tier: L H L H   L H L H

Note: L and H stand for L(ow) and H(igh) tone. The lines link vowels with
    tones. An unbroken line indicates pre-linking and a broken line shows
    the establishment of a fresh link (see [8.4] below).

Observe that when the vowel at the end of *kusa* 'to grind' is deleted, its high tone is preserved and inherited by the *e* of *ebyo* 'those'. If tone were an inherent property of the vowel, the deletion of the vowel would mean the demise of the tone as well. The survival of the tone bears testimony to its independence from the vowel.

The choice of the name **autosegmental phonology** for this model is intended to reflect the fact that phonological representations consist of **segments** like stress, tone, vowels and consonants that appear on autonomous tiers. On their tiers, all phonological elements, be they tones, consonants, vowels, etc., behave as though they were 'segments'. They cannot occupy the same space. They must follow each other in a linear sequence.

The tiers in terms of which phonological representations are organised are independent, but they are not isolated from each other. On the contrary, they are linked in complex hierarchical structures and are capable of interacting with each other (cf. Goldsmith, 1976, 1979, 1990; Clements and Keyser, 1983). In this theory articulation is seen as the bringing together of different autonomous gestures resulting, for instance, in voicing, lip rounding, high pitch that make up a speech sound.

Tone (and stress) are best represented on a separate tier from vowels and consonants which belong to the segmental tier. Mediating between these tiers is the **skeletal tier** (or **CV-tier**). The skeletal tier holds together the different phonological tiers in the same way that the binding holds the pages of a book together. Association lines indirectly link stress or tone to actual vowel and consonant sounds through the skeletal tier, as shown in [8.3] below.

In [8.3a] the skeletal tier is represented by x's (with each x standing for a segment like a consonant or vowel (cf. Hyman, 1985). An alternative representation, with the skeletal tier represented by C and V slots (as in [8.3b], is proposed by Clements and Keyser (1983).

[8.3] **a.** Tonal tier:

	H		L	L		H
	|		|	|		|
Skeletal tier:	x	x	x	x	x	x
	|	|	|	|	|	|
Segmental tier:	a	f	i	a	f	i

b. Tonal tier:

	H		L	L		H
	|		|	|		|
Skeletal tier:	V	C	V	V	C	V
	|	|	|	|	|	|
Segmental tier:	a	f	i	a	f	i

The rest of the notation used in autosegmental phonology is explained in [8.4]:

[8.4] a. | An unbroken **association line indicates pre-linking**, i.e. prior association in the lexicon of elements on separate tiers

     b. ¦ A broken association line indicates **linking**, i.e. the creation of an association line

     c. ⊥ A crossed-through **association line** shows **delinking**, i.e. the sever- † ence by rule of an association line linking elements on different tiers.

     d. [ ] A left bracket shows the left boundary and a right bracket shows a right boundary. (In this book these brackets will only be used where their presence is particularly important for the point under consideration.)

     e. Ⓥ A circle around an item indicates that the item has been deleted.

The theory does not require a one-to-one association of elements on different tiers. Elements at any one tier may be linked one-to-many with elements at another tier. This can be seen in the tonal examples in [8.5a] of **rising tones** or **falling tones**. Such tones are called **contour tones**. They are created when independent high and low tones are simultaneously linked to a single vowel (or to a single tone-bearing consonant).

[8.5] a.

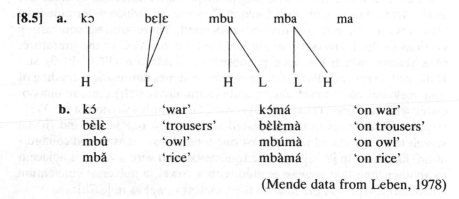

     b.   kɔ́   'war'   kɔ́má   'on war'
            bèlè   'trousers'   bèlèmà   'on trousers'
            mbû   'owl'   mbúmà   'on owl'
            mbǎ   'rice'   mbàmá   'on rice'

(Mende data from Leben, 1978)

Note: H marks a high tone (´)
       L marks a low tone (`)
       HL marks a falling contour tone (ˆ)
       LH marks a rising contour tone (ˇ)
       Blank marks 'no tone'

Conversely, the theory allows elements at one tier to remain unassociated with any item at another tier. Thus, in tone languages, sometimes a syllable representing a particular morpheme may not be associated with any tone in the lexicon. Instead, it may acquire later the tone of an adjacent syllable. In Mende, as seen in [8.5b], the morpheme *-ma* 'on' is underlyingly toneless. It gets its tone when the last tone of the stem to which it is suffixed **spreads** to it.

If phonological representations consist of independent tiers, how are these tiers eventually brought together in the production of speech? Are there any constraints on the ways in which elements on different tiers can be linked? In order to account for the constraints on the linking of elements on various tiers, autosegmental phonology has always incorporated **Mapping Principles**, which are claimed to be part of Universal Grammar. Mapping principles go under the names of **Universal Linking Conventions** (Pulleyblank, 1986) and the **Well-formedness Condition** (-Phonotactic) (WFC) (Goldsmith, 1976, 1990; Clements and Ford, 1979).

The WFC (for tone) is expressed in this fashion by Goldsmith (1990: 319).

[8.6]   1.   All vowels are associated with at least one tone.
        2.   All tones are associated with at least one vowel.
        3.   Association lines do not cross.

The WFC is concerned with safeguarding **phonotactic constraints**. Its role is to ensure that restrictions on the combination of phonological elements are not violated.

As we mentioned above, mapping principles are referred to by some not as the WFC, but as Universal Linking Conventions which determine how elements on different tiers may be associated. There are two contrasting versions of the Universal Linking Conventions or WFC in the literature. One version, which has been proposed by Goldsmith (1976, 1990) and Halle and Vergnaud (1980) among others, assumes automatic spreading of tone melodies on to tone-bearing units (normally vowels) that are unassociated with any tone. Thus, in [8.6] above, Goldsmith's version of the WFC requires (i) all tones to be associated with at least one vowel and (ii) all vowels to be associated with at least one tone. This means that in configurations like those in [8.7a], where a tone associated with a vowel is adjacent to another tone that is unassociated with a vowel, a universal convention *automatically* spreads the tone to the toneless vowel as in [8.7b]:

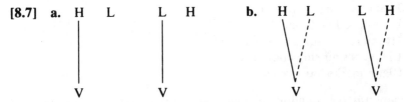

By contrast, some linguists, like Pulleyblank (1986), assume that the spreading of melodies to toneless vowels is *non-automatic*. On this view, spreading only occurs if a specific rule in a given language sanctions it. In many languages, a tone that is underlyingly unassociated with a vowel does not automatically spread to an adjacent vowel. Instead, it is deleted – unless a specific rule links it to a vowel. For reasons that will become clear later in this chapter, this kind of **non-automatic spreading** is what is required for morphological purposes.

We will be adapting the Mapping Principles outlined below for linking tones to tone-bearing units in our morphological analysis. These principles are based on Pulleyblank (1986) and Archangeli (1983).

[8.8]   Universal Linking Convention

(a)   Link a sequence of autosegments (e.g. tones) with a series of elements on the skeletal tier that are capable of bearing them (e.g. link tones with vowels) (see [8.3]);

(b)   Perform the linking going from the beginning to the end of the word. Unless specific instructions are given in the grammar of the language to do otherwise, link autosegments (e.g. tones) with units that are capable of bearing those autosegments (e.g. vowels) in a one-to-one fashion.

(c)   Association lines do not cross in the linking process.

The last point needs some comment because its significance is not obvious. There is a good reason for the ban on crossed association lines. To see why, let us begin by observing that each autosegmental tier contains a linearly ordered sequence of 'segments'. In this context the term segment includes not only consonants and vowels, but also any other elements on autosegmental tiers such as C and V slots (or x's), stress and tone. The segments on separate tiers are linked by an ordered sequence of association lines. Association lines indicate that a given segment, say the vowel /i/, is articulated in a temporal sequence linked to another specified segment, say a high tone (Goldsmith 1976: 28–30). The prohibition of crossed association lines reflects the fact that when an articulatory gesture is interrupted by another distinct gesture, if we want to resume the original gesture, we have to start afresh. This means that a representation like [8.9a] is ruled out because the same H tone is simultaneously associated with the first and third syllables since the association lines linking H and L tones to vowels end up crossing. In circumstances like these, we must posit two separate H tones linked with the first and third vowels, as in [8.9b]:

[8.9]   **a.**

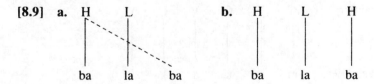

The WFC does not police the system so effectively that it prevents any unlawful representations from ever arising. Violations of the WFC can arise. But if they do, the WFC triggers appropriate repair rules to rectify the situation. It is assumed that the optimal repair strategy in each case will be one that involves the smallest number of rules. This follows the generally accepted principle of economy. (There is no virtue in using five rules to deal with a problem that can be solved using one rule.) The **repair operation** always makes the minimal number of changes required to satisy the WFC-phonotactic in [8.6]. In [8.10a] the repair operation associates unassociated vowels (V) with tone (T) while in [8.10b] it associates unassociated tones with vowels:

[8.10]

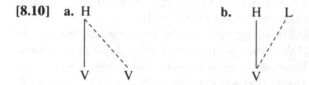

As we have hinted, the Mapping Principles outlined above are applicable not only to tone and tone-bearing units but also to any other elements that are on separate tiers. They regulate the ways in which association lines may link any autosegments to elements that bear them. For a detailed discussion of autosegmental phonology see Goldsmith (1976, 1990), Clements and Ford (1979), Clements and Goldsmith (1984), Pulleyblank (1986), and Katamba (1989).

8.2.2 The Skeletal Tier

The original impetus for autosegmental phonology came from the analysis of tone. But before long phonologists recognised that the essentials of this approach could be extended to other phonological phenomena. For instance, just as tones may be linked in different ways to tone-bearing units, segments may be linked in a variety of ways to the **skeletal tier**. This has important repercussions for morphological theory, as we will see below.

Normally, in the lexicon every vowel segment starts off being associated with a V-slot and every consonant with a C-slot on the skeletal tier (also called the CV-tier) (see [8.3]). We illustrate this with the Luganda word *mukazi* 'woman':

[8.11]

Luganda has both long vowels and geminate ('double') consonants. The geminate consonants are longer and have a more forceful articulation than plain, short consonants. Clements (1985) and Katamba (1985) have shown that in Luganda, the ways in which vowel and consonant segments are linked by association lines to the skeletal tier throws light on their phonological properties.

Geminate consonants arise if, where there are more C (or x) slots on the skeletal tier than there are consonantal segments on the segmental tier, a single consonantal segment is simultaneously associated with two C-slots on the skeletal tier. You can see this in [8.12] where the geminate *tt* of *tta* 'kill' contrasts with the single *t* of *ta* 'let go'.

[8.12]

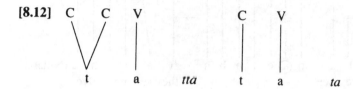

Similarly, although the norm is one-to-one linking of vowels and V-slots on the skeletal tier, it is possible to find more than one vowel segment linked to a V-slot. Vowels simultaneously linked to several V-slots are long. Length is phonemic. The verbs *siiga* 'smear, paint' and *siga* 'sow, plant' are distinguished by the length of their first vowel, as seen in [8.13]:

[8.13] **a.**

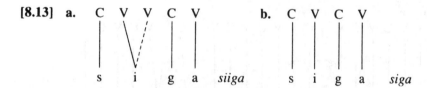

Observe that in [8.13a] **compensatory lengthening** takes place when a single vowel is doubly linked with two V slots in the underlying representation. Thus /i/ surfaces as long [i:] in *siiga*.

Using the notation of autosegmental phonology, account for the 'compensatory lengthening' in the following:

[8.14] **a.** /ba- e- laba/   [be:laba]
              'they-themselves-see'   'they see themselves'
         b. /tu- e -laba/   [twe:laba]
              'we-ourselves-see'   'we see ourselves'

    c.   /bi- e- laba/   [bye:laba]
           'they-themselves-see'   'they (e.g. animals) see themselves'

Observe that in [8.14a], if the first of two adjacent vowels is nonhigh, it is deleted. Then the vacant V slot left behind at the skeletal tier is inherited by the second vowel which is 'lengthened in compensation', as seen in [8.15].

[8.15]

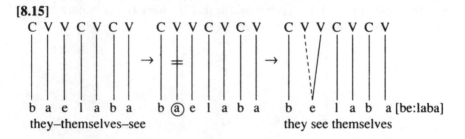

However, if the first vowel is high, i.e. /i/ or /u/ (as in [8.14b] and [8.14c] respectively), it is not deleted. Instead, it is associated with the C slot to its left and surfaces as a nonsyllabic glide /y/ or /w/. Again, the first V-slot which is vacated is taken over by the second vowel, which is 'lengthened in compensation':

[8.16]   **a.**

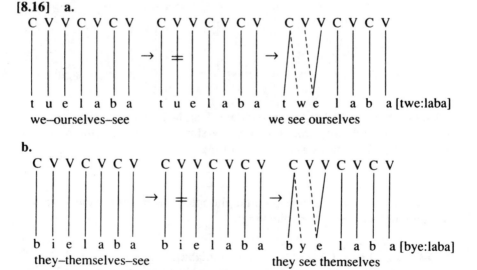

The aim of this phonological interlude has been to give you a grounding in the essentials of autosegmental phonology. The discoveries about the nature and organisation of phonological representations made in this model form the backbone of the prosodic morphology model to which the rest of this chapter is devoted.

8.3 PROSODIC MORPHOLOGY

In chapter 3, in our discussion of typology, we highlighted four principal methods of word-formation namely:

(i)	affixation	(ii)	compounding
(iii)	conversion	(iv)	incorporation

These four morphological types are very common, but they do not exhaust all the possibilities found in the languages of the world. There are other ways of forming words. They include **infixing** (i.e. inserting a word-building element within the root) and **reduplication** (i.e. the full or partial repetition of the base). Traditional morphological approaches offer no satisfactory method of dealing with these processes. But the model of prosodic morphology presented here accounts for them insightfully.

8.3.1 Arabic Binyanim

In Semitic languages, such as Arabic and Hebrew, words may be formed by modifying the root itself internally and not simply by the **concatenation** (i.e. linking together) of affixes and roots as happens in an inflecting, agglutinating or incorporating language (see section (3.6)).

In modern English, with a few exceptions including umlaut words such as *foot* (singular) and *feet* (plural) and ablaut words *ride* ~ *rode*, word-formation does not involve changes in the root vowel. These changes are historical fossils. The typical word-formation methods in English are:

[8.17]   **a.**   Affixation: (i)   prefixation (as in *un-kind*)
           and   (ii)   suffixation (as in *kind-ness-es*)
      b.   Compounding (as in *housewife*)
      c.   Conversion (as in *slow* (Adj) → *slow* (Adv))

As we saw earlier, in English infixing in derivational morphology (as in *Kalama-goddam-zoo*) is marginal to the system.

In Semitic morphology we encounter a very different situation. Much of the word-formation takes place root-internally. Infixing and modification of the root, rather than the stringing together of discrete morphemes, is the norm. Any attempt to segment words into morphemes as we have done for

English in [8.17a] would be misguided. You can verify this by examining the Arabic data below:

[8.18] a. kataba   'he wrote'
        b. kattaba   'he caused to write'
        c. kaataba   'he corresponded' (reciprocal, i.e. letters went to and fro between him and someone else)'
        d. takaatabuu   'they kept up a correspondence'
        e. ktataba   'he wrote, copied'
        f. kitaabun   'book (nomimative)'
        g. kuttaabun   'Koran school (nominative)'
        h. kitaabatun   'act of writing (nominative)'
        i. maktabun   'office (nominative)'
        j. makaatibu   'offices (nominative)'
        k. kutiba   'it was written'
        l. nkatab   'subscribe'

(from McCarthy, 1981)

Traditional scholarship has always recognised the fact that Arabic verbs are structured round a root consisting of consonants only. Verb forms are assigned to one of fifteen derivational classes or **binyanim** (singular **binyan**). Each binyan brings with it its own **vocalism** (i.e. vowels), and sometimes further consonants.

What is the consonantal core around which the words in [8.18] are built? What is the meaning with which it is associated?

The words in [8.18] are built around the consonants *ktb* which are associated with the meaning 'write'.

Often there is clear semantic and morphological evidence (which we will simply take for granted here) that one binyan is derived from the other. Consider these data based on McCarthy (1981: 378) and Wickens (1980):

[8.19] a. DERIVED FORM   DERIVATIONAL SOURCE

           Second Binyan   First Binyan
           kaððab 'to consider a liar'   kaðab 'to lie'
           waqqafa 'to arrest, stop   waqafa 'to stand, come to a
           someone'   halt'
           qattala 'massacre'   qatala 'to kill'

   b. Third Binyan   First Binyan
        kaatab 'to correspond'   katab 'write'
        qaatala 'to fight with'   qatala 'to kill'

What are the meanings associated with the second and third binyanim respectively?

The answer is that the second binyan derivation has a **causative** or intensifying meaning while the third binyan has a **reciprocal** meaning.

    We will use the data in [8.18] and [8.19] for our case study of the treatment of infixing in prosodic morphology, adding to it when the need arises as we go along.

8.3.2 Prosodic Morphology and Nonconcatenative Morphology

Traditional morpheme theory is ideal for the description of word-building process whereby morphemes are concatenated (i.e. are attached one after the other). It is not at all well-suited to the task of describing nonconcatenative morphological processes involving, for example, infixing or the internal modification of the root. So, although it has been recognised for a long time that in Semitic languages the root, usually consisting of three consonants (e.g. *ktb* or *qtl*), serves as the skeleton to which flesh is added in the process of word-formation, before the advent of prosodic morphology there was no theoretically elegant way of describing this method or word-formation.

    Prosodic morphology was initiated by McCarthy (1979, 1981). He noted the similarity in the behaviour of vowels introduced into consonantal roots by morphological processes in Arabic on the one hand, and that of **phonological prosodies** like tone spreading on the other. (See section 8.2.1). He hypothesised that the verb in Arabic has elements arranged on three independent tiers at the underlying level of representation in the lexicon, the three tiers being the **root tier** (also called the **consonantal** tier), the **skeletal tier** and the **vocalic melody tier**:

[8.20]

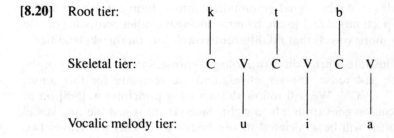

These three tiers are linked together by association lines similar to those we used earlier in our discussion of autosegmental phonology.

We will now look more closely at the three tiers in turn. Our observations will reveal three things:

(i)　　The meaning of a verbal lexeme is signalled at the **root tier** by the consonantal segments. Usually a verb root has three consonants in its underived lexical entry in the lexicon. Thus the root *ktb* represents the lexeme 'write', which is realised by a variety of word-forms. The different word-forms realise different grammatical words depending on the other consonantal and vowel segments introduced by word-formation rules at different strata in the lexicon. If in place of *ktb* we substituted another **tri-consonantal root**, we would get a different lexeme. For example, the root *qtl* represents the lexeme 'kill', *qrʔ* represents the lexeme 'read' and *fʕl* represents the lexeme 'do'.

(ii)　　The **skeletal tier** (which is also called the **CV-tier**) is like a potter's template. So, it is also called the **prosodic template tier**. It provides a canonical shape that is associated with a particular meaning or grammatical function. The template CVCCVCV, for instance, carries the grammatical meaning 'causative'. Hence *kattaba* means 'he caused to write' and *qarraʔa* 'he caused to read'.

　　　　The skeletal tier plays a key role in morphology. It enables us to extract from each derivational and inflectional class of words 'the characteristic pattern assumed by roots and vowel melodies – in fact the basic generalisation underlying Semitic morphological systems' (McCarthy, 1982: 192).

(iii)　　The **vocalic (vowel) melody tier** provides information analogous to that carried in English by inflectional affixes like tense, aspect, number or derivational affixes. It is used to relate, for example *kataba* 'he write' ~ *kitab* 'book' and *maktabun* 'office'. (We are oversimplifying the account somewhat and for the moment overlooking any consonants in prefixes and suffixes that are introduced by inflectional processes).

　　　　Just as tone rules introduce tone melodies consisting of one or more high or low tones which are linked to vowels, morphological rules in Arabic signal grammatical information concerning voice, aspect, mood and so on, by introducing melodies consisting of one or more vowels that fill different vowel slots on the skeletal tier.

Let us illustrate a prosodic morphology representation with the vocalic melody for past tense, the vowel /a/, and the template for this tense, which is CVCVCV. We will follow the mapping principles in [8.8] on p. 159. Association goes from left to right. Since there is just the one vowel segment /a/, it will be associated to the first V slot. This still leaves two unassociated V slots. Spreading is non-automatic. But a rule of Arabic licences the spreading of /a/ to these vacant slots. There are languages such as Yawelmani, with similar morphological patterns, where spreading fails to take place (for details see Archangeli 1983: 375–6). As for the skeletal tier, it contains the root consonants *ktb*.

The language specific vowel spreading melody needed to derive *kataba* 'he wrote' has this form:

[8.21] Root tier:

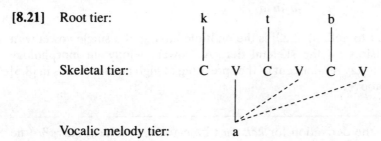

How should the past tense of *qrʔ* 'read' and *fʕl* 'do' be represented?

You already know that the C slots on the skeletal tier are going to be filled respectively by the consonantal segments *qrʔ* 'read' and *fʕl*. You needed to establish two more things: (i) the skeletal tier template for the past tense, and (ii) the vocalic melody for the past tense. As you can see by inspecting the skeletal tier in [8.21] above, the past tense template is CVCVCV and the vocalic melody is /a/. With this information in hand, you should have had no trouble coming up with this solution:

[8.22] a. Root tier:

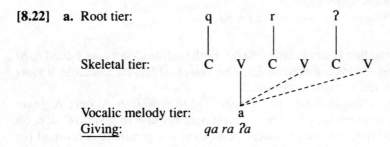

b. Root tier:

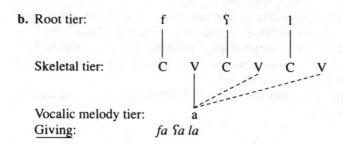

Skeletal tier:

Vocalic melody tier: a
Giving: *fa ʕa la*

A key point to note in [8.22] is the multiple linking of a single vowel with several V slots on the skeletal tier. A 'vowel melody' in morphology behaves in a way reminiscent of the spreading of high and low tones in [8.5] on p. 157 above.

Now show the derivation for *kattaba* ('he caused to write'), *qarraʔa* ('he caused to read') and *faʕʕala* ('he caused to do').

The form *kattaba* is represented as follows.

[8.23]   Root tier:

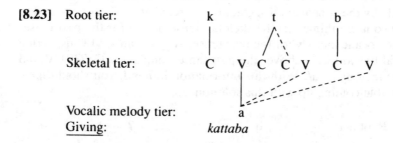

Skeletal tier:

Vocalic melody tier: a
Giving: *kattaba*

In [8.23], the first C slot is linked to /k/, both middle C-slots are linked to /t/ and the last C slot is linked to /b/. The vowel /a/ fills all available V slots going from left to right as before.

As seen, multiple linking is not restricted to vowels in Arabic. A single consonant segment may also be linked simultaneously to several C slots on the skeletal tier. A multiply linked consonant is a geminate consonant (as we have already seen in our discussion of Luganda in [8.12] above). Consonant gemination is a characteristic way of signalling morphological information in Arabic. The causative verb template, which is exemplified by *kattaba*, is CVCCVCV. To form the causative, the second consonant segment of a triconsonantal root is geminated.

Given the roots *qrʔ* 'read' and *fʕl* 'do', for *qarraʔa* 'he caused to read' and *faʕʕala* 'he caused to do', what we need is essentially the same as

[8.23] above – only the consonants at the skeletal tier are different in each case:

[8.24] **a.** Root tier:

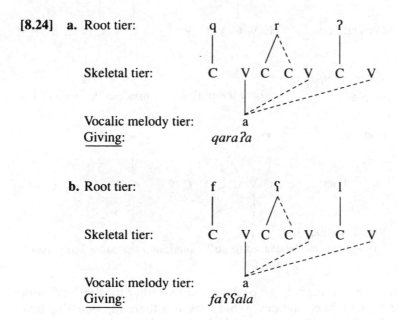

Vocalic melody tier: a
Giving: *qara?a*

  b. Root tier:

Vocalic melody tier: a
Giving: *faʕʕala*

So far we have only considered verb roots such as *ktb*, *qr?* and *fʕl* which contain three consonants. But not all verbal roots are triconsonantal. In [8.25] we see that there exist forms which have just two distinct underlying consonants like *zr* 'pull' and *sm* 'poison'. The challenge now is to find a way of describing these geminate roots. Account for the formation of the past tense and causative forms of the following:

[8.25] **a.** zarara 'he pulled'   **b.** samama 'he poisoned'
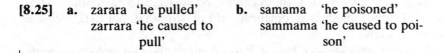
    zarrara 'he caused to   sammama 'he caused to poi-
        pull'   son'

The past tense template is CVCVCV, with /a/ as the vocalic melody while the causative melody is CVCCVCV, again with /a/ as the vocalic melody. The first C-slot is filled by /z/ and the second C-slot is filled by /r/ but this still leaves one vacant C-slot in the case of 'he pulled' and two in the case of 'he caused to pull'. The by now familiar Arabic spreading rule applies and links the remaining vacant C-slot on the skeletal tier with /r/:

[8.26]

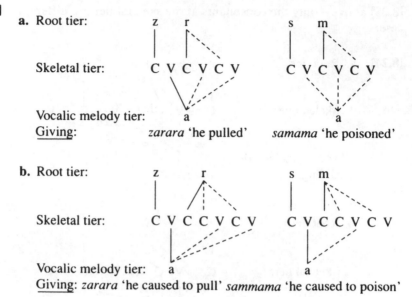

a. Root tier:

Skeletal tier:

Vocalic melody tier:
Giving: *zarara* 'he pulled'   *samama* 'he poisoned'

b. Root tier:

Skeletal tier:

Vocalic melody tier: a   a
Giving: *zarara* 'he caused to pull' *sammama* 'he caused to poison'

You might be wondering why we posit *zr* and *sm* rather than *zrr* and *smm* as the underlying representations. What reason is there for assuming that these roots have two and not three underlying consonants?

McCarthy (1981: 383) argues that verb roots like *zr* 'pull' and *sm* 'poison' contain just two underlying consonants in their lexical entries. If they had three consonants, their underlying representations would be **zrr* and **smm* respectively, with two identical consonants occurring next to each other. But such sequences are disallowed by the **Obligatory Contour Principle** (OCP). This principle states that: 'At the melodic level, adjacent identical elements are prohibited' (McCarthy, 1986: 208).

The OCP is a principle of Universal Grammar which was originally proposed by Leben (1973) in order to account for the fact that in tone languages identical H or L tones cannot be adjacent to each other. Thus, any HLL tone melody is automatically simplified to HL but LHL melodies are unaffected.

McCarthy has extended the OCP to all autosegmental melodies, including those of consonantal segments on the skeletal tier. The OCP either blocks or triggers the repair of representations where identical elements would be identical within a morpheme in a lexical representation (Yip, 1988). In the particular case that we are considering, the OCP blocks *zrr* and *smm*. In such roots the second consonant is geminated by the rule of left to right spreading so that it fills the vacant third slot in the CV template provided by the skeletal tier. Hence, such roots are known as **geminate roots**.

In the light of the foregoing discussion, explain how the causative past tense form of the verb *dahraja* ('caused to roll') is formed. Why is there no consonant gemination? Show the derivation of this form.

The example of *dahraja* shows that in addition to the typical triconsonantal roots like *ktb* and geminate (two-consonant) roots like *zr(r)*, Arabic also has **quadriliteral roots** which consist of four consonants like *dhrj* ('roll'). Quadriliteral roots are not fundamentally different from the more common triconsonantal roots. They only differ in having an extra C-slot in their prosodic template. The rules developed so far are sufficient to account for them. The causative binyan has the template CVCCVCV. With a triconsonantal root (see [8.23]), the middle consonant has to be geminated because there are more C slots than there are consonant segments to fill them. However, a quadriliteral root has enough consonants to allow a one-to-one association. There is no need to resort to gemination. The association simply proceeds following the regular conventions of left-to right association. You can see this in the representation of the causative binyan in [8.27].

[8.27] Root tier:

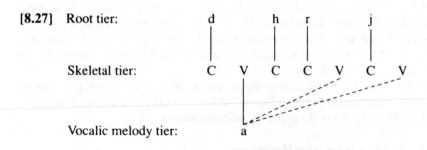

Now show how the Syrian Arabic forms in [8.28b] are derived:

[8.28] **a.** daxala 'entered' **b.** ddaxala 'mediated'
   tajara 'did business'   ttajara 'did his business'

Let us illustrate the answer with *ddaxala*. What we have here again is a template with four C slots and a root that has just three consonants. The first consonant segment is entered in the lexicon **pre-linked** with the second C slot on the skeletal tier as shown in [8.29a]. This is a marked association of C slots with segments. But it is needed to ensure that we do not end up with the incorrect **daxalla*, the form that would result from the normal unmarked left to right association of segments with C slots on the skeletal tier.

[8.29]   **a.** Root tier:

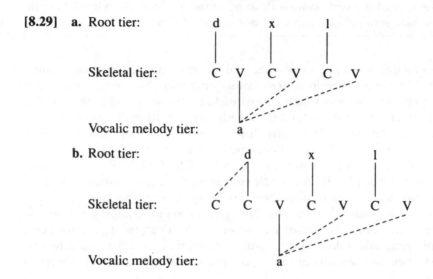

Subsequently, as seen in [8.29b], right-to-left spreading takes place and the empty first C is associated with the nearest free consonantal segment – which is on its left. The situation is parallel to that in [8.26]. The only difference is that here the association goes from right-to-left and not in the usual left-to-right manner.

In this way prosodic morphology accounts not only for the typical triconsonantal roots, but also for the less common geminate and quadriliteral roots without needing any ad hoc stipulations.

8.3.3 The Morpheme Tier Hypothesis

Prosodic morphology also incorporates the **morpheme tier hypothesis**. This is the claim that, in the lexicon, the representation of each morpheme in a word occupies a separate tier (McCarthy, 1981). So lexical representations contain another tier, namely the morpheme tier, which hitherto we have not included in our rules. The morpheme tier is conventionally symbolised by μ.

As we saw in section (2.3), infixing raises insurmountable problems for a theory which assumes that morphemes are made up of morphs. In nonconcatenative morphological systems, where words are not necessarily made up of sequences of morphemes in a row, it is common to find **discontinuous morphemes** which are interrupted by infixes (Harris, 1951; McCarthy, 1981). All the sounds that represent the same morpheme are not necessarily adjacent to each other. As you can see in [8.30] below, which is typical, a root morpheme like *ktb* may be interrupted by occurrences of the vowels that represent a different morpheme.

[8.30] a.

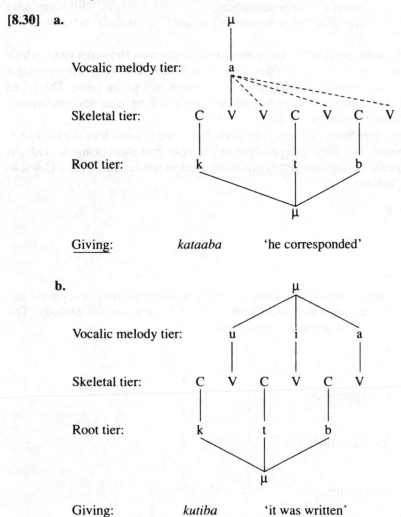

Giving: *kataaba*   'he corresponded'

b.

Giving: *kutiba*   'it was written'

In each example in [8.30] there are two morphemes: the root morpheme, represented by the consonantal elements on the root tier, and the morpheme conveying grammatical information, which is represented by the vocalic melody tier.

Up to this point the only discontinuous morphemes that we have encountered have been represented by intertwining consonants on the root tier with vowels of the vocalic melody tier, as in [8.30]. But infixed segments need not always be realised by vowels. Consonants can also be infixed. In Syrian Arabic, the reflexive morpheme is realised by a /t/ infix:

[8.31] **a.** smʕ 'hear'   **b.** tfʕ 'lift'
     samiʕa 'hear something'   rafaʕa 'lift something'
     stamaʕa 'hear for oneself (listen)' rtafaʕa 'lift oneself'

In such cases, putting all the consonants together on the same tier – which is separate from the vocalic tier – is not sufficient. The infix representing a separate morpheme must occupy a separate tier of its own. The [t] of *rtafaʕa* and *stamaʕa* is an infix representing a 'reflexive or detransitivising' morpheme referred to as the '*t-morpheme*'.

 If we adopt the morpheme tier hypothesis, we assume that the infixed 't-morpheme' is totally independent of the root and the vocalic melody. It appears on its own tier and is pre-associated in the lexicon with a C-slot as shown below:

[8.32] μ
    |
    t
    |
    C

In the case of *stamaʕa* for instance, the 't-morpheme' is on a separate tier from both the root tier consonants *smʕ* and the *a* vocalic melody. The representation of *stamaʕa* is as follows:

[8.33] **a.**

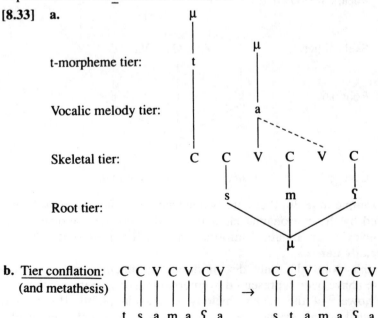

b. Tier conflation:
  (and metathesis)

Provide a morphological derivation of the following:

[8.34] kitaab-un   'book' (nominative)
       maktab-un   'office' (nominative)

These examples show that, in addition to the infixing processes of the kind we have seen above, Arabic morphology does also display affixation. The prosodic template for turning the triconsonantal verb root *ktb* into a noun is CVCVVC and the vocalic melody is /i a/ or /a/. In addition, the analysis needs to include the suffix *-un* as in *kitaab-un*, as well as both the prefix *ma-* and the suffix *-un* as in *ma-ktab-un*.

Kitaab-un can be represented as we have shown in [8.35a] and *ma-ktab-un* as shown in [8.35b], with the suffix *-un* and the prefix *ma-* placed on separate morpheme tiers.

[8.35] a.

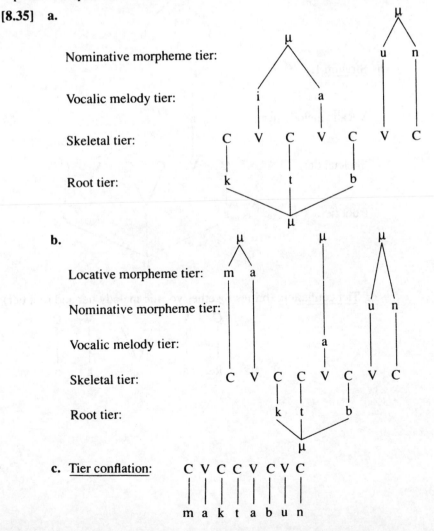

c. Tier conflation:

Nodes dominating morphemes may be regarded as the equivalent of brackets. We will assume that **tier conflation** occurs at the end of each cycle in the lexicon removing M nodes. This is analogous to the way in which the **Bracket Erasure Convention** removes all brackets at the end of each cycle in the lexicon (recall section (6.2.4)). So morphemes which start off on separate tiers at the beginning of a lexical cycle end up on the same tier when all the rules at a given stratum have applied. Effectively, this means that morphological rules can only access information that is available at the stratum at which they apply.

I will illustrate this with the derivation of the hypothetical form *takattab* (fifth *binyan*) reflexive 'cause to write to oneself'. Comparable forms, e.g. *takassab* 'to earn', do exist.

[8.36]   **a.** <u>Underived:</u>

b. <u>Stratum 1:</u>

Vocalic melody tier:

Skeletal tier:

Root tier:

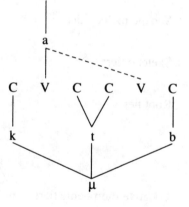

c. <u>Tier conflation:</u> (brings together vocalic melody tier and root tier)

C   V   C   C   V   C
| | \ / | |
k   a   t   a   b

d. <u>Stratum 2:</u>

ta-morpheme tier: t   a

e. <u>Tier conflation:</u> (brings together *ta* and *kattab*)

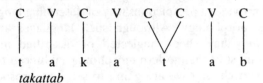

<u>Output:</u>   *takattab*

In this section we have seen the basic motivation for the morpheme tier hypothesis and the mechanisms by which its implications are dealt with. In the process, we have seen its similarity with the Bracket Erasure Convention and noted the relationship between prosodic morphology and lexical morphology.

8.4 CONCLUSION

In languages like Arabic, whose morphology is largely nonconcatenative, very often words are formed, not simply by concatenating affixes and roots, but rather by infixing, gemination and other changes taking place internally within the root. Consequently, many morphemes are discontinuous. Prior to the advent of prosodic morphology, there was no satisfactory mechanisms for analysing such languages.

We have seen how prosodic morphology, by adapting the principles and devices of autosegmental phonology, provides a framework for the analysis of languages of this morphological type. We have seen that the skeletal tier which plays a pivotal role in autosegmental phonology also serves morphological functions. The root consonants and the vocalic melody are represented on independent tiers. Mapping Principles regulate the linking of these independent morphological tiers.

The patterns of infixing in Arabic *binyanim* are treated in terms of the association of morphological CV templates on the skeletal tier with segments. The C slots are associated with consonants of the root morpheme and the V slots are associated with inflectional and derivational morphemes. In all cases the tiers are linked via the skeletal CV-tier which is the spine that holds together the different tiers. The association of the morphological consonantal root tier and the vocalic melody on the one hand, with the skeletal CV-tier on the other, is modelled on the phonological mapping of autonomous tones onto tone-bearing units in tone languages.

Prosodic morphology was conceived in the wider context of a generative grammar that incorporates the theory of the lexicon and morphology encapsulated in lexical morphology. The theory assumes a lexicon where different morphemes are on separate morpheme tiers. At the end of each lexical cycle, tier conflation (which is analogous to Bracket Erasure) takes place.

Prosodic morphology is a significant development in morphological theory. It provides a perspicuous way of describing languages with nonconcatenative morphology. Although such languages are neither rare nor obscure, they have been neglected or described in an unilluminating manner in most of the modern morphological literature.

In the next chapter we are going to see how prosodic morphology has been extended from the analysis of Semitic morphology to the description of other nonconcatenative morphological phenomena such as reduplication and metathesis.

EXERCISES

1. Why are the phonological representations in (a) well formed while those in (b) are not?

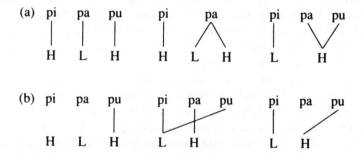

2. In Egyptian Arabic the verbal root for 'sit' is *gls*.
   (a) What do you expect *galasa* to mean?
   (b) Use *galasa* to explain what is meant by a *discontinuous morpheme*.

   (c) Explain why the assumption that morphemes *consist* of morphs runs into difficulties when confronted by words like *galasa*.

   (d) How is *galasa* derived from *gls*?

   (e) Show the derivation of the noun *magaalis* 'councils' from the root *gls*.

3. (a) Given the roots *klm* 'speak', *ksr* 'break', *zr* 'pull' and the information that (i) the passive vocalic melody is *uia* and (ii) the passive skeletal template is CVCVCV, how does one say 'it was spoken', 'it was broken' and 'it was pulled'?

   (b) Using the notation of prosodic morphology, write the derivation of each one of the words which you have formed in 3(a) above.

4. Study the following data from Syrian Colloquial Arabic and answer the questions that follow:

Perfect		Subjunctive	
daraset	'I studied'	tedros	'that she study'
darasna	'we studied'	yedros	'that he study'
darasti	'you (feminine) studied'	tedersi	'that you (masculine) study'
daras	'he studied'		
darset	'she studied'		
darasu	'they studied'		

(from Younis, 1975)

   (a) What is the consonantal root meaning 'study'?

   (b) Use the data above to argue for a prosodic analysis of Syrian Arabic. Show why an attempt to analyse the data purely in terms of prefixation and suffixation would be unsuccessful.

   (c) Provide prosodic morphology derivations of *daras*, *darset*, *tedros* and *tedersi*.

5. Show the derivation of the Standard Arabic forms below:

faʕala	'he did'
tafaaʕala	'to exert mutual pressure'
ʔinfaʕla	'it has been done'
ʔiftaʕala	'he made it up (it was not genuine)'

9 Template and Prosodic Morphology

9.1 WHAT IS REDUPLICATION?

In the last chapter we used the theory of template morphology to analyse Arabic, a language that uses infixing extensively. This chapter extends template morphology to the analysis of two other nonconcatenative phonological phenomena. We begin with a discussion of **reduplication**, a process whereby an affix is realised by phonological material borrowed from the base. Although this process is very widespread, it has tended to be treated as a marginal curiosity by many Eurocentric writers on morphology. After dealing with reduplication, we shall next extend the theory to the analysis of **metathesis** in section (9.4).

Sapir (1921: 76) observed that:

Nothing is more natural than the prevalence of reduplication, in other words, the repetition of all or part of the radical element. The process is generally employed, with self-evident symbolism, to indicate such concepts as distribution, plurality, repetition, customary activity, increase in size, added intensity, continuance.

He noted that, even in English, reduplication is not altogether unknown. He pointed to examples like these:

[9.1]

pooh-pooh	goody-goody	wishy-washy
sing-song	roly-poly	brain-drain
harum-scarum		

The term reduplication is used in a narrower sense in this book than in Sapir's writings. It is restricted to situations where the repeated part of the word serves some derivational or inflectional purpose. Apparent examples of reduplication like Sapir's *harum-scarum* and others like *brain-drain* and *dodo* where the recurring forms do not serve a grammatical or semantic function that could be served by a comparable affix are excluded.

In this section I summarise the common functions served by reduplication. The discussion and data draw on the survey by Moravcsik (1978).

180

How is the plural of nouns formed in Papago?

[9.2] Singular | | Plural
bana | 'coyote' | b<u>aa</u>bana
tini | 'mouth' | t<u>ii</u>tini
kuna | 'husband' | k<u>uu</u>kuna

(based on Langacker, 1972)

Note: Here and elsewhere in this chapter orthographic forms rather than phonemic transcriptions are used for phonological representations except where this might lead to confusion.

In Papago, the plural is formed by prefixing a copy of the first syllable of the singular noun and lengthening its vowel. This is a good example of the use of **nominal reduplication** to signal plural.

Noun reduplication may also express the meanings 'every X' and 'all X'. Thus in Luganda, *babiri* 'two' becomes *babiri-babiri*, with the entire word reduplicated, to express the meaning 'every two'.

Moravcsik (1978: 319) reports similar facts in these unrelated languages:

[9.3] Unreduplicated | | Reduplicated | |
bar | 'two' | barbar | 'all two' | (Tzeltal)
ren | 'man' | renren | 'everybody' | (Mandarin)
anak | 'child' | anakanak | 'various children' | (Malay)

In verbs, reduplication often indicates continuation, frequency or repetition of an event or action. In some cases, the repetition involves the same participants as seen below:

[9.4] Reduplicated

-pik | 'touch it' | -pipik | 'touch it lightly, repeatedly' (Tzeltal)
guyon | 'to jest' | guguyon | 'to jest repeatedly' (Sundanese)

Sometimes, however, reduplication involves repetition of the event or action, but with different participants, as in these Twi examples:

[9.5] Unreduplicated | | Reduplicated |
wu | 'die (of one or several persons)' | wuwu | 'die in numbers'
bu | 'bend/break (in many places)' | bubu | 'bend/break (break many things)'

Often reduplication has an **augmentative meaning**. It signals an increase in size, frequency or intensity. This is illustrated by the following:

[9.6] Unreduplicated	Reduplicated
dolu 'full'	dopdolu 'quite full' (Turkish)
dii 'to be good'	díidii 'to be extremely good' (Thai)

Conversely, reduplication may have a diminutive effect, often with connotations of endearment as in [9.7a] or simply of attenuation as in [9.7b]:

[9.7] **a.** Unreduplicated	Reduplicated
xóyamac 'child'	xoyamacxóyamac 'small child' (Nez Percé)

b. kὲε 'old (of people)' kὲε-kὲε 'elderly' (Thai)
kàw 'old (of things)' kàw-kàw 'oldish' (Thai)

We have seen the range of meanings conveyed by reduplication. Obviously, reduplication involves copying. But what exactly gets copied?

9.2 IS REDUPLICATION CONSTITUENT COPYING?

At first sight, a case could be made for treating reduplication always as nothing more than **constituent copying**. We could say that a rule makes a copy of some constituent, e.g. the word, and attaches it to a pre-existing token of that constituent. This seems to work in some cases. Let us take, for example, Walpiri, a language of Australia, where many nouns referring to humans form their plural by **total reduplication** (i.e. copying the entire word):

[9.8] Singular | | Plural |
---|---|---|---
kurdu | 'child' | kurdukurdu | 'children'
kamina | 'girl' | kaminakamina | 'girls'
mardukuja | 'woman' | mardukujamardukuja | 'women'

(Walpiri data from Nash, 1980)

Total reduplication can also happen in other word-clases. Tagalog uses it with an attenuative meaning in verbs like *sulatsulat* 'write a little' (cf. *sulat* 'to write'). Shi, a Bantu language, on the other hand, uses total reduplication of an adjective to express an intensive meaning: *nyeerunyeeru* 'very white' (cf. *nyeeru* 'white').

Study the following data from Maori (Krupa, 1966) and answer the questions that follow:

[9.9] **a.**

mano	'thousand'	manomano	'innumerable' (nouns)
reo	'voice'	reoreo	'conversation'

b.

ako	'learn'	akoako	'consult together' (verbs)
kimo	'wink, blink'	kimokimo	'wink frequently'
patu	'beat, kill'	patupatu	'kill all'

c.

mate	'sick'	matemate	'sickly (adjective)
wera	'hot'	werawera	'rather hot, warm'

d.

pango	'black'	papango	'somewhat black, dark' (adjectives)
whero	'red'	whewhero	'reddish'

(i) Describe the reduplication process used in each set of examples.
(ii) What is the meaning associated with each reduplication process?

In the noun examples in [9.9a] there is full reduplication: the entire word is copied and the resulting word has an augmentative meaning. Likewise, in [9.9b] the entire verb is copied and the resulting word has a reciprocal, augmentative or frequentative meaning. Again, in [9.9c] full reduplication takes place but this time it has an attenuative meaning. It indicates a reduction in intensity in the derived adjective. Partial reduplication (copying only the first syllable of a word), which is shown in [9.9d], has the same semantic effect of diminution of intensity in adjectives.

Cases like the above, where a constituent like a syllable, a morpheme or a word is copied in reduplication, will not detain us. They are simple and offer no great theoretical challenge. What is less straightforward, though by no means rare, is reduplication that does not involve constituent copying. In Maori, for instance, reduplication does not always copy elements that are identifiable as constituents. Thus in [9.10] we see the reduplication of the first CV of a word. By accident, in [9.10a] that CV happens to constitute a syllable. As you can see (data from Krupa, 1966), in [9.10b] the copying still goes ahead, although it results in splitting a syllable and leaving behind a stranded *i* and *e*:

[9.10] **a.**

tango	'take up, take in hand'	tatango	'snatch from one another'
kimo	'wink, close the eyes'	kikimo	'keep the eyes firmly closed'

b.

nui	'big'	nunui	'big plural'
moe	'sleep, close the eyes'	momoe	'keep the eyes closed, sleep together'

Maori is not unique in this respect. Moravcisk reports that there are other languages that reduplicate a part of a word that does not form a constituent. See the examples below:

[9.11]　**a.**　qa·x̦ 'bone'　　qaqa·x̦ 'bones' (Quileute)

　　　　b.　gen 'to sleep'　　ggen　'to be sleeping' (Shilh)

In Quileute plural-formation in [9.11], a copy is made of the first CV of the input morpheme, taking only the *qa* fragment of the first syllable for reduplication and leaving behind the /x̦/ – even though this splits the syllable. Likewise, in Shilh, the first consonant is copied, leaving behind the rest of the syllable.

The obvious conclusion to draw from this is that although some instances of reduplication can be viewed as constituent copying, it would be wrong to assume that reduplication always requires constituent copying. Sometimes reduplication copies fragments that are not morphemes or even syllables. How should such reduplication be represented? In the next section we present a theoretical approach that offers some insight into these reduplication phenomena that not involve constituent copying.

9.3 CV-TEMPLATES AND REDUPLICATION

McCarthy (1981) and Marantz (1982) have extended McCarthy's CV-template morphology model for the analysis of Arabic to account for what happens in reduplication. Their lead has been followed by Broselow (1983), Archangeli (1983), Broselow and McCarthy (1983) among others. Below we shall see how they have done it.

9.3.1 Underspecification

The essence of reduplication in CV-template morphology is summed up in this way by Borselow and McCarthy (1984: 25):

> reduplication is a special case of ordinary affixational morphology, where the affixes are phonologically underspecified, receiving their full phonetic expression by copying adjacent segments.

In other words, reduplication is essentially the affixation of a **morpheme template** (in the shape of a CV-skeleton) to a stem. Normally, the dictionary entry of a morpheme includes a specification of its semantic, syntactic, morphological and phonological properties. What is odd about reduplication as an affixation process is the fact that the skeletal CV-template introduced by the affix is **underspecified**. The reduplicative morpheme has a phonologically defective dictionary entry. While its syntactic and semantic properties are specified, the phonological part of the entry is left incomplete. It consists merely of a CV-skeletal template unattached to any segments, or it only has some C or V elements linked to segments. In order for that morpheme to be phonologically represented, a phonemic melody must be mapped on to every C and V slot in the morpheme template. This is achieved by copying a portion of the segmental representation of the base, or even the entire base, to which the underspecified morpheme is attached. The sounds copied, as we saw above, may or may not represent a constituent of the base.

The primary objective of underspecification is to enable us to write the most economical grammar possible. This is achieved in part by omitting from individual lexical representations anything that is general and predictable by rule (Kiparsky, 1985; Steriade, 1987; Archangeli, 1988). If we have general rules that copy out bases or portions thereof, we avoid the preposterous situation where one would simply have an indefinitely long list of allomorphs of, say, the plural morpheme in Walpiri. Without assuming a general reduplication principle for Walpiri (see [9.8]), for example, the singular and plural forms of nouns would have to be listed separately, implying that the plural morpheme has hundreds of unpredictable allomorphs.

Marantz (1982) proposes a way of providing a phonemic melody for an underspecified affix via reduplication. He requires the grammar to state the following:

  (i)   the shape of the reduplicative CV-template,
  (ii)   whether the reduplicative CV-template is prefixed, infixed or suffixed,
  (iii) the part of the base copied as the 'melody',
  (iv)   the direction of mapping: is the melody mapped on to the CV-template left-to-right or right-to-left?

Marantz's proposals are broadly accepted by linguists working in this model although there are some relatively minor differences in the way they apply them.

Below we will use the somewhat modified version of the principles for dealing with reduplication proposed by Broselow and McCarthy (1983: 27):

[9.12] Mapping Principles in Reduplication
  (i)   Introduce an underspecified affix (prefix, suffix or infix);
  (ii)   create an unassociated copy of the phonemic melody of the root or stem or base;

(iii) associate the copied phonemic melody on to the CV-skeleton one-
       to-one, with vowels being linked to V-slots and consonants with
       C-slots. In the case of a prefix the association goes from left to
       right while in the case of a suffix it goes from right to left;
(iv)   finally, erase all superfluous phonemic material or any CV slots on
       the skeletal tier that remain unassociated at the end.

In the following sub-sections we will apply these principles to a wide range
of reduplication processes. You may need to refer back to [9.12] frequently. You will also probably need to refer back to (8.2.1). The mapping
principles presented here are specially adapted from the general mapping
principles of autosegmental phonology that were introduced in that
section.

9.3.2 Reduplication as Prefixation

We start our reduplicative CV-template presentation with a consideration
of Agta reduplication.

Study these data from Agta, a language of the Philippines:

[9.13] takki   'leg'   taktakki   'legs'
        uffu   'thigh'   ufuffu   'thighs'

(based on Marantz, 1982)

a. Describe how the plural is formed.
b. Write morpheme CV-templates along the lines we used for Arabic
    nonconcatenative morphology in the last chapter to derive the
    plural of 'leg' and 'thigh'.

Agta reduplication is no respector of constituent structure. In a CVCCV
noun, the first CVC is reduplicated, even where this means splitting a
geminate consonant. Hence *takki* becomes *tak-takki*. However, in a VCCV
only the first VC part is reduplicated. Again, this happens even where a
geminate is split in the process. Hence, *uffu* becomes *uf-uffu*.

Marantz (1982) shows that all this can be simply stated using morpheme
templates if we assume that the plural of nouns is formed by prefixing CVC
on the morpheme template tier (i.e. CV-skeletal tier) where the noun has a
CVCCV root. At the start of the derivation, there are no consonant or
vowel segments attached to the C and V slots. A reduplication rule
subsequently copies the segmental melody in its entirety as you can see in
[9.14b]:

[9.14]   **a.** <u>Singular</u>   **b.** <u>Plural</u>

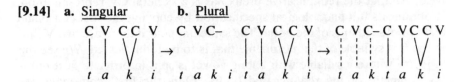

<u>Note</u>: any *k* doubly linked to two C slots is a geminate *kk*. Refer back to the discussion of the WFC and association principles in section (8.2) of the last chapter, and also see [9.12].

Following the mapping principles introduced in [9.12], there is no automatic spreading. Segments associate in a one-to-one fashion starting from the beginning of the word. So, in the reduplicated part of the word, the first three segments *t a k* associate in a one-to-one fashion with the C V C introduced by reduplication. The last segment, i.e. *i*, is left unassociated since a V does not (normally) associated with a C slot. As was stated in section (8.2) the model includes a general pruning convention that deletes at the end of the derivation any tone or segment that is left unassociated with some slot on the skeletal tier. Conversely, any C or V that is not linked to a segment by the end of the derivation is also automatically deleted (cf. [9.12]). These conventions ensure that only *tak* is prefixed by the reduplication process as shown in [9.14c].

For our next example we turn to Tagalog, a language of the Philippines, which has been described by Carrier-Duncan (1984) using CV-template morphology. Tagalog has a number of reduplication processes that are of great interest. In disyllabic stems, the initial CV of the stem is copied. The vowel of the reduplicated affix is short, regardless of whether the original vowel of the base is long or short. And the CV copied need not be a whole syllable:

[9.15]   **a.** kandilah 'candle'

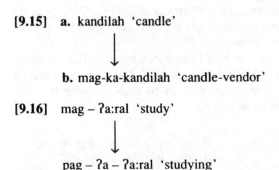

   b. mag-ka-kandilah 'candle-vendor'

[9.16]   mag – ʔaːral 'study'

↓

   pag – ʔa – ʔaːral 'studying'

In [9.15] *mag-* is attached before the base, and a reduplicative prefix with no segmental realisation of its own is interposed between *mag-* and the base. All that the reduplicative prefix has is a skeletal CV representation. It obtains its full phonological specification by copying the first consonant and vowel segments of the noun. The same happens if there are two V slots in the first syllable of the morpheme that is to be reduplicated. We see this in [9.16] where a syllable with a long vowel is split in order to fit it in the CV template of the reduplicated prefix. Only the first, consonant and vowel segments of the verb are copied, although the input has two V slots in the first syllable, since that syllable has a long vowel. Thus, a short vowel and not a long one appears in the reduplicated form *pag ʔa̲ ʔa:ral* (not *pag ʔa̲: ʔa:ral*).

Study the Tagalog data below:

[9.17]   **a.** tahi:mik 'quiet'
$$\downarrow$$
b. tahi:-tahi:mik 'rather quiet'

[9.18]   **a.** baluktot 'crooked'
$$\downarrow$$
b. balu:-baluktot 'variously bent'

<u>Hint</u>: treat long vowels as parallel to geminate consonants.

    a. How does reduplication work where the base contains a trisyllabic morpheme?
    b. Write a CV-template to account for this type of reduplication.

In [9.17], where the base is trisyllabic, reduplication is only carried out up to the second vowel. This is done even if it means splitting a CVC syllable and leaving behind the consonant that closes the syllable. Then the second vowel is lengthened if it is not already long.

In [9.17] and [9.18] reduplication first introduces a prefix that is solely represented by CVCVV template on the skeletal tier. In the case of *tahi:mik* this CV-template happens fortuitously to coincide with the first two syllables of the word as seen in [9.19]:

[9.19]   **a.**   C   V   C   V   V –

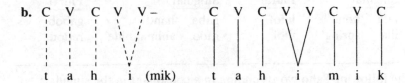

General mapping principles (see [9.12] above) associate the C and V slots on the skeletal tier to the copy of the segmental realisation of the prefix created by reduplication. The surplus segments (*mik*), with no CV slots to map on to, are pruned by convention.

No doubt you will have realised that a very similar analysis is needed for *balu:baluktot*. Your solution should resemble [9.20]:

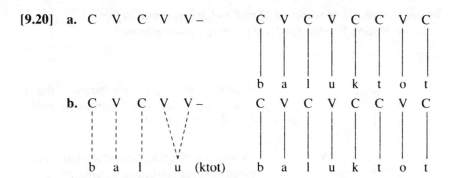

Here again we begin by adding the reduplicative CVCVV template as shown in [9.20a]. Then, in [9.20b] we make a copy of the segmental representation of *baluktot* and associate it to the morpheme CV-template following mapping conventions. The vowel *u* is doubly associated with the two V slots and surfaces as long. The remainder (*ktot*), which has nowhere to dock on the morpheme template, is deleted.

9.3.3 Reduplication as Suffixation

Reduplication may also introduce suffixes. Welmers (1973: 224) reports that in Saho, a typical Cushitic language, one common pattern of forming the plural is to attach a **reduplicative suffix** consisting of a vowel pre-linked in the lexicon to the partially specified reduplicative suffix. This vowel is usually /o/. (But some nouns have a vowel other than /o/ introduced; e.g. *boodo* 'hole' becomes *boodad* (not **boodod*) and *angu* 'breast' becomes *angug* (not **angog*).) Regardless of whether the vowel inserted is /a/ or /o/, it is followed by a copy of the last stem consonant:

[9.21]

Singular		Plural	Singular		Plural
lafa	'bone'	lafof	gaba	'hand'	gabob
illa	'spring'	illol	rado	'animal hide'	radod

Study the following Saho words ending in a consonant in the singular:

[9.22]

Singular		Plural	Singular		Plural
af	'mouth'	?	*nef*	'face'	?

a. What do you expect to be the plural forms of 'mouth' and 'face'?
b. Using morpheme templates, show the derivation of your proposed form for the plural of 'bone'.

The respective plural forms of *af* and *nef* are *afof* and *nefof*.
   The derivation needed for *lafof* ('bones') is as follows:

[9.23]

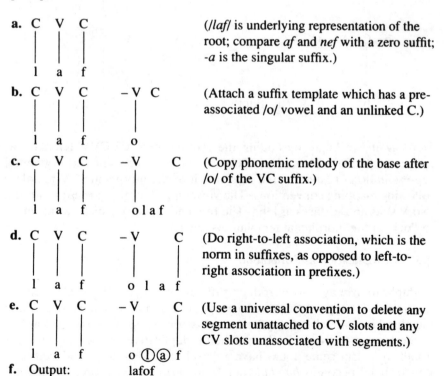

a. C V C
   | | |
   l a f
   (/*laf*/ is underlying representation of the root; compare *af* and *nef* with a zero suffit; -*a* is the singular suffix.)

b. C V C   – V C
   | | | |
   l a f   o
   (Attach a suffix template which has a pre-associated /o/ vowel and an unlinked C.)

c. C V C   – V   C
   | | | |
   l a f   o l a f
   (Copy phonemic melody of the base after /o/ of the VC suffix.)

d. C V C   – V   C
   | | | | |
   l a f   o l a f
   (Do right-to-left association, which is the norm in suffixes, as opposed to left-to-right association in prefixes.)

e. C V C   – V   C
   | | | | |
   l a f   o (l)(a) f
   (Use a universal convention to delete any segment unattached to CV slots and any CV slots unassociated with segments.)

f. Output: lafof

We will now leave reduplicative suffixes and turn to infixes.

9.3.4 Internal Reduplication

The norm in most languages is for affixes to appear as prefixes or suffixes on the edges of the stem or base to which they are attached.

Infixes (i.e. affixes inserted internally in a stem or base) are unusual, as we have already seen. An even greater morphological oddity is the phenomenon of **infixing reduplication**, whereby a copy of part of the base is inserted in the base as an affix. Though not common, this phenomenon is found in a cross-section of the world's languages including, among others, Levantine Arabic, various native American languages of the Salishan family spoken in the Pacific northwest, Washo, a Hokan language, Quileute, several Austronesian languages including Samoan, Chamorro and Nakanai (cf. Broselow and McCarthy, 1983).

Internal reduplication is very similar at a formal level to **infixing** in Arabic. The essential difference is that, unlike Arabic infixes, which are represented in the lexicon by vocalic melodies, reduplicative infix morphemes are devoid of segmental material. They obtain the segments that phonologically manifest them by raiding the bases to which they are attached.

In Samoan, for instance, plurals of verbs are formed by reduplicating a CV sequence. Verb stems may contain one, two or three syllables. Syllables consist of either a lone V or a CV sequence (cf. Broselow and McCarthy, 1983: 30). Consider the following:

[9.24] Disyllabic verbs

no·fo	*no*·no·fo	'sit'
mo·e	*mo*·mo·e	'sleep'

Note: · separates syllables.

As you can see in [9.24], in disyllabic verbs, a CV prefix is attached in the plural and a copy of the first consonant and vowel of the stem is made. The copy is then linked to the prefixed empty CV slots. This procedure is reminiscent of the CV reduplication in Maori [9.10].

More interesting are verbs with three syllables. Their plural is formed, not by adding a CV prefix, but rather by *infixing* an underspecified CV on the skeletal tier before or after the second syllable. (As you can see, there is no way of knowing for sure but we will simply assume it is before the second syllable for the purposes of writing the rule. Since it is an infix, the question of left-to-right or right-to-left attachment to the stem does not arise.)

[9.25]   Three syllable roots

alofa	a*lo*lofa	'love'
savali	sa*va*vali	'walk'
maliu	ma*li*liu	'die'

Represent CV infixation in *savavali* using a morpheme template.

Your solution should be along these lines:

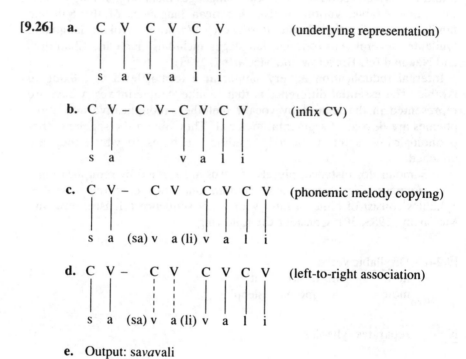

[9.26] a. C V C V C V   (underlying representation)

      s a v a l i

b. C V – C V – C V C V   (infix CV)

      s a   v a l i

c. C V –   C V   C V C V   (phonemic melody copying)

      s a (sa) v a (li) v a l i

d. C V –   C V   C V C V   (left-to-right association)

      s a (sa) v a (li) v a l i

e. Output: sa*va*vali

The derivation in [9.26c] (where a portion taken from the middle of a morpheme is inserted in the middle of that morpheme in reduplication) is highly marked. Left-to-right association starts with the second syllable. The first one is **invisible** (cf. Inkelas, 1989).

9.3.5 Prosodic Morphology

McCarthy and Prince (1988, 1990) have proposed a new theory of **Prosodic Morphology** which makes a number of important claims about the interaction between phonological structure and morphology. Prosodic

Morphology claims that a morphological melody maps directly on to a prosodic phonological template consisting of a genuine prosodic unit such as the syllable the foot, phonological word, etc. They argue in favour of by-passing the CV-skeleton in representing reduplication and instead mapping morphological representations directly onto prosodic units like moras, syllables, feet and phonological words. The evidence for this is that in many cases a reduplicative process supplies a template characterisable in terms of such prosodic units rather than C and V slots.

Their proposal is centred round the three principles below which are set out in McCarthy and Prince (1990: 209–10):

[9.27]   (i)   The Prosodic Morphology Hypothesis
           This requires that templates are stated neither in segmental terms nor in terms of C and V slots on the skeletal tier, but rather in terms of the units of prosody, e.g. the mora, syllable, foot, prosodic word, and so on.

     (ii)   The Template Satisfaction Condition
           This principle makes it obligatory to satisfy all elements in a template. It is forbidden to leave any part of the morphological template unassociated with some prosodic unit. The association follows both universal and language specific principles.

    (iii)   The Prosodic Circumscription of Domains
           This states that the domain in which morphological processes take place may be circumscribed not only by morpho-syntactic factors, but also by prosodic criteria. In particular, morphological operations may target the **minimal word** within a domain rather than the whole domain.

The proposal that reduplication is best described in terms of the copying or transferring of prosodic melodies rather than C and V slots on the skeletal tier has been influential. A number of linguists have taken it on board and applied it not only to Arabic, as McCarthy and Prince themselves have done, but also to a range of other languages which at first sight might not appear ideal for a prosodic treatment. These languages have included Sanskrit (cf. Steriade, 1988) as well as Kinande (cf. Mutaka and Hyman, 1990).

In the remainder of this section I am going to focus on Kinande. I will show how Mutaka and Hyman (1990) have used Prosodic Morphology to describe reduplication in Kinande. The discussion in this section is based on their account of Kinande nominal reduplication.

Before we are in a position to discuss reduplication, we need to outline the structure of nouns in this language. Typically, a Kinande noun has the structure in [9.28]:

[9.28]	augment-	prefix	stem	
	o	ku	gulu	'leg'
	a	ká	tì	'stick'
	e	kí	témbekalį̀	'kind of tree'

Note: The symbol į represents a very high front vowel.

Most Kinande noun stems are bisyllabic, although there are some longer nouns as well as some monosyllabic ones. This fact is important, as we shall soon see.

Kinande is a tone language, with a three-way, underlying phonological contrast between high (´), low (`) and toneless (unmarked) vowels. However, on the surface, there is only a two-way contrast between H and L tone. Any toneless vowel is assigned a tone post-lexically. For simplicity's sake, we will not refer to tone in what follows because it is not relevant.

We are now ready to treat reduplication. A noun can be reduplicated to create a new noun with an intensified meaning that can be glossed as 'a really (good) x'. We will start by presenting reduplication in disyllabic nouns.

[9.29] **a.** Nouns with bisyllabic stems

o.ku-gulu	'leg'	o.ku-gulu.gulu	'a real leg'
o.mu-gógò	'back'	o.mu-góngo.góngò	'a real back'
o.kú-boko	'arm'	o.kú-bokó.boko	'a real arm'
a.káhúkà	'insect'	a.ká-húká.húkà	'a real insect'
o.múkalį̀	'woman'	o.mú-kalį.kalį̀	'a real woman'
o.mu-longò	'village'	o.mu-longo.longò	'a real village'
o.mu-síkaa	'girl'	o.mu-síka.síkaa	'a real girl'

b. Nouns with monosyllabic stems

o.mú-twe	'head'	o.mú-twe.mú-twe	'head'
e.bi-laa	'bowels'	e.bi-laa.bi-laa	'bowels'
a.ká-tì	'stick'	a.ká-tì.ká-tì	'stick'

Note: The augment is separated from the prefix with a dot (.) for reasons that will become clear shortly.

---

a. How does the reduplication of nouns with monosyllabic stems differ from that of nouns with disyllabic stems?

b. Give the reduplicative template that will account for reduplication in both kinds of noun.

I hope you have discovered that if a noun has a monosyllabic stem, the stem syllable, together with the prefix, is reduplicated. However, if a noun has a disyllabic stem, both stem syllables are reduplicated but the augment is excluded. The **minimal word** targeted by reduplication (cf. [9.27]) is that portion of the word that includes the prefix and the stem but not the augment. On the basis of this analysis, Hyman and Mutaka propose that Kinande reduplication should be accounted for by the statements in [9.30].

[9.30] **a.** The reduplicative template is a prosodic unit consisting of *two syllables*.
  b. The template is suffixed.
  c. Copy the entire melody of the noun (except for the augment).
  d. Map the melody to the template right-to-left.

When attempting the exercises below, you should see also the discussion of mapping principles in [9.12] and in section (8.2.1) above.
  Following these principles they provide these derivations of *okugulugulu* and *a.ká-tí ká-tì*:

[9.31] **a.**

  b.

  c.

  d.

Ø (i.e. is deleted)

Note: σ stands for 'syllable'.

Study the following nouns and answer the questions that follow:

[9.32] **a.** e.n-daa 'belly' e.n-da. n-da. n-daa
       b. é.m-bwa 'dog' é.m-bwá. m-bwá. mbwa
       c. e.n-dwa 'wedding' e.n-dwa. n-dwa. n-dwa

a. What is the shape of the reduplicative template?
b. Is the template prefixed or suffixed?
c. What part of the base is copied as the melody?
d. What is the direction of mapping of the melody?

The answers we are looking for are:

a. There is a bisyllabic reduplicative affix template.
b. The template is suffixed.
c. The *nasal* prefix is copied together with the monosyllabic stem.
d. The melody is mapped from right-to-left.

Now provide a Prosodic Morphology derivation for:
  e.n-dwa. n-dwa. n-dwa

The expected answer is given in [9.33].

[9.33] **a.**   σ   (<u>Input</u> base noun including
                  ⟋⎥⟍   prefix and stem)
          e.n-d w a

     b.   σ   σ   σ   (Suffixation of bi-syllabic
            ⟋⎥⟍   prosodic template)
         e.n-d w a

     c.   σ   σ   σ   (First copy of base noun
         ⟋⎥⟍ ⟋⎥⟍   melody with right-to-left)
       e.n-d w a . n-d w a   mapping)

d. σ   σ   σ   (Second copy of base noun
    ∕⋏∖ ∕⋏∖ ∕⋏∖   melody with right-to-left)
    e.n-d w a . n-d w a . n-d w a mapping)

e.   Output: e.n-d w a . n-d w a .   n-d w a

In [9.33] we have clear evidence of the fact that the reduplicated suffix must be a bi-syllabic foot. Here the base, including the nasal prefix, gives us just one syllable. The requirements of the **Template Satisfaction Condition** (cf. [9.27]) are not met. So the reduplication process demands a double reduplication to provide enough syllables in [9.33d] in order for every syllable node to be associated with a syllable.

By now you might be wondering what happens to nouns with stems of more than two syllables. How do they get reduplicated? The answer is that they do not. Hyman and Mutaka show that Kinande puts a **morpheme integrity condition** on reduplication. A reduplicative process must use up all segments representing the base morpheme, or none at all.

As you can see in [9.34], reduplicating a word whose stem contains more than two syllables would mean using some of the sounds representing a morpheme as a melody and discarding the rest. This would violate the morpheme integrity condition. So it is disallowed.

[9.34] **a.** o.tu-gotseri   'sleepiness'   (no reduplication)
       b. e.bínyurúgúnzù 'butterflies'   (no reduplication)

Languages differ typologically in this regard. Whereas morpheme integrity is a determinant of reduplication in a language of the Kinande type, it is totally irrelevant in languages such as Maori (cf. [9.10]) and Quileute (cf. [9.11]).

With this we will leave Prosodic Morphology and reduplication and turn to metathesis.

9.4 METATHESIS

Metathesis is the process whereby segments are switched round in a word. This process is somewhat rare but it occurs in a range of languages, including Hebrew, Arabic and Hanunoo. The theory of template morphology presented earlier in this chapter has been successfully extended to the analysis of this process.

Being a morphological process that does not involve a simple concatenation of morphemes, metathesis has been difficult to describe using traditional theories. However, it lends itself naturally to a prosodic template approach. We will show this using an example from Hanunoo, a language of the Philippines.

Gleason (1955) reports that in Hanunoo, the morpheme *ka-* is prefixed to a numeral like *lima* or *pitu* to create a new word meaning 'x times'.

[9.35]   lima   'five'   kalima   'five times'
         pitu   'seven'   kapitu   'seven times'

However, when *ka-* is prefixed to a numeral with CVCV structure where the first vowel is /u/, that vowel is deleted by the **syncope** rule as in the examples:

[9.36]   duwa   'two'   kadwa   (*kaduwa)   'twice'
         tulu   'three'   katlu   (*katulu   'three times'

The /u/ syncope rule can be represented as in [9.37]:

[9.37]   k   a

         C   V   - C   V   C   V
                      ‖
                      ⓤ

The /u/-syncope rule **feeds** another rule and it is that other rule that will be our primary concern. First observe the effect of the syncope in column B in [9.38] below. You will see that, after syncope has applied, if the first consonant of the base is a glottal stop as in column A, when *ka-* is prefixed the metathesis rule applies. It re-arranges the consonants of column B, to yield the output in column C.

[9.38]   A   B   C
         ʔusa   'one'   (→ kaʔsa)   kasʔa   'once'
         ʔupat   'four'   (→ kaʔpat)   kapʔat   'four times'
         ʔunum   'six'   (→ kaʔnum)   kanʔum   'six times'

Work out a derivation for *kasʔa* ('once') using the notation of template morphology.

Your derivation should be like [9.39]:

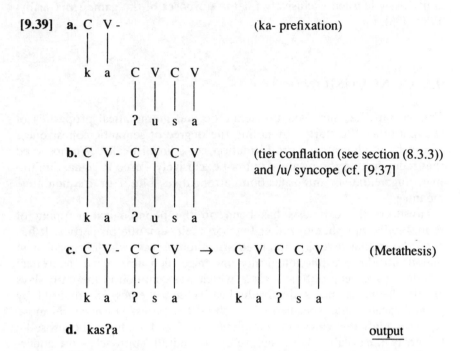

[9.39]   **a.** C V -   (ka- prefixation)

k   a   C V C V

ʔ   u   s   a

b. C V - C V C V   (tier conflation (see section (8.3.3))

                         and /u/ syncope (cf. [9.37]

k   a   ʔ   u   s   a

c. C V - C C V   →   C V C C V   (Metathesis)

k   a   ʔ   s   a   k   a   ʔ   s   a

d. kasʔa   output

Predictably, when a C on the skeletal tier which is already associated to a segment is moved by metathesis, the segment (in this case /ʔ/) also moves with it.

Metathesis is not unique in re-ordering C and V elements of the skeletal tier. An analogous process is found in 'secret languages' and **transposition word games** mostly used by adolescents in various languages. McCarthy (1982) has given an account of a word game in the Arabic dialect spoken by the Hijazi Bedouins. This game is played by transposing the consonants on the root tier, leaving the vocalic melody intact as seen below:

[9.40]   Normal form: kattab
            Game forms: battak   takkab   tabbak   bakkat   kabbat   'wrote'

The fact that speakers are able to manipulate consonants of root morphemes *independently* of the rest of the sounds in a word gives credence (if not perhaps absolute proof) to the key assumption of template morphology that different morphemes – e.g. the root as opposed to inflectional and derivational morphemes – are initially on separate tiers in the lexicon.

The hesitant note that I have struck is due to a recognition that, since word games are peripheral to the linguistic system, evidence obtained from studying them must be treated with a degree of circumspection. This is because they might involve specialised non-linguistic principles that are consciously learned exclusively for the purposes of the game (McCarthy, 1982: 196).

9.5 CONCLUSION

This chapter has surveyed the semantic and grammatical properties of reduplication. We started by noting the degree of semantic cohesiveness shown by reduplicative word-formation cross-linguistically. We observed that reduplication is typically – but not exclusively – used to signal diminutive, augmentative, intensification, attenuative, plural or frequentative meanings.

However, the emphasis has been on the phonological mapping of reduplicative morphemes rather than on their semantic properties. It has been shown that reduplication is an affixation process which is peculiar in that it introduces a phonologically underspecified affix. Two theoretical positions have been outlined. One, which was common in the early days of the theory, assumes that reduplicative morphemes are realised by phonologically underspecified CV-skeletal templates. However, in more recent studies the view that templates consist of units of the prosodic hierarchy like syllables has been preferred. In both approaches the underspecified templates are not linked to any consonant and vowel segments. They get their full phonological representation by copying the melody of the base.

The chapter has ended with a brief discussion of metathesis and transposition word games (see section (9.4) which illustrate ways in which the theory can handle other nonconcatenative morphological processes. All these processes are reminiscent of semitic infixing which gave Template and Prosodic Morphology its original impetus.

With this we end the discussion of the interplay between phonology, morphology and the lexicon. The next and final part of the book will focus on the relationship between syntax, morphology and the lexicon.

EXERCISES

1. Write a rule to account for the reduplication shown by the Latin data below:

Present		Perfect	
pend-o:	'I hang (intr.)'	pe-pe-nd-i:	'I have hanged'
mord-eo:	'I bite'	mo-mo-rd-i:	'I have bitten'
tond-eo:	'I shear'	to-to-nd-i:	'I have shorn'

   (from Kennedy, 1948)

2. Study carefully the patterns of reduplication exemplified by the following Luganda words:

(i)	mu-too	'young'	mu-toototoo	'fairly young'
	mu-bii	'bad'	mu-biibibii	'rather bad'
	mu-tii	'cowardly	mu-tiititii	'rather cowardly'

(ii)	ki-bisi	'wet'	ki-bisibisi	'rather wet'
	mu-gezi	'clever'	mu-gezigezi	'rather clever'
	ki-lebevu	'slack'	ki-lebevulebevu	'rather slack'
	mu-gayaavu	'lazy'	mu-gayaavugayaavu	'rather lazy'

   (Note: doubling of vowels shows length; *mu-* and *ki-* are noun class prefixes.)

   (a) What is the meaning contributed by reduplication?
   (b) Describe verbally the patterns of reduplication in the above data.
   (c) Write prosodic derivations (similar to those we used for Kinande) that show the derivation of *mutoototoo*, *mubisibisi* and *mugayaavugayaavu*.

3. Study the following data from Ateso (Hilders & Lawrence, 1957).

aiduk	'to build'	aituduk	'to cause to build'
ailel	'to be glad'	aitelel	'to gladden (i.e. to cause to be glad)'
ainyam	'to eat'	aitanyam	'to feed (i.e. to cause to eat)'
aiwadik	'to write'	aitawadik	'to cause to write'
aicak	'to throw'	aitacak	'to cause to throw'

   (a) Identify the infinitive morpheme.
   (b) Write down the root of each verb.
   (c) Give a formal statement of the derivation of the causative.

4. Write derivations of the diminutive forms in the following Agta data (from Wilbur, 1973):

wer	'creek'	wala-wer	'small creek'
talobag	'beetle'	tala-tálobag	'lady bird'
pirák	'money'	pala-pirák	'a little money'

5. Study the Hanunoo transposition word-game data below:

A Base form		B Word game form
rignuk	'tame'	nugrik
bi:ŋaw	'nick'	ŋa:biw
ʔusah	'one'	saʔuh
balaynun	'domesticated'	nulayban
katagbuʔ	(no gloss)	kabugtaʔ

(based on Conklin, 1959; McCarthy, 1982)

(a) Identify the segments moved in the transposition game.
(b) How is vowel length affected by the transposition process?
(c) Using templates, show how the game forms *ŋabi:w*, *saʔuh* and *kabugtaʔ* are formed.

Part III
Morphology and its Relation to the Lexicon and Syntax

10 Inflectional Morphology

10.1 INTRODUCTION

The main aim of this portion of the book is to examine the interaction between morphology and syntax. A question that will recur at several points is whether there is a clear difference between the structure of words, which is the domain of morphology, and the structure of sentences, which is the domain of syntax. Are the rules that regulate sentence structure different in kind from the rules that govern the internal structure of words? In answering this question we will see that, although morphology interacts with other components of the grammar (in particular syntax) and shares some of their rules, it nevertheless has a degree of internal coherence which makes it merit separate treatment as a distinct component of the linguistic model.

The investigations begin in this chapter with an exploration of the nature of **inflectional** morphology. First, the theoretical basis of the inflection-derivation dichotomy is scrutinised. This is followed in the second half of the chapter by a survey of phenomena marked using inflection in the languages of the world. In the next chapter we examine in detail the role of syntactic structure, at the core of which is the verb, in determining the form of words when they appear in sentences. That chapter is essentially an elaboration of the theory of how case is assigned and how it is mapped on words. The book concludes with an analysis of idioms and compounds which highlights the similarities, as well as differences, between lexical items and syntactic phrases.

10.2 INFLECTION AND DERIVATION

What is inflection? The standard intuition among linguists is that inflectional morphology is concerned with syntactically driven word-formation. Inflectional morphology deals with syntactically determined affixation processes while drivational morphology is used to create new lexical items (cf. section (3.2)).

In practice, however, there is not always unanimity in the classification of processes as inflectional or derivational. Grammarians working on the same language may not agree as to which processes are to be treated as inflectional and which ones are to be regarded as derivational. Across languages there can be even greater confusion. As we shall see shortly, a process classified as inflectional in one language may be analogous to a process regarded as derivational in another. Clearly, there is a need for a principled way of determining whether a given process is inflectional or derivational. Below we will examine ways in which that need might be met.

10.2.1 Differentiating between Inflection and Derivation

In this subsection I outline a number of criteria that have been proposed in order to put the dichotomy between inflection and derivation on a firmer theoretical footing. This is important since much morphological theorising is based on the assumption that morphological processes fall into two broad categories: inflection and derivation.

10.2.1.1 Obligatoriness
Greenberg (1954) proposed the criterion of **obligatoriness** to characterise inflection. He argued that inflection occurs when, at different points in a sentence, syntax imposes obligatory choices from a menu of affixes. If the right choice is not made, an ungrammatical sentence results.

This can be seen from the behaviour of the inflectional category of number in the demonstrative in English. The demonstrative must always have the same number category as the noun it modifies, as seen in [10.1a]:

[10.1] a $D_{sing} N_{sing}$ $D_{plur} N_{plur}$ b $D_{plur} N_{sing}$ $D_{sing} N_{plur}$
           this book   these books   *these book   *this books
           that book   those books   *those book   *that books

If the demonstrative has a different marking for number from that of the noun it modifies (cf. [10.1b]), the result is ungrammatical.

According to Greenberg, no such obligatoriness exists in the case of derivation. Syntax *per se* does not force the choice of a specific derivationally derived lexical item in order to ensure that ungrammaticality is avoided. To take a simple example, an English noun does not have to be affixed with *-er* in any syntactic position. Hence *-er* is a derivational suffix.

We see in [10.2] that a subject NP need not contain a noun with the agentive nominaliser *-er*. All kinds of nouns which do not have that suffix can freely substitute for each other (cf. [10.2b]) as subject NPs without affecting grammatical well-formedness:

[10.2] a. The farmer is in the barn.   b. The cow is in the barn.
            The teacher is in the barn.   The pig is in the barn.
            The baker is in the barn.   The man is in the barn.

By contrast, where, as in Latin, a noun does have to be affixed with a particular suffix when it is in a subject or object NP etc. the suffixation is inflectional. Contrast *Agricola$_{[SUBj]}$videt* 'The farmer sees' *Agricolam$_{[OBJ]}$videt* 'He sees the farmer'. (See section 10.4.3 below.)

Unfortunately, the criterion of obligatoriness cannot always successfully distinguish between inflection and derivation. There are cases where syntactic well-formedness requires the selection of a form with a particular derivational suffix. Compare the following:

[10.3]   a.   I opened it *awkwardly*   **b.**   The *teacher* is in the office.
         *I opened it *awkward*.   *The *teach* is in the office.

To get a well-formed sentence in [10.3a] we must apply the derivational rule that suffixes *-ly* to the adjective *awkward* and turns it into the adverb *awkwardly*. Similarly, in [10.3b], the italicised noun phrase is ill-formed unless a derived noun (with the *-er* suffix) appears after the determiner.

Evidently, such transcategorial derivation is a problem for a definition of inflection in terms of syntactic obligatoriness. The use of the words *teacher* and of *awkwardly*, which are formed by derivation (rather than *teach* and *awkward*), is essential in order to ensure grammaticality.

Nonetheless, the claim that some affixes are syntactically more pertinent than others is well-founded. Furthermore, the more syntactically pertinent affixes tend to be the ones that are obligatory. Thus, for example, inflectional *-s* in verbs, which realises the syntactically pertinent properties of third person, present tense and singular number, is obligatory. But the derivational prefix *ex-*, as in *ex-wife*, which is not syntactically pertinent, is not obligatory in nouns appearing in any sentence position. We will return to this important issue of the correlation between obligatoriness and syntactic pertinence in (10.2.1.3) below.

10.2.1.2 Productivity

Productivity (or generality) is another property that is often said to distinguish inflection from derivation. It is claimed that derivational processes tend to be sporadic while inflectional processes tend to apply automatically across the board to forms belonging to the appropriate paradigm (cf. section (4.3)).

A good illustration of this is tense marking in verbs. Every verb in English takes the inflectional category of past tense (usually realised as *-ed*). By contrast, it is very much a hit or miss affair whether a verb will take the *-ant* derivational agentive nominal forming suffix. We have *apply* ~ *applicant* but not *donate* ~ *donant* (see [4.24] on p. 80 above).

Unfortunately, the generality criterion often runs into trouble because (i) there exist very regular derivational processes (such as the suffixation of the English adverb-forming *-ly* suffix, as in *quick-ly*, to adjectives to form adverbs) which are every bit as predictable as any inflectional process, and (ii) there exist exception-ridden inflectional processes (cf. section 4.3).

The Russian verb system illustrates the latter problem in a very striking way. Halle reports that about 100 Russian verbs belonging to the so-called inflectional second-conjugation are defective. (See the discussion of conjugation in section (10.3.1) below). For no apparent reason, these verbs lack first person singular present tense forms as you can see from this selected list:

[10.4]   *lažu   'I climb'
         *pobežu (or pobeždu)   'I conquer'
         *deržu   'I talk rudely'
         *muču   'I stir up'
         *erunžu   'I behave foolishly'

Note: [ž] = IPA [ʒ]   č = IPA [tʃ].

(from Halle, 1973: 7)

Without a doubt first person, singular number and present tense are inflectional categories in Russian. Nevertheless this large sub-class of verbs is not inflected for these categories. From this we conclude that belonging to a regular paradigm is an important tendency in inflectional morphology. But it is not an essential characteristic that unfailingly separates inflection from derivation.

10.2.1.3 Inflection is syntactically motivated
Earlier, in section (3.2) and also later in (10.2), we observed that, typically, affixes serving a syntactic function are inflectional while those which are used to create new lexical items are derivational. Unfortunately, the line between what is or is not syntactically motivated is often blurred. You can see the nature of the problem if you consider English verbal forms ending in *-en* or *-ed* (usually abbreviated to *V-en/V-ed*).

Matthews (1974: 53–4) draws attention to *V-en/V-ed* forms which, in some cases, are borderline between **verbal past participles** and **participial adjectives**. He observes that, in the phrase *a crowded room*, the word *crowded* can be correctly classified as a 'participial adjective' since it can appear as an adjective in the frame 'a very $-_{Adj}$N'. *A very crowded room* is syntactically parallel to *a very small/clean/cold room*. This contrasts with the V-ed form *heated* in *a well heated room* where *heated* is a verbal past participle. Putting *heated* in a position which can only be filled by an adjective, for instance following *very* in **a very heated room*, results in ungrammaticality.

Thus, the same formatives *-en/-ed* can represent either an inflectional morpheme (when they mark the past participle of a verb as in *He has heated the room* (not **He has heat/heats/heating the room*), or a derivational morpheme (when they mark the change of a verb into a participial adjective, as in *a very crowded room*).

Not surprisingly, the specific problem of *-en/-ed* in English has received a good deal of attention. In addition to Matthews (1974) see also Allen (1978: 262–5), Lieber (1980: 29–30) and Scalise (1984: 127–9) for further discussion.

It is obvious that one of the major problems in separating inflection from derivation is finding a way of characterising properly the notion 'syntactically determined', since this notion is crucial in defining inflection. Following S. R. Anderson (1982, 1985, 1988), we will distinguish between inflection and derivation in this way:

[10.5]   Inflectional morphology deals with whatever information about word-structure that is relevant to the syntax. Inflectional properties of words are assigned by the syntax and depend on how a word interacts with other words in a phrase, clause or sentence.

S. R. Anderson (1988a: 167) identifies four kinds of morphological properties or categories that characterise inflection:

[10.6]
  (i)   **Configurational properties.** These are so called because the choice of a particular inflection is determined by the place occupied by a word in a syntactic configuration, i.e. its position and function as a constituent of a phrase, or some other syntactic structure. (E.g. in some languages, a noun that is the object of a preceding preposition must receive accusative case marking; the direct object of a verb must be in the accusative case; a verb in a subordinate clause must have a special form such as the subjunctive mood, etc.)
  (ii) **Agreement properties.** These are determined by the characteristics of another word or words in the same construction. (E.g., if an adjective modifies a singular noun, it must be assigned a singular affix whose form depends on the form of the affix in the noun it is modifying.)
  (iii) **Inherent properties,** such as the gender of a noun, that must be accessed by agreement rules. (E.g. the gender of a French or German noun determines the gender of the adjective that modifies it, etc.).
  (iv) **Phrasal properties,** which belong to an entire syntactic phrase but are morphologically realised in one of the words of that phrase. (E.g. genitive *'s* marking in English phrases like *the Mayor of Lancaster's limousine*, where, although the *mayor* is the possessor of the *limousine*, the *'s* inflection is attached to *Lancaster*. (This type is problematic. See section (10.5) below.)

In a given language, whether a particular process is to be viewed as inflectional or derivational always depends on the extent to which it is determined by purely syntactic processes. That is why, as we shall see, the same category may be derivational in one language and inflectional in another. This is true even of processes, such as number, that seem to be indisputably inflectional. In some languages, such as Chinese and Vietnamese, plural is only optionally marked in nouns. It is not a syntactically obligatory property of nouns (see section. (10.4.1)).

In the next few paragraphs, I shall use the cross-linguistic treatment of **diminutives** and **augmentatives** to show that whether a given process is regarded as inflectional or derivational depends largely on the extent to which it is syntactically determined. (cf. S. R. Anderson, 1982, 1990).

(a) Find one fresh example of each diminutive suffix that occurs in [10.7].

(b) Should these suffixes be treated as inflectional or derivational?

[10.7]	duck	~	duckling	goose	~	gosling
	pig	~	piglet	brace	~	bracelet
	pig	~	piggy	dog	~	doggy
	kitchen	~	kitchenette	laundry	~	launderette

In English, it is uncontroversial to view *-ling*, *-y*, *-let* and *-ette* as derivational suffixes, because diminutive formation is not part of any general, syntactically driven paradigm. One good argument for this view is that no syntactic rule of English needs to make reference to the property 'diminutive'. But the reverse is the case in some other languages (cf. S. R. Anderson, 1990). For instance, in many African languages, diminutives and augmentatives are marked by affixes that are at the heart of the inflectional system. Arnott (1970) has shown that in Fula a noun belongs to a neutral (i.e. unmarked) class to which it is assigned either arbitrarily or on the basis of some tenuous semantic considerations. But a noun can be taken out of its neutral class and assigned to a number of other classes which are marked by other prefixes. Typically, the pattern constitutes a paradigm in which the noun appears in one of the following classes:

[10.8] a. singular neutral classes (1, 9–23) and plural neutral classes (2, 24, or 25);

b. singular diminutive classes (3, 5) and plural diminutive class (6);

c. singular augmentative class (7) and plural augmentative class (8).

This is exemplified by the **full noun paradigms** of the noun stems *laam-* 'chief' and *dem-* 'tongue' which are listed as follows in Arnott (1970: 79):

[10.9]   **a.**   Neutral Classes

1	laam-ɗo	'chief'	16 ɗem-ŋgal	'big tongue'
2	laam-ɓe	'chiefs'	24 ɗem-ɗe	'big tongues'

b.   Diminutive Classes

3	laam-ŋgel	'petty chief'	3 ɗem-ŋgel	'small tongue'
5	laam-ŋgum	'worthless little chief'	5 ɗem-ŋgum	'puny little tongue'
6	laam-kon	'petty chiefs'	6 ɗem-kon	'small tongues'

c.   Augmentative Classes

7	laam-ŋga	'mighty chief'	7 ɗem-ŋga	'big tongue'
8	laam-ko	'mighty chiefs'	8 ɗem-ko	'big tongues'

Study the Luganda data below and show why it is unclear whether the diminutive and augmentative suffixes are inflectional or derivational:

[10.10]   **a.**

		(Unmarked) Class	Diminutive	Augmentative
	Sg	mu-kazi (class 1) 'woman'	ka-kazi (class 12) 'little woman'	gu-kazi (class 20) 'enormous woman'
	Pl	ba-kazi (class 2) 'women'	bu-kazi (class 14) 'little women'	ga-kazi (class 22) 'enormous women'
	Sg	ki-sero (class 7) 'basket'	ka-sero (class 12) 'little basket'	gu-sero (class 20) 'enormous basket'
	Pl	bi-sero (class 7) 'baskets'	busero (class 14) 'little baskets'	ga-sero (class 22) 'enormous baskets'
b.	*Sg*	kasolya (class 12) 'roof'	kamyu (class 12) 'hare'	kasera (class 12) 'porridge'
	Pl	busolya (class 14) 'roofs'	bumyu (class 14) 'hares'	busera (class 14) 'porridges'
c.	*Sg*	kikazi (class 7) 'large woman'	kisajja (class 7) 'large man'	kigatto (class 7) 'large (ugly) shoe'
	Pl	bikazi (class 8) 'large women'	bisajja (class 8) 'large men'	bigatto (class 8) 'large (ugly) shoes'

The situation in Luganda and many other Bantu languages is similar to that in Fula. Nouns inherently belong to a morpho-syntatic noun class and take their prefixes in an unmarked way.

Interestingly, the standard noun class system used to divide nouns into syntactic sub-classes is also used to mark diminutives (often with ameliorative connotations) and augmentatives (often with pejorative connotations). This derivational use of inflectional affixes is seen when we compare *ka-/bu-* in [10.10a] and [10.10b]. Classes 12/14 function as unmarked class markers in [10.10b] but in [10.10a] they are used as diminutive prefixes. Likewise, in [10.10c], class 7/8 *ki-/bi-* prefixes are augmentative but in [10.10a] they are unmarked class prefixes. However, some prefixes, e.g. *gu-* (class 20) and *ga-* (class 22) in [10.10a] can only be used as augmentatives. Such prefixes can be appropriately classed as derivational.

Cross-linguistic comparisons are thus rendered difficult since the same category (e.g. diminutive) may be inflectional in one language but derivational in another. We cannot assume that, if a category is treated as an inflection in one language, it will be inflectional in the next language we encounter. To complicate matters further, within the same language the same affix may have both inflectional and derivational uses.

10.2.2 Relevance and Generality

Would a semantics-driven approach to the problem assist us significantly? Unfortunately, the answer has to be that this is unlikely. Semantic properties in themselves do not determine whether a given concept is going to be expressed using inflection or derivation in a particular language. Thus, as we have seen, 'diminutive' is derivational in English but inflectional in Fula (cf. Sapir, 1921: 107; S. R. Anderson, 1982, 1990).

However, some linguists disagree. Bybee (1985), for instance, gives semantics a major role in determining whether a category is realised inflectionally or derivationally. According to her, there are two factors that play a role in determining whether a particular concept is expressed lexically, morphologically (usually by inflection) or syntactically. They are the criteria of **relevance** and **generality** (in the sense explained earlier in Chapter 4). Bybee (1985: 13) defines relevance in these words:

> A meaning element is relevant to another meaning element if the semantic content of the first directly affects or modifies the semantic content of the second.

She hypothesises that, if two semantic elements are highly relevant to each other, they are likely to be expressed lexically or by morphological inflection. But if they are not highly relevant to each other, syntactic expression is more probable.

Let us explore the problem by looking at causatives in English.

[10.11]   **a.**   In English sometimes the causative idea is expressed lexically, as in *teach* (i.e. cause to learn or know). Find another
morphologically simple lexeme of this kind.

        b.   Find one more morphologically complex lexeme like *deaf-en*
where the causative is marked morphologically by affixation.

        c.   Finally, give one example where the causative is expressed
syntactically using a construction that contains *make* or
cause.

Some possible answers are listed below:

[10.12]   **a.**   <u>Lexical causatives</u>
drop 'cause to fall'
kill 'cause to die'

        b.   <u>Morphological causatives</u>
widen 'make wide'
shorten 'cause to become short'

        c.   <u>Syntactic causatives</u>
make someone happy
cause to riot

The three classes of causatives in [10.12] reflect the three ways in which
syntactic-semantic notions and categories may be expressed: (i) lexically,
by underived lexicalised forms like *drop* which simply have to be listed in
the dictionary, (ii) morphologically, typically by inflectional morphemes
like plural and occasionally by derivational affixation (e.g. causative *-en* in
shorten (i.e. cause to become short), and (iii) syntactically, by separate
words in phrases and clauses (e.g. *make someone happy* (not **happy-en
someone* or *happy someone)). So, it appears, the nature of the concept
per se does not determine how it is expressed.

Against this, Bybee argues that the choice whether to represent a
concept lexically, morphologically or semantically is not arbitrary.
Relevance is always a major factor in opting for **lexicalisation** (i.e. expressing a concept using a lexical item). But what counts as 'semantically
relevant content' is partly cognitively and partly culturally determined.

Bybee's argument can be illustrated with the Luganda examples below:

[10.13]   kulya   'to ingest solid food'
          kuwuuta   'to ingest semi-solid food (e.g. porridge)'
          kunywa   'to ingest liquid food (i.e. to drink)'

Here the first meaning element is 'the action of ingesting' and the second is 'the food substance'. The nature of the food ingested is obviously highly relevant to the action of ingesting. Hence the two semantic elements are expressed lexically. A different lexical item is used depending on whether the food that is consumed is solid, semi-solid or liquid.

Whereas Luganda makes a three-way lexical distinction in [10.13], English only makes a two-way distinction between 'consuming solid food', i.e. *eating*, and 'consuming non-solid food', i.e. *drinking*. What is crucial is that in both languages the same semantic parameters are mutually relevant (the act of ingesting food and the nature of the food that is consumed). It is extremely unlikely that a language would lexicalise semantic concepts like 'to ingest food sitting down' and 'to ingest food standing up on one leg' because posture is not strictly relevant to the process of ingesting food.

What differences in meaning or syntactic function are considered sufficiently important to be expressed by morphological inflection is something that seems to be open to subjective interpretation and to depend on culture.

To overcome these problems some linguists have sought a formal principle that enables the linguist to avoid subjectivity in characterising syntactic relevance in inflection. Di Sciullo and Williams (1987: 70) propose that 'affixes more relevant to syntax appear outside affixes less relevant to syntax'. Earlier, Williams (1981) had claimed that features that are syntactically pertinent are at the margins of words because, given the theory of how features percolate (i.e. spread) from the head to the rest of the constituent, it is only from that vantage point that features of the affix can affect the entire word. (See the discussion of affix ordering in section (6.2.1) and the discussion of headedness in (12.4.1)).

If a given concept is not lexicalised but is morphologically expressed, on what basis is the choice between inflection and derivation made? I argued in section (10.2.1.3) above, that semantics does not determine whether a given property will be inflectional or derivational.

Bybee takes the opposite view, as we have seen. She claims that there is a semantic basis for the distinction between inflection and derivation. She contends that 'the content of morphological categories determines whether they will appear in inflectional or derivational expression' (Bybee, 1985: 98) The more relevant a given category is in a given situation, the more likely it is to be inflectional rather than derivational.

Bybee uses the relevance principle to predict that certain categories are more likely to be inflectionally or derivationally expressed. In verbal morphology, for example, prime candidates for derivational expression are **grammatical function changing rules**. These are processes that alter the number or nature of noun phrase arguments (subject and object(s) which a verb can take: cf. section 11.5).

Passivisation is a good example of such a process. It changes the object into the subject and may allow the deletion of the original subject, thus reducing the number of noun phrases that are in construction with the verb from two to one. This is seen in the Luganda sentences in [10.14]. Compare the active sentence in [10.14a], where a subject and an object noun phrase are both present, with [10.14b], where the object noun phrase is missing and the verb is marked with the passive derivational suffix -*ebwa*:

[10.14]   **a.**   Active   **b.**   Passive

Subj.	Verb	Obj.		Subj.	Verb
Petero *a*-li-yoza		engoye.		Engoye	*zi*-li-yoz-*ebw*a.
Peter he-*fut*-wash		clothes		The clothes	they-*fut*-wash-passive
'Peter will wash the clothes'				'The clothes will be washed.'	

In the light of the foregoing discussion it will not come as a surprise to see that passivisation is regarded as a derivational process by Bantuists on the grounds that it brings about a drastic change in the syntactic properties of the verb.

Bybee's relevance theory also correctly predicts that agreement categories such as the pronominal subject markers *a*- and *zi*- in [10.14], which index the arguments of the verb, are inflectional.

But this still leaves us with a problem when the same category is expressed both inflectionally and derivationally, as in the case of the diminutive with which the discussion started in section (10.2.1.3). Bybee argues that some difference in meaning should be expected in such cases. She appeals to the principle of **generality**: where similar conceptual content is expressed by both inflection and derivation, the inflectional expression displays a fully general and totally predictable meaning while the derivational one tends to be less general and more idiosyncratic.

She illustrates this principle with number. In many languages, inflectional marking of plurality in the verb indicates number of the subject or object noun phrase. In English, a verb is inflected to agree in number with the subject:

[10.15]   **a.**   Margaret dances (=singular, third person, present tense)
         b.   The girls dance (=plural, third person, present tense)

By contrast, in some Native American languages (e.g. Diegueno and Kwakwala, also called Kwakiutl), verbs are not inflected for number agreement. In these languages number is a derivational category and plurality has a different meaning from the one it has in a language like English where it is realised inflectionally. In Kwakwala number marking may be used to express three distinct kinds of meaning. Plurality may indicate (i) repeated action (that is the meaning of [10.16a]), (ii) several subjects (see [10.16b]) or (iii) an action occurring simultaneously in different parts of a unit (see [10.16c]). (See Boas, 1947: 246; S. R. Anderson, 1981, 1982: 589; Bybee, 1985: 103.)

[10.16] a. mɛdɛ'lqwɛla   'it is boiling'
         b. meʔmɛdɛ'lqwɛla   'many are boiling'
         c. maʔɛ'mɛdɛ'lqwɛla   'is boiling in all of its parts'

Bybee claims that number illustrates the difference in generality between inflection and derivation. An inflectional category is ordinarily obligatorily applicable to most stems occurring in the appropriate syntactic environment which have the requisite phonological, morphological and syntactic characteristics.

In order for an affixational morpheme to be so general it must have minimum semantic content so that it can be attached to a very broad class of stems. The concept 'plural' meaning 'more than one' is a good candidate for an inflectional morpheme since it is not very specific. Contrast it with a putative English morpheme -od with the specific meaning 'while hopping on the toes of the left foot and scratching one's back', which could be attached to verb stems. Not much mileage would be got out of using an inflectional morpheme with such a restricted meaning. According to Bybee, that is the lesson we learn from the real examples of the various types of verbal plurality in Kwakwala in [10.16]: if a semantic concept is reasonably fleshed out, it becomes more narrow and hence more likely to be expressed derivationally (see also section (10.4)).

Let me summarise Bybee's position. First, the relevance hypothesis predicts that semantic elements that are mutually highly relevant to each other are likely to be expressed lexically or by inflection. In the event of their being expressed morphologically, semantic elements which are less relevant to each other tend to be expressed derivationally. However, where there is a very low degree of mutual relevance, semantic elements are most likely to be expressed syntactically.

Second, a semantic notion is more likely to be expressed using an inflectional morpheme if it has a non-specific meaning which allows it to be applicable to a wide range of stems. Conversely, a semantic notion is more likely to be expressed using a derivational morpheme if it has a rather specific meaning.

Third, semantic notions that are both highly relevant and specific tend to be lexicalised as in the case of 'cause' and 'fall' which is expressed by the lexical item *fell* (as in *to fell a tree*). Semantic notions which might be general but are not highly relevant tend to be syntactically expressed (e.g. *make someone happy*). It is the area in between where a significant degree of relevance and generality coincide that is the domain of inflection.

Bybee's arguments are quite plausible. But in the light of the evidence presented earlier in section (10.2.1.3), showing that we cannot predict whether inflection or derivation will be used to signal a particular concept (e.g. the diminutive, plurality etc.), we must remain sceptical about semantically-based criteria for separating inflection from derivation.

To sum up, the difference between inflection and derivation is a *cline* rather than a dichotomy. **Prototypical inflectional morphemes** (e.g. verbal affixes in English) are very strongly syntactically determined while prototypical derivational morphemes (e.g. *-er* as in *worker*) are very weakly syntactically determined. In between there is a continuum of syntactic determination. The terms 'inflection' and 'derivation' simply indicate the *degree* of syntactic relevance. Just as generality cannot be treated as the litmus test that separates inflection from derivation, syntactic or semantic relevance cannot either.

10.2.3 Is Morphology Necessary?

If certain morphological properties of words must be 'seen' by syntactic rules and if there are inflectional properties of words that are determined by syntax, the question arises whether there is a need for a separate morphological component in our grammar. Might all morphology not be reduced to syntax and phonology (cf. Lees, 1960)?

A defence of the separate existence of a morphological component has been mounted by Chomsky (1970), Jackendoff (1975) and S. R. Anderson (1988a) and many other proponents of what is referred to as the **lexicalist hypothesis**. These linguists claim that the principles that regulate the internal structure of words are quite different from those that govern sentence structure, the domain of syntax. Some of the major differences are summarised in [10.17]:

[10.17]	Word structure rules	Sentence structure rules
(i)	Lexical rules may relate items from different word-classes (e.g. derive adverbs from adjectives).	Syntactic rules do not change word-classes.
(ii)	Lexical rules apply to the output of other lexical rules but not to the output of syntactic rules] (e.g. [develop$_V$] → [develop$_V$ + ment$_N$] → [development$_N$ + al$_{ADJ}$] → [developmental]$_{ADJ}$).	Syntactic rules have access to the output of both lexical and syntactic rules.
(iii)	Lexical rules may have arbitrary lexical exceptions (e.g. normally an agentive noun can be derived from a verb by suffixing *-er*. However, this is	Syntactic rules do not have arbitrary lexical expressions (such as the passivisation of a sentence being blocked if the subject NP contains the noun *safari* be-

	not allowed if the verb is *spy* (**spier*)).	cause that word is borrowed from Swahili).
(iv)	Lexical rules use recursion only to a limited extent. (Words like *great-great-great-great-great-great-great-aunt* while not ruled out, are decidedly odd. See section (3.3)). The only exception is compounding (see Chapter 12).	Recursion is extensively used in syntax.
(v)	Morphology has paradigms. (See the paragraph below.)	Syntax has no paradigms.

As we have already noted, the problem of distinguishing between morphology and syntax is especially pressing in the case of inflectional morphology. What are the properties of inflectional morphology that set it apart from syntax?

An interesting proposal for distinguishing between inflectional morphology and syntax has been made by Carstairs (1987). He has expounded the last point in [10.17] by arguing that one of the key differences between syntax and inflectional morphology is that the latter is subject to the **Inflectional Parsimony Principle**, while the former is not subject to a comparable syntactic parsimony principle. The parsimony principle states that two inflections cannot be functionally identical unless they are in complementary distribution. What this means is that where, in inflectional morphology, one form selects a given formative to realise a certain morphosyntactic property or properties (e.g. gender, number, tense, aspect etc.), that selection precludes the selection of other formatives which realise the same morpheme(s). This is the insight traditionally expressed in terms of **inflectional paradigms**.

Let us consider a French example. In written French, the third person singular present tense can be realised by *-e* or by *-it*, as in *elle donne* 'she gives' and *elle finit* 'she finishes'. So-called *-er* verbs (whose infinite form ends in *-er*) belong to the paradigm of *donner* 'to give' while so-called *-ir* verbs belong to the paradigm of *finir* 'to finish'. The paradigms are mutually exclusive. *Donner* cannot take the suffix *-it* as in **donnit*. Nor can *finir* take the suffix *-e* as in **fine*.

Carstairs points out that the existence of paradigms means that the potential number of inflectional forms of a word is restricted. This is a highly desirable result, given the fact that, even with this restriction in force, a stem in an inflecting language can co-occur in different word-forms with hundreds or even thousands of different affixes.

The **Inflectional Parsimony Principle** is similar to Aronoff's blocking

principle in derivational morphology. Because of blocking, the prior existence of a word with a particular meaning usually inhibits the derivation of
other words with the same meaning (cf. section 4.2.1). Paradigms are the
implementation of blocking in inflectional morphology. In our French
example, the existence of *finit* blocks the creation of **fine* (see section
(10.3.1)).

Carstairs argues, then, that the Inflectional Parsimony Principle is one of
the properties that distinguishes inflection from syntax. Inflectional morphology has paradigms as a consequence of the Inflectional Parsimony
Principle. But syntax has nothing comparable. Hence syntax has no paradigms. We do not find patterns of syntactic distribution where (nearly)
synonymous words are divided in complementary sets for the purpose of
applying syntactic rules. The claim is that the notion of complementary
distribution which plays a key role in morphology does not apply in syntax.

Let us take the concrete example of two sets of (roughly) synonymous
English nouns. We shall put *tummy*, *pail*, *posterior* and *truck* in set A and
belly, *bucket*, *backside* and *lorry* in set B.

Syntax cannot have a constraint that says, for instance, that set A nouns
only occur in subject noun phrases while set B nouns occur in object noun
phrases. The nearest that syntax comes to being paradigmatic is in cases of
rare, sporadic **lexically governed rules**. Carstairs illustrates this with the
behaviour of a number of adjectives.

[10.18] **a.** Kim is sick. **b.** Kim is ill.
              (sick = predicate adj.)   (ill = predicate adj.)

          c. The judge is dead. **d.** *The judge is late.
              (dead = predicate adj.)   (meaning 'not alive')
                                            (late = predicate adj.)

We observe that in [10.18a] and [10.18c] both *sick* and *dead* can be used as
predicate adjectives. This contrasts with [10.19] where in [10.19a] *sick* can
be used as an attributive adjective but in [10.19b] *ill* cannot. Thus, syntactically, the two synonyms appear to be in complementary distribution.

[10.19] **a.** Kim is a sick child. **b.** *Kim is an ill child.
              (sick = attributive adj.)   (ill = attributive adj.)

          c. The dead judge was ancient. **d.** The late judge was ancient.
              (dead = attributive adj.)   (late = attributive adj.)
                                                    (meaning 'not alive')

The situation with *late* and *dead* is similar. *Late* is allowed as an attributive
adjective in [10.19d]; it is disallowed as a predicate adjective in [10.18d].
However both *dead* and *late* are allowed as attributive adjectives in [10.19c]

and [10.19d]. Notwithstanding this complementarity, the evidence does not warrant setting up syntactic paradigms. Marginal and rare distributional facts of this kind do not constitute the basis of a paradigm.

We shall assume that the statements in [10.17] in general, and the Inflectional Parsimony Principle in particular, provide between them a proper basis for establishing a separate morphology.

Let us now examine in detail the nature of inflectional morphology. The discussion in the rest of the chapter is based on S. R. Anderson's work. We shall examine in turn some inflectional properties of verbs and nouns.

10.3 VERBAL INFLECTIONAL CATEGORIES

We will start our discussion with the morphology of the verb. In most languages the verb shows greater morphological complexity than any other word class. I will consider in turn inherent, agreement and configurational properties of verbs (see [10.6]).

10.3.1 Inherent Verbal Properties

In this sub-section I describe inflection for tense, aspect, mood and conjugation classes. These categories all add further specification to the event, state, process or action, which we will refer to as the **predication**, indicated by the verb. This is not meant to be an exhaustive list. It only represents the commonest inherent morphological categories that recur generally in the verbal inflection of languages.

Normally, **tense** indicates the time of the predication in relation to some particular moment. This moment is typically the moment of speaking or writing (e.g. tense indicates whether the event happened prior to the moment of speaking (past tense), is contemporaneous with it (present tense) or subsequent to it (future tense). We will provisionally regard tense as an inherent property of the verb (until we refine our analysis further in section (10.3.3)).

In the morphology of many languages at least three tenses are distinguished: past, present and future. The Swahili examples in [10.20] are typical:

[10.20]	Past	Present	Future
	(Before now)	(Now)	(After now)
	ni-li-leta	ni-na-leta	ni-ta-leta
	I-*past*-bring	I-*pres*-bring	I-*fut*-bring
	'I brought'	'I bring'	'I shall bring'

But tripartite tense distinctions are by no means universal. In some systems only a two way distinction is made. Sometimes, the distinction is drawn between **past** vs **non-past** (i.e. present-future) as in Yidiɲ where, for example, *gali:ŋ* 'went (past)' is used if the action happened in the past while *galiŋ* 'go (non-past)' is used if the action is happening now or is to happen in the future (cf. Dixon: 1977: 211–3).

Sometimes it is future and non-future that are distinguished. The present and the past are marked in the same way, but are differentiated from the future. That is the situation in many native American, Australian and Papuan languages of New Guinea. In such systems, temporal reference, which is normally the function of tense, is still present but it is subordinate to the task of indicating the **status** of an event with regard to the degree of certainty of its taking place. The basic temporal dichotomy is between the past and the present, which are **real** and the future which is **unreal** (we know what the past and present hold but we are not at all certain about the future). Usually the technical term used for 'real' in this sense is **realis** while the technical term for 'unreal' is **irrealis**.

Dani is a good example of this type of language. It makes three status distinctions: real, likely and possible (Bromley, 1981; Foley, 1986):

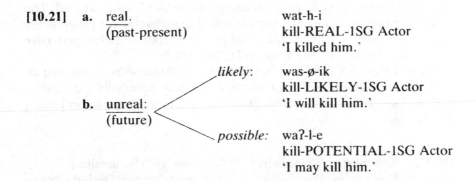

[10.21] **a.** real.   wat-h-i
                (past-present)   kill-REAL-1SG Actor
                                    'I killed him.'

                        likely: was-ø-ik
                                    kill-LIKELY-1SG Actor
          b. unreal: 'I will kill him.'
                (future)
                        possible: waʔ-l-e
                                    kill-POTENTIAL-1SG Actor
                                    'I may kill him.'

As seen in [10.21], tense, which has to do with temporal reference, intersects with status, which has to do with probability and certainty (some of the concepts to be discussed on pp. 222–4, under mood).

Aspect is another common inherent verbal category. Its function is to highlight the internal temporal unfolding of the predication. Essentially, aspect indicates whether an event, state, process or action that is denoted by a verb is completed or in progress. Aspect will be approached through a brief examination of the opposition between the simple past and imperfect in literary French:

[10.22]   **a.**   Simple past tense
                     Il chant*a*.
                     'He sang.'

          b.   Imperfective/progressive aspect
                     Il chant*ait* quand Yvonne arriva.
                     'He was singing when Yvonne arrived.'

          c.   Perfective aspect
                     Il *avait* chanté quand Yvonne arriva.
                     He had sung when Yvonne arrived.'

In all three cases the event of singing took place in the past (this is shown
by the presence of the -*a* simple past tense suffix of *chanta* and *arriva*). The
inflectional differences in [10.22b and c] in the verb *chanter* 'sing' do not
show when the singing occurred. Rather, they reveal something about the
evolution of the singing. In [10.22a] all we know is that the event took
place in the past. No aspect is shown. However, in [10.22b] we know not
only that the singing happened in the past but also that it was still in
progress when something else, Yvonne's arrival, happened. This contrasts
with [10.22c] where the singing was completed before Yvonne arrived. The
aspect used for incomplete actions is called **imperfective** (or progressive)
aspect and that used to indicate completed actions is called **perfective**
aspect (see Comrie, 1976; Lyons, 1977: 703–18).

   Mood is also an inherent verbal category. Its function is to describe an
event in terms of whether it is necessary, possible, permissible, desirable,
and the like. The semantic concepts expressed in the examples in [10.23]
are typical.

[10.23]   You must go. (Necessity)   You may go. (Permission)
          You can go. (Possibility)   You ought to go. (Desirability)

Many languages, including English, indicate modality syntactically using
auxiliary modal verbs like *must, can, may* and *ought*.   However, there are
languages in which such concepts are signalled by inflecting the verb. For
instance, in Greenlandic Eskimo, inflection is used to mark mood.
Inflectional affixes include a **potential mood** marker, indicating that something is possible, an **epistemic mood**, showing extent of the speaker's
certainty, an **evidential mood**, used in hearsay reports where the speaker
cannot personally vouch for the truthfulness of a statement, and a **debitive
mood**, used to express physical or moral obligation. Here are some
examples (from Fortescue, 1984):

[10.24]   **a.** Potential mood

        timmi-*sinnaa*-vuq
        fly   can 3PS_I
        'It can fly'

      b. Epistemic mood

        Nuum-mi   api-*nnguatsiar*-puq
        Nuuk-locative snow presumably 3PS_I
        'It's presumably snowing in Nuuk'
        nilli-*runar*-puq
        be-cold undoubtedly 3PS_I
        'It (the water) is undoubtedly cold' (e.g. from looking at it)

      c. Evidential mood

        nalunaaqutaq pingasut tuqu-*sima*-vuq
        clock   three   die apparently 3PS_I
        'He died at three o'clock'

      d. Debitive mood

        imir-niru-*sariaqar*-putit
        drink more must 2PS_I
        'You must drink more.'

Note: 2PS = 2nd person singular; 3PS = 3rd person singular; I = indicative.

Conceptually, the categories of mood, tense and aspect are not entirely independent of each other. So they are often simultaneously signalled by the same form. For instance, an event that is already completed at the moment of speaking is also in the past tense. Consequently, in many languages the difference between past tense and perfective aspect is blurred. This is true of Latin, for example, where *amāvī* can be glossed either as 'I loved' or 'I have loved'.

Paralleling this, the distinction between future tense and potential (or irrealis i.e. non-actual) mood is not always strictly maintained. This was hinted at in [10.21] with reference to Dani. Further exemplification comes from Ngiyambaa:

[10.25] yurun-gu nidja-l-aga

Donaldson (1980: 160–1) points out that this sentence can be appropriately glossed as 'It might rain' or 'It will rain'. In the former case, an irrealis (non-actual) meaning would be selected while in the latter case it is the future meaning that would be most relevant. Of course, Ngiyambaa and

Dani are not unique. In most languages the fact that statements about the future are no more than (uncertain) predictions is reflected in the morphology. Even an apparently confident future tense statement like 'I will come tomorrow' is ultimately no more than an optimistic prediction.

Turning to the present tense, we note that an event that is incomplete at the moment of speaking can naturally be indicated by the present tense. This results in the conflating of the opposition between progressive aspect and present tense in many languages. In Luganda, for example, *bakola* can mean either 'they work' or 'they are working'.

Conjugation class is the last inherent category of the verb that we will deal with. In many languages verbs belong to a number of distinct morphological classes called **conjugations** and verbal paradigms are normally described in terms of conjugations. The particular subset of inflectional affixes that a verb can take depends on the conjugation that it belongs to. Conjugations are a good example of the parsimony principle and blocking in inflectional morphology as we have already noted (cf. [4.18] on p. 77).

Examine closely the Latin data below:

[10.26]	Conjug.	Infinitive	2SG Present	ISG Future	Past
	1st	amāre	amās	amābō	amāvi
		('to love')	('you love')	('I shall love')	('I loved')
	2nd	monēre	monēs	monēbō	monuī
		('to advise')	('you advise')	('I shall advise')	('I advised';
	3rd	regere	regis	regam	rēksī
		('to rule')	('you rule')	('I shall rule')	('I ruled')
	4th	audīre	audīs	audiam	audīvī
		('to hear')	('you hear')	('I shall hear')	('I heard')

(based on Kennedy, 1948)

Regular verbs belong to one of these four conjugations. Alternative realisations of various inflectional categories of the verb are selected, depending on the conjugation that verb belongs to. A verb like *vidēre* 'to see', which belongs to the same conjugation as *monēre* will take the inflections that *monēre* takes.

What morphemes are represented by the different inflectional suffixes in [10.26]? Do not overlook the fact that some of the morphemes are cumulatively realised (i.e. a given form may represent several different morphemes simultaneously).

Your answer should look like this:

[10.27]

	1st Conjug.	2nd Conjug.	3rd Conjug.	4th Conjug.
Infinitive	-āre	-ēre	-ere	-īre
2SG Present	-ās	-ēs	-is	-īs
ISG Future	-bō	-bō	-am	-am
ISG Past	-āvī	-ī	-ī	-vī

In Latin the same formative may simultaneously represent several morphemes – in this instance conjugation class, person, number and tense (e.g. in *amās*, the form -*ās* represents 1st conjugation, 2nd person, singular, present tense).

10.3.2 Agreement Properties of Verbs

In many languages the verb has **agreement markers** which are determined by the characteristics of some other word in the same construction. Such markers may indicate properties such as person, gender and number (cf. [10.6] on p. 209).

In English the verb agrees with the subject in number. Hence the difference between *she walks* (*she walk*) and *they walk* (*they walks*). However, except for present tense verbs whose subject is a third person singular noun phrase or pronoun, normally verbs carry no visible agreement markers. The only verb that shows extensive agreement is *be* (cf. *I am*, *he/she/it is* and *you/they are*).

Study the data in [10.28] from Runyankore, a Bantu language of Uganda.

[10.28] Class a

Omushomesa aza.	'The teacher is coming.'
Abashomesa baza.	'The teachers are coming.'
Omukazi aza.	'The woman is coming.'
Abakazi baza.	'The women are coming.'

Class b

Ekicuncu kiza.	'The lion is coming.'
Ebicuncu biza.	'The lions are coming.'
Ekisegyesi kiza.	'The porcupine is coming.'
Ebisegyesi biza.	'The porcupines are coming.'

Class c

Embuzi eza.	'The goat is coming.'
Embuzi ziza.	'The goats are coming.'
Ente eza.	'The cow is coming.'
Ente ziza.	'The cows are coming.'

a. Identify the noun stems and prefixes.
b. Describe the verb agreement phenomena which you observe.

The class a noun stems -*shomesha* and -*kazi* take the prefix *omu*- in the singular and *aba*- in the plural; class b stems -*cuncu* and -*segyesi* take the prefix *eki*- in the singular and *ebi*- in the plural; the class c noun stems -*buzi* and -*te* take a prefix with the form *eN*- in both the singular and the plural (with *N* always realised as a nasal homorganic with the first stem consonant).

The verb -*za* 'come' agrees with the subject. It carries a subject marker whose form is selected depending on the number and noun class of the subject noun phrase. If the subject belongs to class (a), the subject marker is *a*- in the singular and *ba*- in the plural; if it belongs to class (b) the subject marker is *ki*- in the singular and *bi*- in the plural; and if the subject belongs to class (c), the subject marker is *e*- in the singular and *zi*- in the plural.

What modifications do we need to make to the above statement in order to accommodate these additional data? (Underlining indicates emphasis.)

[10.29] a. Tureeba omushomesa.   'We see the teacher.'
            Omushomesa tumureeba. 'We see the teacher.'
            Tureeba abashomesa.   'We see the teachers.'
            Abashomesa tubareeba. 'We see the teachers.'

         b. Tureeba ekicuncu.   'We see the lion.'
             Ekicuncu tukireeba.   'We see the lion.'
             Tureeba ebicuncu.   'We see the lions.'
             Ebicuncu tubireeba.   'We see the lions.'

         c. Tureeba embuzi   'We see the goat.'
             Embuzi tugireeba.   'We see the goat.'
             Tureeba embuzi.   'We see the goats.'
             Embuzi tuzireeba.   'We see the goats.'

In [10.29], in addition to the subject marker *tu-* 'we', the verb stem *-reeba* 'see' contains an **object marker** which agrees in noun class and number with the object noun if the object is put before the verb for emphasis. The object marker is *mu-* in the singular and *ba-* in the plural if the object noun belongs to class (a); it is *ki* in the singular and *bi* in the plural if the noun belongs to class (b); and it is *gi-* in the singular and *zi-* in the plural if the object belongs to class (c). What is given here is only a small sample of verb agreements in Runyankore. As is the norm in Bantu, Runyankore has a very extensive system of noun classes and agreement.

Although verb-object agreement might seem exotic to the English speaker, it is not uncommon in the languages of the world. Besides Bantu and other African languages, some Native American languages also display the same phenomenon.

10.3.3. Configurational Properties of Verbs

Recall that configurational properties are assigned to a word when it occurs in some specified larger syntactic context (see [10.6]). We will distinguish between two kinds of configurational properties of verbs, those assigned in a single clause and those assigned in a larger domain. Many of the properties assigned clausally indicate the kind of relationship that a verb has with some noun phrase or phrases that are in construction with it.

We begin by treating configurational properties assigned within a single clause. In section (10.3.1) we provisionally dealt with tense as an inherent property of the verb. We were simply following the traditional view. However, as we shall now see, recent scholarship has shown that tense is a configurational category of the verb (Chomsky, 1973, 1980).

In Chomskyan Government and Binding theory, tense is regarded as a constituent of the inflectional component of **AUX** (auxiliary). AUX contains modal auxiliaries (symbolised by *M* in [10.30]) like *may*, *can*, *will* etc., and other auxiliary verbs (e.g. *have*, *be*) as well as those features such as person and number, in which a finite verb must agree with its subject. These features are commonly referred to by the label **INFL** (which is short for inflection). INFL is introduced by the phrase structure rule in [10.30]:

[10.30] AUX → INFL (M(*(have + en)* *(be + ing)*)

As seen in [10.31], there are two manifestations of INFL:

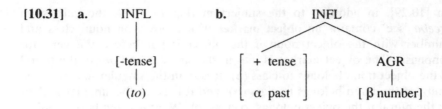

[10.31] a.   INFL   **b.**   INFL

[-tense] $\begin{bmatrix} + \text{ tense} \\ \alpha \text{ past} \end{bmatrix}$ AGR

(*to*)   [β number]

Note: INFL = inflection; AGR = agreement. The Greek letter β is a variable indicating that the subject will share whatever value of the agreeing feature number that the verb has. The variable α in [α past] ranges over [+] (i.e. past) and [−] (i.e. non-past) for English.

INFL is the syntactic head of the sentence, the element that must be obligatorily present in any complete, finite sentence for it to be well-formed. Moreover, if tense is indicated in a sentence, a subject must also be obligatorily assigned to the verb, and *vice versa*. You can see this by comparing the ungrammatical sentences in [10.32a], which have a subject but lack tense, with the equally ill-formed sentences in [10.32b], which have tense but lack a subject:

[10.32] a.   *The girl   to walk.   [+subject   −tense]
                  *The girl   walking.

       b.   *walks   [−subject   +tense]
                  *walked

For them to be well-formed, these sentences must have both subjects and tensed verbs as shown below:

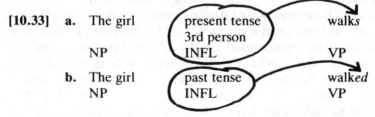

[10.33] a.   The girl   present tense   walk*s*
                                 3rd person
                  NP   INFL   VP

       b.   The girl   past tense   walk*ed*
                  NP   INFL   VP

As seen in [10.33], in English tense is usually moved out of INFL and appears as inflection on the verb, although it is a property that belongs to the sentence as a whole rather than to the verb alone. But if an auxiliary verb is present, as in [10.34], tense must be left within AUX, in which case it appears as inflection on the first (or only) auxiliary:

[10.34] a.

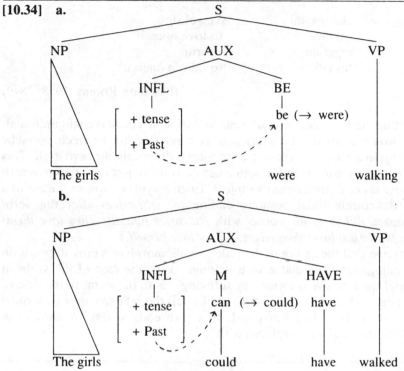

Observe that the presence of tense in INFL also triggers number agreement which only shows up in the present tense if the subject is in the third person. We distinguish between *The girl plays every Monday* and *The girls play every Monday*. Contrast this with the situation shown above in [10.31a] where INFL is tenseless. Untensed INFL is only permissible in non-finite clauses like *She wants to read that novel* where the infinitival clause *to read that novel* lacks both a subject and tense. In this case INFL is realised by the infinitive marker *to*. Hence the ungrammaticality of **She wants to reads that novel*, which shows agreement of an infinitival verb form with *to* in INFL.

Besides tense, there are a number of other common single clause inflectional categories. The Yurok examples in [10.35] illustrate one of them, namely **reflexivisation**:

[10.35] a.   skuyk-   skuykep-
        'to treat well'   'to treat oneself well'
        megetołkʷ.   megetołkʷ. ep
        'to look after'   'to look after oneself'

b.   nɹgɹyk   nɹgɹykep or nɹgɹykɹp
        'to help'   'to help oneself'
        sɹmɹt-   sɹmɹtep or sɹmɹtɹp-
        'to beat'   'to beat or kill oneself'

    c. skewoksim-   skewoksip-
        'to love'   'to love oneself'
        tegerum-   tegerup
        'to talk'   'to talk to oneself'

(based on Robins, 1958: 78–9)

Inflecting the verb itself to mark reflexivisation in Yurok is configurational: what would otherwise be expressed as a relationship between syntactic elements in a clause is signalled morphologically within the verb itself. This is clear if you compare Yurok with English. If the object of a transitive verb like *love* is not distinct from its subject, English syntax requires the use of a separate co-referential pronoun ending in *-self/-selves* after the verb. (Compare *Bill loves his teacher* with *Bill loves himself*; *They love themselves* with **Bill loves themselves*, **They love herself*.)

Observe that the choice of the reflexive allomorph in Yurok depends on the conjugation class that a verb belongs to. In the case of the verbs in [10.35a] the reflexive is formed by suffixing *-ep-* to the stem. In the case of the verbs in [10.35b] (whose root vowel is [ɪ]) the reflexive may be formed either by suffixing [-ep-] or [-ɪp-]. If a verb ends in *-im-* or *um-*, as in [10.35c], the final *-m* is replaced with *-p*.

Study the following Luganda data and explain why reflexivisation is a configurational property in this language as well.

[10.36] **a.** Abakinjaagi ba-li-sala ennyama
           butchers they-*fut*-cut meat
           'The butchers will cut the meat'

        b. Abakinjaagi ba-li-*ee*-sala
           butchers they-*fut*-themselves-cut
           'The butchers will cut themselves'

The verb *-sala* 'cut' is transitive and is typically followed by an object noun phrase. The object noun phrase normally refers to an individual or entity which is different from the one referred to by the subject noun phrase. That is the case in the first sentence in [10.36a]. However, it is possible for the object noun phrase to refer to the same individual as the subject noun phrase. Whereas in English this latter situation is signalled using a separate word, i.e. a reflexive pronoun (which ends in *-self*, e.g. *myself*, *yourself* etc.) in the singular and *-selves* (*ourselves*, *yourselves*) in the plural, in Luganda it is the verb itself that is modified as in [10.36b] by placing the reflexive pronoun prefix *-ee-* between the tense prefix and the verb stem.

In addition to inflectional configurational properties that are assigned within a single clause, which we have just discussed, there are others which are determined by the position of a verb in a sequence of clauses. Frequently, languages distinguish between verbs in main clauses and those in subordinate clauses. So grammarians traditionally recognise the so-called **subjunctive mood**, which has the typical inflection that marks a tensed verb appearing in a subordinate clause.

French will provide our example of the subjunctive. Observe the use of *écrit* and *écrive* in the following:

[10.37] a. $_s$[Elle leur écrit]$_s$
$(*_s$[Elle leur écrive]$_s)$
'She writes to them'

b. $_s[[_{s_1}$[Je souhaite/regrette]$_{s_1}$ que$_{s_2}$[elle leur écrive]$_{s_2}]_s$
$(*_s[[_{s_1}$[Je souhaite/regrette]$_{s_1}$ que$_{s_2}$ [elle leur écrit.]$_{s_2}]_s$
'I wish/am sorry she would write to them.'

A verb which occurs in a subordinate clause (S_2) is in the subjunctive mood if the subordinate clause is dependent on a main clause (S_1) with a verb expressing emotion or wish. The subjunctive form in this case is *écrive*.

Although there are other subordinate clauses that require the subjunctive, we will not consider them here. The above examples are sufficient for our present purposes.

What are the morphological differences between sentences expressing direct and indirect statements in the Latin examples below?

[10.38] a. Ego amāvī   puellam.
I   love-past-I girl
'I loved the girl.'
Tū   amāv*istī*   puellam.
You (sg.) love-past-you girl
'You loved the girl.'

b. Dixit   me   puellam amav*isse*. (*Dixit ego puellam amavī)
say-past-he me   girl   to-have-loved
'He said that I loved the girl.'
Dixit   tē   peullam amav*isse*. (*Dixit tū puellam amāv*istī*)
say-past-he you girl   to-have-loved
'He said that you (sing.) loved the girl.'

Note: In the glosses, the morphemes are linearly ordered within words for convenience. This has no theoretical significance whatsoever. In Latin morphemes are normally cumulatively realised.

What we observe here is another example of configurationally determined inflection. The construction in question is called the **oratio obliqua**. If the verb meaning 'love' is in the main clause, *amāvī* and *amāvistī* are chosen in the first and second person singular perfect/past tense respectively (as in [10.38a]). But if the same verb is in a subordinate clause, following a verb of saying in reported speech (oratio obliqua), the perfect infinitive form *amavisse* 'to have loved' must be used as in [10.38b]. In addition, the subject of the verb in the subordinate clause is put in the accusative case. Hence, the first and second person singular pronouns are realised by the accusative *me* and *tē* forms (rather than the nominative *ego* and *tū*). (See the discussion of case in section (10.4.3) below.)

Switch reference is another kind of inflectional phenomenon that is determined by the position of a verb in a wider discourse context that goes beyond a single clause. In many native American, Papuan New Guinea and Australian languages, when two co-ordinate clauses are joined together in a sentence, the verbs receive different inflectional suffixes depending on whether they have the same subject or not.

We will illustrate switch-reference with data from Choctaw, a native American language. Davies (1984) reports that this language has the affix -*cha* which is suffixed to the first verb if the verbs in the two co-ordinate clauses have the same subject:

[10.39] a. ofi   poshohli-li-*cha* tamaha ia-li-tok
             dog   rub-1 nom-SS   town   go-1 nom-pst
             'I patted the dog and went to town.'

         b. *ofi   poshohli-li-*na*   tamaha ia-li-tok
             dog   rub-1 nom-DS   town   go-1 nom-pst

Notes on abbreviations: SS = same subject; DS = different subject; nom = nominative; pst = past tense; 1 = first person; 2 = second person.

However, where the verbs in co-ordinate clauses have different subjects -*na* is attached to the verb in the first clause:

[10.40] a. tobi   apa-li-*na*   tãchi ish-pa-tok
             beans eat-1 nom-DS corn 2 nom-eat-pst
             'I ate beans and you ate corn.'

         b. *tobi   apa-li-*cha*   tãchi ish-pa-tok
             beans eat-1 nom-SS corn 2 nom-eat-pst

In this section we have surveyed a wide range of inflectional categories of the verb. We have considered in turn the inherent properties, agreement properties and configurational properties of verbs. In the next section I provide a similar overview of the phenomena encompassed by nominal inflection.

10.4 INFLECTIONAL CATEGORIES OF NOUNS

We will first examine inherent inflectional categories before going on to discuss agreement and configurational inflectional properties.

10.4.1 Inherent Categories of Nouns

We shall only consider three of the commonest inherent categories of nouns, namely number and gender/class.

Of these three categories, **number** seems to be the most widespread. All speech communities have ways of encoding the notion of countability. Many languages distinguish by inflection between one and more than one, e.g. English distinguishes between (one) *woman* and *women* (i.e. more than one woman). A few languages, including Sanskrit and Greek make a tripartite distinction between one, dual (two) and more than two. In English there is a hint of the dual in the use of *both* as in *both women* (only used when referring to two) vs *many women* and in the use of <u>*between*</u> *two* versus <u>*among*</u> *many*.

Number is an obligatory category in English nouns. Nouns have to carry inflection showing whether they are singular or plural. However, as we briefly noted in (10.2.1.3), number is not universally an obligatory inflectional category. In East Asian languages like Chinese, Vietnamese and Japanese for instance, number is not obligatorily indicated. There is normally no inflection for number. Thus, in Japanese *hón* can mean either *book* or *books*.

Interestingly, there is a strong correlation between animacy and number. Thus, in Japanese, absence of number specification is especially common if a noun is inanimate, like *hón*. Nouns referring to animates, especially humans, are likely to have explicit plural marking. For instance, *hitó* means 'person or people' but *hitó-tati* (with the plural marker *-tati*) is unambiguously plural.

If a language does not obligatorily mark number numerals (e.g. one, two etc.), enumerator or measure or classifier words equivalent to 'several' or 'many' are used in situations where number is relevant in an inanimate noun. Often, forms signalling number do also simultaneously reflect animacy or some other salient property of an entity. In Japanese, the enumerator or classifier selected may depend on whether one is counting people, cars, animals, etc. Compare the enumerator classifier for 'three' which is *san-nin* for people and *san-dai* for cars in the following:

[10.41]   a.   Yuuzin o   san-*nin*   syootaisita
               friend   3- person   invited
               'I invited three friends.'

           b.   San-*dai* no   sidoosya de   ryookoosita
               3-car   car by   travelled
               'We travelled in three cars.'

(based on Kuno, 1978)

Recall the examples of number in Kwakwala in [10.16]. In that language nouns can be marked for number yet number is a derivational and not an inflectional category (S. R. Anderson, 1982). Plural marking is neither assigned on the basis of the syntactic configuration (e.g. to a noun that is the head of some NP in a main clause), nor on the basis of agreement (e.g. between a verb and its subject or between a noun and its modifiers in an NP). Rather, plural has three properties in Kwakwala: (i) only some of the nouns have a plural form; (ii) even if a noun has a plural form, marking the plural is optional; (iii) the meaning of 'plural' is unpredictable.

Languages also often classify nouns by **gender** or **class** (Corbett, 1990). This kind of classification tends to have a residual semantic basis. Thus, in many European languages, nouns referring to animate individuals are usually masculine or feminine, depending on whether the individuals in question are male or female. Nouns referring to inanimate entities are usually neuter. For instance, in German *der Onkel* 'the uncle' is masculine and *die Tante* 'the aunt' is feminine.

However, the importance of the semantic basis of the gender system (where the gender of a word reflects the sex of its denotation) must not be exaggerated. Linguistic gender is essentially a grammatical rather than semantic classification of nouns. It is true that the sex gender of the entity referred to by an expression may play a role and may indeed have been the original motivation for the classification. But meaning is not always paramount. Thus there a few German nouns referring to humans, e.g. *das Mädchen* 'the girl' and *das Kind* 'the child' which are neuter. And there are many nouns which refer to inanimate things which are masculine, e.g. *der Alkohol* 'alcohol', or feminine, e.g. *die Bratpfanne* 'frying pan'. (The gender system of Old English was equally semantically arbitrary. For instance, *wīfmann* 'woman' was neuter while *bōc* 'book' was feminine and *bāt* 'boat' was masculine.) Clearly, grammatical gender does not necessarily reflect sex gender.

The inherent grammatical gender of the noun dictates the choice of agreeing morphemes such as the definite article in German. With a masculine noun *der* is chosen, with feminine nouns *die* is chosen and with neuter nouns *das* is chosen, e.g. *der Onkel* 'the uncle', *die Tante* 'the aunt' and *das Haus* 'house'.

The phenomenon of noun classification is also found in numerous sub-Saharan African languages, native American languages and Australasian languages. But in many of these languages the classification system does not have even the most tenuous semantic (sex) gender basis. Nouns are assigned to different classes, often on a minimally semantic basis, depending on whether they refer to a human/animate individual, or on the basis of salient properties of the entity denoted by a noun such as its shape and size. We can see this in the partial list of Swahili noun classes in [10.42]:

[10.42]		Singular		Plural
	a.	mtu	'person'	watu
		mgeni	'guest'	wageni
		mke	'woman'	wake
	b.	mti	'tree'	miti
		mganda	'bundle'	miganda
		mfereji	'ditch'	mifereji
		mto	'river'	mito
		mtego	'trap'	mitego
	c.	kikapu	'basket'	vikapu
		kitabu	'book'	vitabu
		kioo	'mirror'	vioo
	d.	mbuzi	'goat'	mbuzi
		ndege	'bird'	ndege
		ŋguruwe	'pig'	ŋguruwe

---

a. For each set of nouns in [10.42], identify the noun class marker.
b. What semantic factors play a role in the classification of nouns?

Noun classes are marked using the following prefixes:

[10.43]   Singular   Plural
    a. m- (class 1)   wa- (class 2)
    b. m- (class 3)   mi- (class 4)
    c. ki- (class 7)   vi- (class 8)
    d. N (class 9)   N (class 10)

Notes: N stands for a nasal with the same place of articulation as the following consonant.

In Bantu linguistics the classes are each given a number. The nouns in [10.42a] take the class 1 *m-* prefix in the singular and class 2 *wa-* in the plural, those in [10.42b] take the class 3 prefix *m-* in the singular and *mi-* in the plural, and so on.

Semantics has a rather limited role in the Swahili classifier system. Nouns referring to humans are normally in classes 1/2 and those referring to animals tend to be in classes 9/10. But other noun classes have a very heterogeneous membership. For instance, rivers, bundles, trees and ditches all belong to classes 3/4, although they do not constitute a semantically coherent group. Besides, some human nouns are found outside the human classes 1 and 2. For example, *kipofu* 'blind person' and *kijana* 'youth' are in classes 7/8, which mostly contain nouns referring to inanimate things like *kikapu* 'basket' and *kitabu* 'book'. (However, semantics re-asserts itself in agreement: these words have human classes 1/2 agreements with adjectives, subject and object markers in verbs, and so on.) Finally, classes 9/10 typically contain nouns referring to animals. But they also contain nouns like *ndizi* 'banana', *ndugu* 'brother, sister' and *ngoma* 'drum'.

10.4.2 Agreement Categories of Nouns

It is often the case that, within a noun phrase, agreement rules copy an inherent feature of the noun (e.g. gender and/or number) on to other words such as articles, numerals and adjectives in construction with it.

Describe the exact nature of the agreement displayed in the following Swahili data:

[10.44] **a.** *m*-t-oto   *m*-toto *m*-moja *m*-dogo
            'child' (class 1, sing.)   'one small child'
            wa-toto   *wa*-toto *wa*-tatu *wa*-dogo
            'children' (class 2, pl.)   'three small children'

b. *ki*-tabu 'book'   *ki*-tabu *ki*-moja *ki*-dogo
(class 7, sing.)   'one small book'
vi-tabu 'books'   *vi*-tabu *vi*-tatu *vi*-kubwa
(class 8, pl.)   'three big books'

These Swahili data are a classic example of noun agreement. The adjective and numeral get a class and number agreement marker taken from the menu of prefixes available to forms modifying the noun that is the head of their noun phrase.

Describe the agreement shown in the following French sentences:

[10.45] **a.** La petite fille chante   Le petit garçon chante
'The little girl sings'   'The little boy sings'
     b. Les petites filles chantent   Les petits garçons chantent
'The little girls sing'   'The little boys sing'

In [10.45a] we see the article agreeing in gender and number with the noun. In addition, before a singular noun, the definite article has the form *la* if the noun is feminine and the form *le* if it is masculine. But a plural noun requires the selection of *les*. Gender distinctions are neutralised in the article in the plural.

Furthermore, the adjective agrees with the noun in number and gender. In the singular, a feminine noun conditions the suffixation of *-e* to the adjective while a masculine noun does not require the selection of any overt suffix. In the plural an *-s* suffix is added after the gender suffix.

Finally, there is also subject-verb agreement in number. If the head of the subject NP is singular, the verb must also be singular. Conversely, if the head of the subject NP is plural, then the verb must be plural. So, the singular subject NPs *la petite fille* and *le petit garçon* co-occur with a verb ending in *-e* (*chant-e*) but the plural subject NPs *les petites filles* and *les petits garçons* go with a verb with the *-ent* plural ending (*chant-ent*). (Refer back to section (10.3.2) for a discussion of verb agreement.)

10.4.3 Configurational Categories of Nouns

In this section we will consider properties that nouns receive when they appear in certain larger syntactic configurations. The property that we will focus on is **case**, because of its great importance.

The term case is used in at least two different senses in the literature:

[10.46]   (i)   the **grammatical case**, where case is used to mark the function of a noun (or rather, more precisely, a noun phrase) e.g. as subject or object, depending on its position in relation to the verb in the sentence. Example: in Latin the nominative case marks the subject, while the accusative case marks the object. (Word-order in a sentence is relatively free because the function of an NP is always indicated by its case inflection.)

     a.   *Agricol-a*$_{(nom.)}$ puell-*am*$_{(accus.)}$ *videt*
          subject   object
          Farmer   girl   sees
          'The farmer sees the girl.'

     b.   *Agricol-a*$_{(nom.)}$ *videt*   puell-*am*$_{(accus.)}$
          subject   object
          Farmer   sees   girl
          'The farmer sees the girl.'

    (ii)   the **oblique case** marks not the syntactic function of a noun but rather its semantic function. Typically, oblique cases mark location or direction. (See also (section 11.3).)
        Example: in Latin, with certain verbs of motion, the accusative marks the destination.
        Romam$_{(accus.)}$ *rediit*   'He returned to Rome.'

We will begin the discussion with grammatical case, focusing on a limited range of common grammatical cases involving the relationship between nouns and verbs. (Prepositions too may require nouns in construction with them to be in certain cases. Thus in Latin *ad* 'to' governs the accusative case as in *ad urbem* 'to the town', while *ab* 'from' governs the ablative case as in *ab urbe* 'from the town'.)

Case agreement systems imposed by verbs fall in one of two categories: **nominative-accusative case systems** and **ergative-absolutive case systems**. Consider the following:

[10.47]   **a.**   Fido   barked.   **b.**   Mary   saw   Fido.
             S   Vintr   S   Vtr   O
             She   slept.   She   saw   him.
             S   Vintr   S   Vtr   O

Note on abbreviations: S = subject; O = object; Vtr = transitive verb; Vintr = intransitive verb.

What the examples in [10.47] show is that in English case is not generally overtly marked by inflection in nouns. A noun has the same form regardless of whether it is subject or object. Only pronouns are inflected for case. The subject pronoun of an intransitive verb like *sleep* and the subject of a transitive verb like *see* must both be in the nominative case. Nominative is realised as *she* in [10.47]. A pronoun functioning as the object NP of a transitive verb like *see* must be in the accusative case (which is realised as *her*). The only situation where lexical nouns are morphologically marked to indicate their grammatical function is when they are in the genitive case. (See pp. 240–1 and section 10.5.)

As seen in the table of the forms of the first person personal pronoun in [10.48], in Latin pronouns are extensively inflected for case. (The general meaning of each case can be inferred from the English gloss.)

[10.48]	1st person	Singular	Plural
	Nominative	ego 'I'	nōs 'we'
	Accusative	mē 'me'	nōs 'us'
	Genitive	meī 'my'	nostrī/nostrum 'of us'
	Dative	mihi 'to me'	mōbis 'to us'
	Ablative	mē 'from me'	nōbis 'from us'

However, as we have already noted, unlike English, Latin also typically inflects nouns for case to show their grammatical function:

[10.49] *Agricola*   *venit*   'The farmer comes.'
<Subj>$_{[nom.]}$

Agricola   *amat*   *puellam*   'The farmer loves the girl.'
<Subj>$_{[nom.]}$ <Obj>$_{[accus.]}$

Puella
<Subj>$_{[nom.]}$ *amat*   *agricolam*   'The girl loves the farmer.'
              <Obj>$_{[accus.]}$

Latin is a prime example of a **nominative-accusative** language, i.e. a language where subjects are in the nominative case and objects in the accusative case. As we saw above, English pronouns are also marked for case on the same basis. So, English too is a nominative-accusative language.

Study the data below from Dyirbal and describe exactly how Dyribal case marking differs from that of English and Latin.

[10.50]   **a.**   *Subject*   *Verb*
            ŋuma   banaga+nʸu   'Father returned'
            Father   return+non-future

b.	*Subject*	*Verb*		'Mother returned'
	yabu	banaga+nʸu		
	Mother	return+non-future		

c.	*Object*	*Subject*	*Verb*	'Mother saw father'
	ŋuma	yabu+ ŋgu	buɽa+n	
	Father	mother	see+non-future	

d.	*Object*	*Subject*	*Verb*	'Father saw mother'
	yabu	ŋuma+ŋgu	buɽa+n	
	Mother	father	see+non-future	

(Dyirbal data from Dixon, 1979: 61)

I hope you have discovered that the object of a transitive verb and the subject of an intransitive verb share the same inflectional marking. This particular correlation of case with syntactic function is quite common among the languages of the world. Clearly, such languages are not nominative-accusative. They are called **ergative-absolutive** (or sometimes simply **ergative**). In such languages, the case adopted by the subject of an intransitive verb and the object of a transitive verb is called the **absolute** (or **absolutive**) case. And the case adopted by the subject of a transitive verb is called the **ergative**.

Thus, the object in (c) *ŋuma* 'father' has the same zero marking of case as the subject of the intransitive verb 'returned' in (a). But the subject of a transitive verb is given a different inflectional agreement, namely the suffix *-ŋgu* in ŋuma+ŋgu in (d).

To sum up, **ergative-absolutive** case marking systems group noun phrases differently from nominative-accusative systems. The absolutive case is assigned to both the subject of an intransitive verb and the object of a transitive verb. A different case, the ergative, is assigned to the object of a transitive verb. The case systems of many languages, including Native American languages like Eskimo, most Australian languages, many Papua-New Guinea languages, some Asian languages like Tibetan and one European language, Basque, follow the ergative-absolutive pattern.

Let us now turn to another common grammatical case, the **genitive**. This case is used to signal the fact that one noun is subordinate to the other, i.e. one noun is the head and the other noun is the modifier which adds some further specification to the head. One typical but by no means the only use of the genitive is illustrated in [10.51].

[10.51]   Mary's dress

Here the genitive is probably used to mark possession. The possessor noun, *Mary*, is inflected with *'s* and the head noun, *dress*, appears after it without any inflection. The presence of the word *Mary's* makes the referent of *dress* more specific than say, *a dress*.

What is the function of the genitive [10.52]?

[10.52] **a.** Fido's vet has retired.
 b. Peter's playing thrilled the audience.

Here the genitive is used without a possessive meaning. Obviously, in [10.52a] the *vet* is not owned by *Fido*, rather, the *vet* is associated with *Fido* in a doctor-patient relationship. In [10.52b] *Peter* does not own the *playing*. We could paraphrase the sentence as 'The manner in which Peter played thrilled the audience.' The semantic relationship between the head and its modifier in a genitive construction is vague. That is why I have treated genitive primarily as a syntactic rather than a semantic notion, i.e. in terms of the *syntactic* notions of modification and subordination. I will return to the genitive in English from a somewhat different angle in section (10.5).

Let us now turn to oblique cases. As we observed earlier in [10.46], oblique cases tend to mark functions that are to some degree semantic. Many languages mark such cases not by inflecting nouns, but rather by using prepositions or postpositions for this purpose.

A common oblique case is the **instrumental**, which, as its name implies, marks a noun phrase denoting some entity which is used to perform the action indicated by the verb. In English the preposition *with* is commonly used for this purpose as in *He chopped the tree down with an axe* (i.e. *he used an axe to chop down the tree*). The use of a preposition to mark the instrument is common.

But also common is the use of inflection. In some languages, a noun is inflected with an instrumental affix to indicate this role. This is seen in the Dyirbal examples below from Dixon (1972: 93):

[10.53] **a.** *balan*   *ɖugumbil*   *baŋgul*   *yaṛaŋgu*
 THERE-NOM-II   woman-NOM   THERE-ERG-I   man-ERG
 ba-ŋgu   *yugu-ŋgu*   *balgan*
 THERE-INST-IV   stick-INST   hit-PRES/PAST
 man is hitting woman with stick

     b. *bayi*   *ɖaban*   *baŋgul*   *yaṛaŋgu*
 THERE-NOM-I   eel-NOM   THERE-ERG-I   man-ERG
 ba-ŋgu-l   *ɖirga-ŋgu*   *ɖurgaɲu*
 THERE-INST-I   spear-INST   spear-PRES/PAST
 man is spearing eel with multi-prong spear

c. *balam*　　　　　*dugut̪*　　　　*baŋgugaragu*　　　*ba-ŋgu*
　　THERE-NOM-III　yam-NOM　　two people-ERG　THERE-INST-IV
　　gaḏin-du　　　*bagan*
　　yamstick-INST　dig-PRES/PAST
　　the two [women] are digging yams with a yamstick

Note: Nominative (NOM) is the cases of the intransitive subject and transitive object. It is what is called absolute in the terminology of this book.

Even more general is the spatial use of oblique cases to express **location** or **direction**. (These are often extended metaphorically to concepts of time, beneficiary etc.) In English prepositions rather than case inflections on nouns are used to mark these concepts:

[10.54]　Location:　　　　Mary is *at the station*.
　　　　　Direction:　　　　John is travelling *to the station*.
　　　　　Beneficiary:　　　John sent a ticket *to Mary*.
　　　　　Temporal use:　　Mary arrives *at noon*.

However, in many other languages case inflections fulfil similar functions. This is illustrated by the Ngiyambaa data below (from Donaldson, 1980):

[10.55]　a.　*-ga*　Location without motion
　　　　　　　　　(adessive = 'at, on, by, next/to')
　　　　　　　　　gabada:　　　　　balima-*ga*　wamba-nha
　　　　　　　　　The moon+ABS　sky-LOC　be up-PRES
　　　　　　　　　'The moon is in (literally on) the sky'

　　　　　b.　*-gu*　Motion to (allative function of dative *-gu*)
　　　　　　　　　nadhu　　　yana-nha　galin-*gu*
　　　　　　　　　I+NOM　go-PRES　water-ALLATIVE
　　　　　　　　　'I am going to the water'

Furthermore, Donaldson reports that in Ngiyambaa a locative case ending may be used to signal not only concrete spatial location but also abstract or metaphorical location:

[10.56] **a.** Concrete spatial location

balima-ga	'in the sky'
sky-LOC	

b. Metaphorical location

in time	dhuni-ga	'in the day'
	sun-LOC	(i.e. 'the day')
in states	wagayma-ga	'in jest'
	play-LOC	
	gari-ga	'in earnest'
	truth-LOC	

Ngiyambaa is by no means unusual in its use of the same case marking suffix to locate entities in space, in time, or metaphorically in states. Many other languages do the same. Why should that be the case?

Proponents of the **localist hypothesis** believe that they have an explanation. They argue that:

> Spatial expressions are more basic, grammatically and semantically, than various kinds of non-spatial expressions.   (Lyons, 1977: 718)

This, they claim, is attributable to the fact that concrete location serves as a model for the conceptualisation of various abstract concepts and metaphors (Miller and Johnson-Laird, 1976). Many of the spatial metaphors occur cross-linguistically and are responsible for the striking similarity in the case marking of certain relationships (see Jackendoff, 1976; J. M. Anderson, 1971, 1977; Lyons, 1977; Miller, 1985 and references cited there).

As we have already seen, in English the locative is marked not by inflection on nouns but by prepositions (e.g. *on the table*, *at Lancaster*, *over the moon* etc.). Its most basic function is indicating location or direction. But it may also be used in a metaphorical sense to indicate temporal relationships (e.g. *on Monday*, *at 2.00a.m.*, *in July* etc.). Furthermore, in English, as in Ngiyambaa, the locative has a number of other metaphorical uses that indicate metaphorical (abstract) location or motion (e.g. *He is in a terrible state*; *she sent greetings to the children*; *I gave the book to David*; *Ivan sold the car to Sandra*; *Sandra bought the car from Ivan* etc.) The main difference is that, whereas Ngiyambaa uses case inflection, English uses prepositions to mark these relationships.

I have shown is that spatial concepts appear to play a role in the representation of case. I will call this the **Weak Localist Hypothesis**. Anderson (1977) goes further. He proposes what I will refer to as the **Strong Localist Hypothesis**. He makes the audacious claim that all case relations found in the languages of the world are reducible to the two notions of 'place', i.e. location in space, and (ii) 'source', i.e. movement of entities. These notions are treated as 'features' that may be present or absent. The combinations of these features give four logical possibilities (Anderson, 1977: 116):

[10.57]   ABS = case   LOC = case   ERG = case   ABL = case
                                place   source   place
                                                                              source

Combinations of 'place' and 'source' give the case relations of 'absolutive', 'locative', 'ergative' and 'ablative' (mention of 'case' is obviously redundant here).

Anderson contends that all underlying case relations are transparently reducible to the four patterns he identifies. And he proceeds to demonstrate this with examples from Tibetan and from the Pacific language Colville.

Hurford (1981) rejects Anderson's proposal. He argues that the reduction of all case relations to the four simple patterns above loses its attractiveness once we realise that an unconstrained battery of transformations is needed to translate simple underlying case relations into surface case markings because, within and across languages, the mapping of underlying case on to surface representations often displays byzantine complexity (cf. section (11.3)).

Hurford further criticises the lack of explicitness in the way in which Anderson's approach deals with semantic entailment relations between sentences. For instance, Anderson claims that the tokens of the verb *moved* represent different underlying verbs, with different case specifications for NPs in construction with them, in sentences like *John moved the stone* and *the stone moved*. He rightly recognises the implicational relation between the two sentences and refers to the second one as the 'non-causative congener' of the first. But, unfortunately, he remains silent about the way in which his model captures such implicational relations.

The upshot of this discussion is that localism, in its weak form, provides a potentially useful heuristic for accounting for the fact that location and movement are basic cognitive concepts which provide the basis of much of the conceptual framework for language. However, the manner in which spatial concepts are translated into linguistic concepts in general, and case in particular, is far from straightforward. Hence, any attempt to reduce all underlying cases to a handful of spatial relations in the manner proposed by Anderson is unlikely to succeed.

10.5 CLITICS

The problem which we address in this section has been succinctly stated by Zwicky (1977: 1):

> Most languages – very possibly, all except for the most rigidly isolating type – have morphemes that present analytic difficulties because they are neither clearly independent words nor clearly affixes.

In addition to inflectional affixes, there is another class of bound morphemes called **clitics**, which may be appended to independent words by syntactically motivated rules. Words to which clitics are attached are called **hosts** (or **anchors**). *Mary*, *Tonga* and *newspaper* are the hosts of the genitive clitic *-s* in [10.58]:

[10.58] a. Mary's car
         b. The Queen of Tonga's tiara
         c. The editor of the *Manchester Guardian* newspaper's car

Clitics may be appended at the beginning or at the end of their hosts. A clitic attached to the beginning of a host is called a **proclitic**, and one attached at the end is called an **enclitic**. The *-s*, marking genitive case in English, is an example of an enclitic, while the contracted singular definite article *l'* in French is an example of a **proclitic**. The singular definite article is realised as a separate word (*le* or *la*) if it appears before a noun or adjective commencing with a consonant like *fille* as in *la fille* 'the girl', *le chien* 'the dog'. But it is appended to a host as a proclitic when it occurs before a vowel-commencing noun like *idée* as in *l'idée* (* *la idée*) 'the idea' and *l'ami* (* *le ami*).

A further distinction can be drawn between **simple clitics** and **special clitics** (Zwicky, 1977; Pullum and Zwicky, 1988). A simple clitic belongs to the same word-class as some independent word of the language that could substitute for it in that syntactic position. The French singular definite article exemplifies this. As we have seen above, it is realised as the proclitic *l'* before vowel-commencing hosts but appears as an independent word (i.e. *le* or *la*) before consonant-commencing hosts.

Similarly, English has auxiliary verbs like *have*, *is* and *has* which, when contracted, can be appended as simple clitics to the last word in the subject NP immediately preceding them as in:

[10.59] a. They have eaten   b. They've eaten
             She has eaten   She's eaten
             The big bag is empty   The big bag's empty

The simple clitics *'ve*, *'s* and *'s* occupy the same syntactic position and have the same role as the corresponding full words *have*, *has* and *is*.

Simple clitics only differ from fully-fledged words in that they must be phonologically attached to a host because they are *prosodically deficient* and hence they are incapable of occurring in isolation. In English, simple clitics lack main word stress (as a rule this is because they lack a vowel). So post-lexical phonological rules must ensure that they are attached to a host. When this is done, they become pronounceable as the unstressed part of a bigger phonological domain which is referred to as the **clitic group** (see S. R. Anderson, 1988a; Kaisse, 1985; Klavans, 1985).

Unlike simple clitics, special clitics are not contracted forms of self-standing words. Rather, they are forms that can only occur as bound morphemes appended to hosts in certain syntactic contexts. The genitive *'s* in English is a good example of a special clitic. It never occurs without being attached to a host. (See [10.58] above and [10.64] below.)

Given the existence of two types of bound morphemes with a syntactic role – clitics and inflectional affixes – criteria are needed for distinguishing between them. Zwicky and Pullum, on whose work the following account is based, propose the six criteria below for separating inflecitonal affixes from clitics:

[10.60]   a.   Clitics can exhibit a low degree of selection with respect to their hosts, while affixes exhibit a high degree of selection with respect to their stems.

b.   Arbitrary gaps in the set of combinations are more characteristic of affixed words than of clitic groups.

c.   Morphological idiosyncrasies are more characteristic of affixed words than of clitic groups.

d.   Semantic idiosyncrasies are more characteristic of affixed words than of clitic groups.

e.   Syntactic rules can affect affixed words, but cannot affect clitic groups.

f.   Clitics can attach to material already containing clitics, but affixes cannot.

(from Zwicky and Pullum, 1983: 503–4)

Let us now consider each of these criteria in turn.

Criterion A means that affixes tend to be very selective in their choice of stems, while clitics are usually much less discriminating. As a rule, affixes only get attached to stems belonging to a specific word-class. For instance, in English *-est* attaches to adjectives, *-ing* attaches to verbs, plural *-s* attaches to nouns, and so on.

Some clitics also appear to be somewhat selective. For example, in writing normally only subject nouns and pronouns can be hosts to the contracted auxiliary like the *'s* enclitic form of *is* or *has* as in:

[10.61] Margaret's coming (= Margaret is coming)
  Jack's left (= Jack is/has left)
  He/she'd left (= He/she had left)

However, the contracted auxiliaries may also be cliticised (in speech) to prepositions, verbs, adjectives and adverbs as well:

[10.62] **a.** Preposition
  The house Marie was born *in's* (= in has) been demolished.
  b. Verb
  The jug she *sent's* (= sent is) lovely.
  c. Adjective
  Any minister that is *corrupt's* (= corrupt is) going to be sacked.
  d. Adverb
  All the drivers who are paid *weekly've* (= weekly have) been given a pay rise.

Normally, clitics attach indiscriminately to words in the relevant domain, without any regard to word-class (as seen in [10.62]) or meaning (see also [10.64] and [10.65] below).

Criterion B states that clitics are attached automatically across the board to any word in the relevant domain which has the necessary syntactic and phonological characteristics. But affixation is prone to having inexplicable gaps. Thus, any subject noun or pronoun in sentences similar to those in [10.58], and any preposition, verb, adjective or adverb in the appropriate position in [10.61] would allow the cliticisation of a contracted auxiliary. By contrast, in the combination of affixes with stems, there are usually many gaps, e.g. in English most count nouns have a plural form (usually with an *-s* suffix) which is distinct from the singular, but, for no apparent reason, a few nouns like *deer* and *sheep* lack overt number inflection.

Criterion C indicates that, whereas the combination of a particular affix with a stem may yield idiosyncratic morphophonological results, cliticisation works regularly, with no quirks. For instance, in inflectional morphology the combination of the plural morpheme with a noun stem may result in a regular plural, e.g. *roots*, *ounces*, *boxes*, or in a morphophonologically irregular one like *feet*, *mice* and *oxen*. Similarly, the combination of past tense with a verb stem may give a regular output as in *talked*, *mended* and *fished* or an idiosyncratic one as in *brought*, *put* and *shook*. Cliticisation, however, is not subject to morphophonological idiosyncrasies. All contracted auxiliaries, for instance, follow general rules, e.g. they are subject to the regular rule of voice assimilation (which we discussed in section 2.2)).

Criterion D indicates that the semantic contribution to a sentence made by a simple clitic in a clitic group is identical to the semantic contribution made by a corresponding full word to the meaning of an equivalent sentence. For instance, clitic groups containing contracted *-ve* will always be identical in meaning with constructions where the full form containing *have*. (*I've seen her* has the same meaning as *I have seen her*.)

By contrast, inflection sometimes displays idiosyncratic properties. Thus, whereas *best* is normally simply the superlative form of *good*, meaning 'surpassing all others in quality' (as in *the best man for the job*), in the sentence *He was the best man at their wedding* the superlative *best* has no such meaning. In this latter case the expression *best man* has an idiosyncratic meaning, namely the 'groomsman at a wedding'.

Criterion E. Syntactic operations may affect words such as inflected verbs, nouns, pronouns and adjectives as syntactic units. But no syntactic operation treats a host-plus-clitic (e.g. *I've, corrupt's, to's, you'd*) as a single unit. This is to be expected given the disparate nature of host-plus-clitic combinations (cf. [10.62]).

Criterion F. Whereas clitics can be appended to forms containing clitics, no inflectional morphemes can be attached to forms containing clitics. Hence, [10.63a] is allowed, but [10.63b] is not.

[10.63] **a.** *I'd've* brought some for you, if I'd known.
 b. **I'd've-ing* brought some for you, if I'd known.

We shall now turn to another important feature of special clitics, that has been highlighted by Klavans (1985: 104), which separates them from inflections. It is the tendency for them to have **'dual citizenship'**. By this Klavans means that the 'clitic is structurally a member of one constituent but phonologically a member of another'. We have already encountered it in the examples in [10.58] on p. 245 which are repeated below as [10.64] for convenience:

[10.64] **a.** Mary's car
 b. The Queen of Tonga's tiara
 c. The editor of the *Manchester Guardian* newspaper's car

In some cases, e.g. *Mary's car*, the word that the clitic *'s* leans on to phonologically also happens to be the word that it leans on syntactically and semantically. However, if you compare [10.64a] with [10.64b], you will see that, in the latter example, the syntactic and phonological hosts are not identical. The phonological host is *Tonga* but the syntactic/semantic host is the entire NP headed by *the Queen*. *Mutatis mutandi*, the same analysis

applies to [10.64a], where the NP headed by the *editor* is the syntactic/
semantic host and *newspaper* is the phonological host.

Clearly, the placement of the enclitic is done regardless of meaning.
Although it is the *Queen* and not *Tonga* that possesses the *tiara* and the
editor rather than the *newspaper* that owns the car, the *'s* is attached
neither to *Queen* nor to *editor*. The requirement is to attach *'s* to the last
word of the genitive noun phrase.

Now identify the phonological host and the syntactic host of *'s* in the
following Glasgow dialect examples:

[10.65]   a.   The boy I stayed with's granny
          b.   The man that he robbed's sister

These data underline the fact that the phonological and structural hosts of
a clitic may be distinct. In informal colloquial Glasgow English the *'s*
genitive marker may attach to almost any word that occurs at the end of the
construction modifying a preceding possessor noun phrase in a genitive
NP. Thus *'s* attaches to the preposition *with* in [10.65a] and to the verb
robbed in [10.65b], which function as the phonological hosts. Syntactically
(and semantically), however, it is the entire noun phrases that have *the boy*
and *the man* as their heads that function as the hosts.

The freedom of the *'s* genitive to be attached to the end of whatever
word precedes the last noun of a genitive NP shows that the *'s* genitive is a
case enclitic and not a case inflection suffix. It is involved in building a
phonological domain which we can refer to as a clitic group rather than a
simple word-form containing an inflected word. A true inflectional morpheme (like plural -*s*) becomes an integral part of the base to which it is
suffixed, phonologically, syntactically and semantically. By contrast, a
clitic is a semi-detached element of the word of which it is a part. So it is
capable being part of the word-form that hosts it for phonological purposes
and at the same time to belong either to a different word or to the phrase as
a whole for syntactic and semantic purposes. In other words, whereas
suffixes exemplify *lexical inflection* by introducing a morpheme that becomes fully integrated in a word, clitics involve what could, by analogy, be
loosely termed '*phrasal inflection*', for the morpheme they introduce is
integrated at phrase rather than word level.

Following Klavans (1985), we will assume that in order to account for
the realisation of special clitics, cliticisation rules must specify the
following:

[10.66]   (i)   the class of phrase to which the clitic is attached (e.g. NP,
                 VP).
          (ii)   the location of the host within the phrasal domain in which

the clitic is attached, e.g. is the clitic appended to the first N in NP (or V in VP), the last N in NP (or V in VP)?

(iii)   whether the clitic precedes (i.e. is a proclitic) or follows the host (i.e. is an enclitic). This applies to both simple and special clitics.

(iv)   whether the syntactic host and the phonological host of the clitic are the same or different.

In the case of genitive *'s*, as we have seen, the grammar specifies that:

[10.67]   (i)   the clitic attaches to NPs,

(ii)   the phonological host is the word before the last noun of the NP,

(iii)   the clitic is an enclitic,

(iv)   the phonological and syntactic host are not necessarily the same. The phonological host is the word before the last noun in the genitive NP while the syntactic host is the entire NP headed by the possessor noun.

To conclude, this chapter has dealt with an old chestnut, namely the nature of inflectional morphology and the criteria for distinguishing between inflection and derivation. First, I examined the criteria of obligatoriness, syntactic relevance and generality which have been proposed for distinguishing between inflection and derivation. I showed that these criteria are useful, and they may serve to characterise prototypical inflectional and derivational processes. But they do not always succeed in separating inflection from derivation, for, in reality, the boundary between inflection and derivation is often fuzzy. In the course of the discussion I emphasised the interaction between syntax and inflectional morphology without ignoring the differences between them.

In the second part of the chapter I surveyed a wide range of inflectional processes using S. R. Anderson's theory of inflectional morphology. I focused on the inflectional categories that are either inherently present in verbs and nouns or acquired by verbs and nouns when they occur in certain syntactic contexts.

In the final section I distinguished between two types of bound morphemes whose distribution in syntactically determined, namely clitics and inflectional affixes. I surveyed criteria for separating clitics from inflectional affixes and highlighted the problems involving clitics with dual citizenship, i.e. clitics which phonologically belong to one word and structurally belong to another unit.

EXERCISES

1. (a) Identify the root and inflectional morphemes in the data below from the Ugandan language Ateso and state how they are realised.

   (b) Comment on the problems these data pose for a theory of morphology that assumes that there is a direct, one-to-one relation between *morphemes* and the *morphs* that represent them. (Refer back to section (2.2.2).)

   (c) Use the data to show the role of blocking in inflectional morphology (cf. section (10.2.3)).

   (i) asuba   'I make myself'   enoma   'I beat myself'
        isuba   'you make yourself'   inoma   'you beat yourself'
        esuba   'he makes himself'   inoma   'he beats himself'
        kisuba   'we make ourselves'   kinoma   'we beat ourselves'
        isubas   'you make yourselves'   inomas   'you beat yourselves'
        esubas   'they make themselves'   inomas   'they beat themselves'

   (ii) alemar   'I take myself out'   etemokin   'I prepare myself'
        ilemar   'you take yourself out'   itemokin   'you prepare yourself'
        elemar   'he takes himself out'   itemokin   'he prepares himself'
        kilemar   'we take ourselves out'   kitemokin   'we prepare ourselves'
        ilemaros 'you take yourselves out'   itemokinos 'you prepare yourselves'
        elemaros 'they take themselves out' itemokinos 'they prepare themselves'

   (Ateso data from Hilders and Lawrence, 1957: 52–3)

   (d) Study the additional data below and determine whether the morphosyntactic property signalled by the verbal affix is inherent, configurational or assigned by agreement.

   (i)   Esubi   etelepat   emesa
         he-makes   boy   table
         'The boy makes the table'

   (ii) Inomi   etelepat   emoŋ
         he-beats   boy   bull
         'The boy beats the bull'

   (iii) Elemari   etelepat   emesa
         he-takes out   boy   table
         'The boy takes out the table'

   (iv) Itemokini   etelepat   etogo
         he-prepares   boy   house
         'The boy prepares the house'

2. (a) Which case is realised by each one of the italicised particles in the Japanese sentences below?

   (b) What are the main formal differences in the marking of grammatical case in Japanese, English (cf. [10.47]) and Latin (cf. [10.48, 10.40])? Be as exhaustive and as explicit as you can.

   (i) Taroo *ga*   zidoosya *de* Hanako *to*   Tookyoo *kara*
       Taroo NOM   car   by Hanako with Tokyo *from*
       Hiroshima   *made* ryokoosita.
       Hiroshima   up-to   travelled.
       'Taro travelled with Hanako by car from Tokyo to
   Hiroshima.'

   (ii) Taroo *no*   otoosan *ga*   Amerika *e*   itta
        Taroo's GEN   father NOM   America to   went
        'Taroo's father went to America.'

   (data from Kuno, 1978: 79)

3. (a) List all the morphemes in the Luganda verbs below.
   (b) What syntactic factors account for the alternation between -*a* and -*e* occurring as the final vowel of the verb?

   (i) balaba   'they see'   musoma   'you (pl.) read'
       tugenda   'we go'   bagenda   'they go'
       tugamba   'we tell'   tubagamba   'we tell them'
       mugamba   'you (pl.) tell' tubalaba   'we see them'

   (ii) musome   'you (pl.) read!' mugende mulabe   'go (pl.) and
                                                      see (pl.)!'
        balabe   'let them see!' tubagamba balabe 'we   tell
                                                 them to see'
        bagende   'let them go!'   bagambe bagende basome 'tell
                                                 them to go and read'
        mugambe   'you (pl.) tell!' tugende tubagambe basome 'let
                                                 us go and tell them to read'

4. (a) Analyse the examples below and identify the case inflections for subject and objects.
   (b) Identify subject–verb agreement markers, where appropriate.
   (c) In the light of your analysis, determine which ones of these languages are nominative-accusative and which ones are ergative-absolutive.

   (i) Iatmul (Papua New Guinea) (data from Foley, 1986: 102–3)

       *nti*w   y*i*-nt*i*
       man   go-3SG MASC
       'the man went'

takwə   kɨya-lɨ
woman   die-3SG FEM
'the woman died'

ntɨw   takwə   vɨ-ntɨ
man   woman   see-3SG MASC
'the man saw the woman'

takwə   ntɨw   vɨ-lɨ
woman   man   see-3SG FEM
'the woman saw the man'

(ii) Warrgamay (Australia) (based on Dixon, 1980: 287)

ŋulmburu   gaga+ma
woman   go+FUT
'The woman will go'

maal   gaga+ma
man   go+FUT
'The man will go'

ŋulmburuŋgu   maal   ŋunda+lma
woman   man   see+FUT
'The woman will see the man'

maaldu   ŋulmburu   ŋunda+lma
man   woman   see+FUT
'The man will see the woman'

(Note: the choice of *-lma* or *-ma* future inflection depends
  on the conjugation of the verb.)

(iii) Barai (Papua New Guinea) (data from Foley, 1986: 103–4;
citing Olson, 1981):

e   ije   ruo
man the   come
'the man came'

e   ije   barone
man the   die
'the man died'

e   ije   ame   kan-ia
man the   child hit-3PL
'the man hit some children'
* 'the men hit a child'

> *e   ije   ame   kan-i*
> man   the   child hit-3SG
> 'the man hit a child'

(iv) Yidiɲ (Australia) (based on Dixon, 1977)

> *buɲa:ŋ   wagu:ḍa   wawa:l*
> woman   man   see-PAST
> 'the woman saw the man'

> *wagu:ḍa   buɲa:y   galiŋ*
> man   woman-with   go-PRES
> 'the man is going with the woman'

> *wagu:ḍaŋgu buɲa   gali:ŋal*
> man   woman   go-with-PRES
> 'The man is taking the woman'

> *ŋayu   buɲa   giba:l*
> I   woman   scratch-PAST
> 'I scratched the woman'

> *wagu:ḍaŋgu   buɲa   giba:l*
> man   woman   scratch-PAST
> 'The man scratched the woman'

> *wagu:ḍaŋgu   guda:ga   wawa:l*
> man   dog   see-PAST
> 'The man saw the dog'

> *wagu:ḍa   dungu   buɲa:ŋ   ḍina: baṛa:l*
> man   head   woman   with-foot   strike-PAST
> 'The woman kicked the man's head'

5. Discussion this question: Morphology can be separated in a straightforward way from syntax: morphology deals with word-structure and
   syntax with sentence structure. Do you agree?

11 Morphological Mapping of Grammatical Functions

11.1 INTRODUCTION

In the last chapter we saw that a distinction between morphology and syntax is both legitimate and useful. But this is not to say that there is no interaction between the two. Many morphological processes are syntactically motivated. Such processes fall in the province of inflection. A survey of inflectional phenomena was presented and we examined ways in which morphological theory throws lights on them. The present chapter will take the investigations further by addressing this central question: how are grammatical functions mapped onto morphological representations?

I will begin with a preliminary discussion intended to provide the necessary background for understanding the key syntactic and semantic concepts that are involved. First, I will introduce the notions of thematic (semantic) roles, grammatical relations and the theory of case assignment. (All these terms are explained below.) Next I will explore the morphological effects of syntactic rules that change the canonical pairing of thematic roles with grammatical functions. This is crucially important because these pairings affect the way in which grammatical functions are ultimately represented on the surface by morphology. The final part of the chapter will take the discussion further afield through an investigation of the phenomenon of **incorporation** whereby the syntax requires the inclusion of one word within another. We will see how this method of word-formation raises in an acute manner the issue of the nature of the difference between syntax and morphology.

11.2 PREDICATES, ARGUMENTS AND LEXICAL ENTRIES

Normally, sentences are constructed in such a way that some constituents identify particular individuals or things (or more abstract entities like ideas) and other constituents say things about those individuals or entities. Constituents which indicate individuals or entities are called **referring expressions**, while those which attribute to them properties, processes, actions, relations or states are called **predicates**. Look at the examples in [11.1]:

[11.1] **a.** Buffy growled.
       b. Mary will come.
       c. The stick broke.

Sentence [11.1a] asserts that there is an individual named *Buffy* who *growled*. The sentence **refers** to an individual and **predicates** (or assigns) the property growling of that individual at some time in the past. In sentence [11.1b] on the other hand, the property of *coming* is predicated of the individual *Mary* (at some time in the future). Finally, in [11.1c] the property of *breaking* is attributed to an entity, *the stick* (at some time in the past). Predicates take referring expressions as their **arguments**. An argument, in this sense, is an individual, entity or item (e.g. *Buffy, Mary, the stick* etc.) about which a predicate says something.

11.3 THETA-ROLES AND LEXICAL ENTRIES

Languages use syntax and inflectional morphology to encode some of the semantic relations which obtain in a sentence between a predicate and its arguments. However, before we are in a position to see how the encoding is done, we need to have a clearer idea of the semantic relations that are encoded. We will use the term **theta-roles** (θ-roles) for these semantic relations. (They are also called (abstract) **case relations** or **thematic relations** in the literature.)

Recognition of θ-roles is essentially based on the intuition which is widely shared among linguists that there is a relatively small number of syntactically relevant semantic properties that play a role in the **transitivity systems** of languages. Traditionally, the term transitivity is used to refer to an action initiated by an actor carrying over and affecting another individual, the patient (cf. Lyons, 1968: 350–71).

Below I present definitions of the θ-roles that we will be using which are based on Gruber (1965, 1976) and Fillmore (1968).

[11.2]   **Agent** is the case of the individual (usually animate) that instigates the action identified by the verb.
        (e.g., *Peter* kicked the ball.)
        Instrumental is the case of the inanimate instrument used to bring about the state of affairs described by the verb
        (e.g., *She wrote with a pencil*.)
        Patient is the case of the entity or individual that undergoes the process or action described by the verb.
        (e.g., *Peter kicked the ball*.)

Benefactive (or **goal** or **recipient**) is the case of the individual who gains from the action or process described by the verb. (e.g., *Mary gave the officer a bribe*.)
('Gain' is interpreted liberally to include situations like *The court gave the officer a ten-year gaol sentence* (for accepting a bribe).)
Theme is semantically the most neutral case. It is very intimately related to the meaning of the verb but it is not easy to be precise about its exact meaning. With verbs of motion, for example, the entity that moves is the theme. Thus, *the plank* is the theme in *The plank floated away*. Likewise, with verbs of giving the thing given is the theme. So, in *Andrew offered Helen the car* the NP *car* is the theme. With verbs describing states, the entity that experiences the state is the theme. So, *Mary* is the theme in *Mary rested*, and so on.
Locative is the case that indicates the location, direction or spatial orientation of the event, state or action identified by the verb (see [10.46] section (10.4.3)).

Explain why the two sentences in [11.3] differ in meaning although they contain exactly the same words.

[11.3]   a.   The children tickled the clown.
       b.   The clown tickled the children.

The difference in meaning is due to the fact that the two NPs that occur in both sentences are associated with different θ-roles in each instance. In [11.3a], the NP *the children* refers to the **agent** that does the action and *the clown* is the **patient** that undergoes the action indicated by the verb *tickled*. In [11.3b], the θ-roles associated with the two NPs are reversed so that now the NP *the clown* is the agent and the NP *the children* is the patient.

Evidently, θ-roles are not an inherent aspect of the meaning of individual words. Sentences [11.3a] and [11.3b] contain exactly the same words but have different meanings. The difference in meaning is directly attributable to the different θ-roles that the NPs *the children* and *the clown* are assigned.

Theta-roles are essentially used to characterise **transitivity**. They specify the parts played by the arguments representing different participants in the action, state or process indicated by the verb. Verbs (the most typical predicates) can occur in frames where there are one, two or three arguments. Accordingly, they are classified on the basis of the type and number of arguments with which they can occur.

Intransitive verbs are **one-place predicates**. They occur in frames with one argument, such as:

[11.4]   Mary smiled.
        agent

Transitive verbs are **two-place predicates**. They occur in frames with two arguments, such as:

[11.5]   Mary kicked   the ball.
        agent   patient

Ditransitive verbs are **three-place predicates**. There are several different types of ditransitive verbs which can be distinguished on the basis of their θ-roles.

    (i)   Verbs of *giving* (e.g. *give*, *send*, *lend*, *post*, etc.) have an agent, a goal and a theme:

         [11.6]   Andrew   sent   Helen   flowers.
                 agent   goal   theme

   (ii)   Verbs of *placing* (e.g. *place*, *put*, *position*, *deposit*, etc.) require an agent, a theme and a locative:

         [11.7]   Dad   put   the cake   in the oven.
                 agent   theme   locative

  (iii)   *Instrumental* verbs require an agent, a patient and an instrument:

         [11.8]   The woman   cut   the tree   with a saw.
                 agent   patient   instrument

                 (Notice that in [11.8] the instrument is optional. We can simply say *The woman cut the tree*. Only the agent and patient are obligatory θ-roles with the verb *cut*.)

In order to ensure that a verb appears in the right syntactic frames, the **lexicon** must specify the θ-roles which it requires. The discussion in this section will therefore focus on the nature of **lexical entries** for verbs.

    The lexical entries for the verbs in [11.4]–[11.8] must contain the following information:

[11.9]   SMILE   V (agent)
           KICK   V (agent   patient)
           SEND   V (agent   goal   theme)
           CUT   V (agent   patient   instrument (optional))

There is still one important question that we have not addressed. How are θ-roles encoded by the rules of grammar? A possible answer is that θ-roles are associated directly with NPs by phrase structure rules, as shown in [11.10]:

[11.10] **a.** S →   NP   VP
                         \<agent\>

         b. VP →   V   NP
                                          \<patient\>

         c. NP →   Det N

         d. N →   Nsg, Npl

         e. Det   → the

         f. V →   Vtrns (i.e. transitive verb)

Given the PS rules in [11.10], draw a tree diagram to represent *The clown tickled the children*.

The tree should look like this:

[11.11]

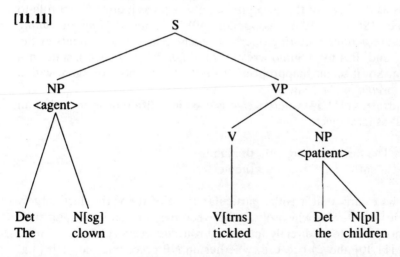

Unfortunately, the idea of θ-roles being directly encoded by syntactic rules does not work. The sentence *The clown tickled the children* has a (nearly) identical syntactic structure to that of:

[11.12]   The children received a tickle.

and so would be assigned the same phrase marker as [11.11] by the PS rules in [11.10].

But this is wrong. It is inappropriate to call *the children* the agent and *a tickle* the patient of *received* in [11.12]. Rather, *the children* is patient and *a tickle* is **theme** (the thing received). The **agent** (doer of the action of giving) is left unspecified. This shows that the NP that goes before the verb is not necessarily the agent and that the NP following the verb is not always the patient.

The sentence in [11.12] is by no means unusual. There are many other English sentences, such as the following, which do not fit the *agent – action verb – patient* pattern generated by the PS rules in [11.10]:

[11.13]　**a.**　That Sumo wrestler weighs 353 pounds.
　　　　　　b.　This curtain washes well.
　　　　　　c.　The door shut.
　　　　　　d.　Roses smell nice.
　　　　　　e.　Janet is a linguist.
　　　　　　f.　Their school is near the bridge.

The examples in [11.13] show that one major determinant of the θ-roles which hold between a verb and its arguments is the meaning of the verb. The verb directly assigns θ-roles to all its arguments. The main verb, which functions as the head of the verb phrase, also assigns indirectly a θ-role to its subject (Stowell, 1981: Koopman, 1984). Thus *tickling* or *hitting someone* are actions requiring doers, as agents, as well as patients at the receiving end. But for a Sumo wrestler to *weigh 353 pounds* is a state, not an action. So, it seems inappropriate to talk about a doer or agent with a verb like *weigh* in [11.13a].

By contrast, in [11.14] where *weigh* is used in a different sense, it has an agent NP as its subject:

[11.14]　The farmer　weighed　the lambs
　　　　　　agent　　　　　　　　　　theme

This shows clearly that it is the particular sense of the verb which largely determines the θ-roles which obtain between the verb and its arguments. We cannot, therefore, directly indicate θ-roles using PS rules (as we tried to do in [11.10] above) because, whether an NP preceding the verb (e.g. *weigh*) is an agent or not depends on the sense of the verb and PS rules are not sensitive to this.

Instead, we will require each lexical entry for a verb to include the θ-roles which that verb assigns to its arguments. To this end, a well-formedness principle called the **Theta-Criterion** will be incorporated in the grammar and given the task of ensuring that:

(i) a verb is only used in frames where the requisite arguments are present, and

(ii) those arguments all have the prescribed θ-roles.

In this way, we can represent the fact that θ-roles change from verb to verb, or rather from verb sense to verb sense. They are not inherently linked with NPs.

What needs to be captured is the fact that arguments are independent variables. Various NP's can fit in the slots for arguments of a given predicate. So, the symbols X? and Y? are used as variables to represent any entity or individual (e.g. *the clown*, *the children*, *a tickle*, and so on) that can function as arguments of these predicates with the θ-roles of agent, patient, theme etc.

The entry for a verb in the lexicon will include a **subcategorisation template** showing its argument structure requirements. An argument is marked with an asterisk (*). Particular predicates manifesting particular verb senses are preceded by #, written in capitals, and numbered. V stands for verb. All the predicates in these examples are verbs. Some illustrative examples of lexical entries are shown below:

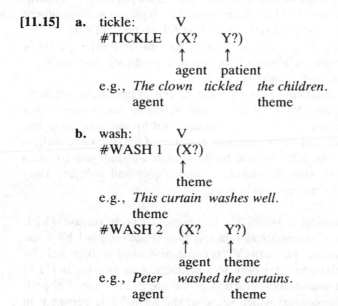

[11.15]　a.　tickle:　　V
　　　　　　　#TICKLE　(X?　　Y?)
　　　　　　　　　　　　　↑　　　↑
　　　　　　　　　　　　agent　patient
　　　　　e.g., *The clown　tickled　the children.*
　　　　　　　　agent　　　　　　　theme

　　　　b.　wash:　　V
　　　　　　　#WASH 1　(X?)
　　　　　　　　　　　　　↑
　　　　　　　　　　　theme
　　　　　e.g., *This curtain　washes well.*
　　　　　　　theme
　　　　　　　#WASH 2　(X?　　Y?)
　　　　　　　　　　　　　↑　　　↑
　　　　　　　　　　　agent　theme
　　　　　e.g., *Peter　washed the curtains.*
　　　　　　　agent　　　　　　theme

Even after the refinement we have introduced, there is still an unsolved problem. The theory does not show how θ-roles get associated with the NP representing the right argument in a particular sentence. For instance, the lexical entries in [11.15] tell us which senses of *wash* and *tickle* require two arguments, and which particular θ-roles hold between those arguments in a particular sense of the verb. But they do not tell us how to work out, for example, that *the curtains*, rather than *Peter*, is the theme in the second sense of *wash*. So, our grammar could generate **the curtains washed Peter*, which is very bizarre in the world as we know it.

To solve this problem we need to add a further dimension to the model of grammatical analysis, namely that of **grammatical relations**.

11.4 GRAMMATICAL RELATIONS

Before we are in a position to discuss grammatical relations, we will need to clarify the distinction between **syntactic categories**, **grammatical relations** (also called **grammatical functions** or **syntactic functions**) and θ-**roles**.

[11.16] (i) **Syntactic categories** like noun phrase (NP) and verb phrase (VP) specify the syntactic type of particular constituents. The syntactic type of a constituent is determined by the category of the **head** of that constituent. A noun phrase is a constituent whose head is a noun while a verb phrase is a constituent whose head is a verb, and so on (cf. section (12.4.4)).

(ii) θ-**roles** (e.g., agent, patient or instrument), specify a semantic relationship between a predicate and its arguments (see section (11.3)).

(iii) **Grammatical relations** indicate the grammatical relationship that holds between two syntactic constituents in a sentence. They are determined, not by semantic considerations, but by the *syntactic* position of a particular constituent. The grammatical relations that we shall use are verb phrase, subject, object, second object and oblique. They are defined in turn below.

The easiest grammatical relation to recognise is **verb phrase** (VP). Unfortunately, it is a grammatical relation that is surrounded by some terminological confusion. The term VP is commonly used ambiguously by generative grammarians to refer both to a syntactic category (as in [11.6] (i) above), and to a grammatical relation as we are doing here. I hope it will be clear from the context which sense of the term VP is pertinent in what follows. (To avoid this ambiguity, some writers reserve the term VP for the syntactic category and use the term **predicate** to describe the grammatical relation that constituents with the syntactic category VP have in a sentence.)

The grammatical relation VP (i.e. predicate) has a verb as its syntactic head. In the examples below, the VP contains only the verb *came* in

[11.17a]. But, in [11.17b], the VP contains the phrase *chased Esmeralda* which is headed by the verb *chased*:

[11.17]   a.   Fay   came
                    S   VP

          b.   The children   chased   Esmeralda
                    S   Obj
                                 ⌣‿‿‿‿‿‿‿‿‿‿‿‿‿⌣
                                      VP

All declarative sentences in English must have a subject (S). In [11.17] *Fay* and *the children* are the subjects of their respective sentences. The archetypal **subject** can be identified using these rules of thumb:

[11.18]   a.   the subject is the topic about which the rest of the sentence says something;
          b.   the subject is the NP that has the θ-role of agent, if that role is present;
          c.   the subject is the NP that precedes the VP, and with which the verb agrees in number.

In reality, however, many subjects do not have all these properties, as we shall soon see.

The **object** (Obj) is the NP that immediately follows the verb. So *Esmeralda* is the object in [11.17b].

In [11.19] there is more than one NP following the verb. The NP that immediately follows the verb is the *object* and the NP that comes after that (first) object NP is called the **second object** (2nd Obj). This is very roughly equivalent to what is called the indirect object in traditional grammar):

[11.19]   The children   gave   Esmeralda   hay
          S   Obj   2nd Obj
                         ⌣‿‿‿‿‿‿‿‿‿‿‿‿‿‿‿‿‿‿‿‿‿‿‿⌣
                                      VP

Finally, any argument of the verb that is realised by a prepositional phrase is an **oblique** NP (Obl):

[11.20]   David   put   the baby   in the cot
          S   Obj   Obj
                         ⌣‿‿‿‿‿‿‿‿‿‿‿‿‿‿‿‿‿‿‿‿‿‿‿⌣
                                      VP

What is the job of **grammatical relations**? Although the precise nature of grammatical relations is surrounded by a degree of theoretical controversy, there is widespread agreement about the purpose which they serve. The mainstream view is that grammatical functions mediate between (semantic) theta-roles that hold between predicates and their arguments on the one hand, and surface manifestations of the grammatical relationships between phrases representing those predicates and arguments on the other (Bresnan, 1982b; Perlmutter, 1983; Marantz, 1984).

Once the need for grammatical relations is recognised, grammars must perform the two tasks:

(i)   They must state how θ-roles are mapped on to grammatical functions in the grammar of a particular language. For example, English mapping principles may take this form:

[11.21]	θ-role	corresponds to	grammatical function
	agent		subject
	patient		object of verb
	locative		oblique NP

(ii) They must state how grammatical functions are marked on the surface, e.g. by word order, prepositions or case inflection. For instance, in English grammatical relations are marked by word order and prepositions (see [11.3]). Many languages, e.g. Latin and Ngiyambaa, use case inflection to mark grammatical relations (see section (10.4.3)).

What we are about to see is that often the mapping of grammatical relations is far from straightforward. This is because there are various syntactic rules which may mask the grammatical function of a particular NP. Much of the morphological complexity found in languages arises from the marking of such masked grammatical functions.

11.5 GRAMMATICAL FUNCTION CHANGING RULES

In the discussion of lexical morphology, we emphasised the point that the lexicon is not simply a long, unstructured list of words. It is not the case that each word in a language is unique. Rather, many phonological, morphological and semantic properties are shared by numerous words.

Following Levin (1985), I will show that words that have similar meanings also tend to have some syntactic properties in common. This is evident in the way that rules that map semantic notions onto syntactic constructions operate. Levin illustrates this using verbs such as *slide*, *roll* and *bounce* which indicate change of position. These verbs show very similar grammatical and semantic properties.

[11.22] **a.** Causative

Fiona bounced   the ball   (across the room)   to Barbara.
agent   theme   (path)   goal

    b. Non-causative

The ball bounced   (across the room)   to Barbara
theme   (path)   goal
(*Path* is a type of locative that indicates the way or course that an entity travels along, through, over, across, etc.)

Both sentences in [11.22] involve motion along a path (which may be optionally mentioned). But they differ in that *bounce* is used as a causative verb in [11.22a], where the instigator of the motion is indicated, while in [11.22b] it is not a causative verb. Whatever makes the ball bounce is not mentioned.

    This is not a peculiarity of the verb *bounce*. Other change of position verbs too can occur in sentences where the agent is mentioned or not mentioned. The arguments of such verbs are assigned the same θ-roles, following the same general principles as those in used in [12.22]. The assigning of θ-roles to verbs with similar meanings is not done ad hoc for each verb but rather follows general principles.

Re-write the following sentences, omitting the agent who initiates the motion. Label the θ-roles:

[11.23] **a.** Mrs Goode sank the knife (through the coat) into his chest.
        b. He dropped the stone (through the water) onto her toe.

Sink and *drop* are motion verbs. They behave in the same way as *bounce*. Without hesitation you will have written down the corresponding sentences in [11.24]:

[11.24] **a.** The knife sank   (through the coat)   into his chest.
             theme   (path)   goal
        b. The stone dropped   (through the water)   onto her toe.
             theme   (path)   goal

A comparison of [11.23] with [11.24] reveals that the grammatical relations that NP arguments have with their predicate are variable. Reasons for the variation may have to do with discourse factors such as wanting to highlight (or play down) the role of a given participant in an event or action (like murdering someone in Mrs Goode's case).

Many verbs can be used transitively, in which case they have an object, or intransitively, in which case they have no object. Above we have seen that a change of position verb like *bounce* can be used transitively as in [11.22a] or intransitively as in [11.22b]. When it is used transitively, it is always causative (there is some individual making some other individual or entity do something). But when used noncausatively, it is intransitive (some entity does something but we are not told what brings this about). All causative verbs whose meaning involves a change of position assign to their arguments the θ-roles of goal, theme, and agent. The corresponding noncausatives only assign the θ-role of theme, and optionally goal (cf. [11.24]). This is a general regularity shown by change of position verbs. It should expressed by a rule. It need not be listed as a special fact about each transitive change of position verb.

What is true of change of position verbs is true of other classes of verbs with similar meanings. They too tend to take the same θ-roles.

What are the θ-roles taken by verbs of placing such as *put*, *place*, *sit*, *stand*, *insert*, *position*, *deposit*, *set*, *rest*, etc.) in:

Fiona *deposited/put/stood/placed*, etc. the book on the table.

The answer is that verbs of placing require an agent, a theme and a location as their arguments, as seen in [11.25]:

[11.25]   Fiona   *deposited/put/stood/placed*, etc.   the book   on the table
         agent   theme   locative

This discussion has introduced us to the notion of changes in the **valency** of verbs, i.e. changes in the number of arguments in the syntactic frame in which the verb occurs which are brought about by grammatical function changing rules. For instance, a transitive verb may occur with an agent, a theme and a goal as in [11.22a], or with just a theme and a goal as in [11.22b]. Usually, changes in the valency of a verb are signalled morphologically.

Grammatical functions are hierarchically ordered across languages. The hierarchy depends on the relative likelihood of NPs associated with particular grammatical functions being affected by certain syntactic rules (e.g. being selected as subject of a sentence or being available to rules that form relative clauses). Keenan and Comrie (1977, 1979) have established this hierarchy:

[11.26]   subject > direct object > non-direct object > possessor

Languages have rules (like causativisation) that change the grammatical functions of NPs by moving them up or down the hierarchy in [10.26]. Such rules mask the relationship between the surface manifestation of **grammatical function** (GF), which is often marked by case or word order, and the semantic role (or θ-role) of an argument.

GF-changing rules tend to have significant morphological repercussions which typically affect verbs more than other word-classes. Now we are in a position to begin examining some of these repercussions.

11.5.1 Passive

Often sentences expressing the same **proposition** (i.e. a statement that may be true or false) can be realised in a variety of ways, depending on how grammatical relations are encoded using the syntax and morphology. Normally, where such choice exists, one way of expressing a proposition is **marked** (i.e. non-neutral) and the other is **unmarked** (i.e. neutral).

[11.27] **a.** **Active voice**

Agent/Subject nominative		*Patient/Object* accusative
The vet	examined	Esmeralda.
She	examined	her.

  b. **Passive voice**

Patient/Subject nominative		*Agent* oblique NP
Esmeralda	was examined	by the vet.
She	was examined	by her.

The sentence in [11.27a], with the subject as agent preceding the verb and the object, who is the patient, following the verb, are unmarked. The morphological case marking also follows the unmarked pattern. The agent, who is also the subject, receives **nominative case** and the patient, who is object receives **accusative case**.

In the passives in [11.27b], the arguments are re-ordered. The patient functions as the subject and the agent is placed after the verb and is put in an oblique by-NP phrase. This change of grammatical function (GF) is marked morphologically in the verb, which now appears in its past participial *V-en / V-ed* form and follows a form of the auxiliary verb *be*. Morphological case marking in nouns is virtually non-existent in English. So, not surprisingly, nouns are not given special inflected forms in passive sentences. However, if pronouns are present, they are duly inflected. The nominative forms (*she*, *he* etc.) are selected for the subject and the objective case forms (*her*, *him* etc.) are used in the oblique by-NP phrase.

Passive is a probably the most familiar GF-changing rule (cf. Perlmutter and Postal, 1977; Bresnan, 1982a; and Baker, 1988). It can be semi-formally stated as in [11.28]:

[11.28]   **a.**   subject → oblique (or null)
        b.   object → subject

We have already seen English examples of passivisation in [11.27]. The active and passive sentences in the two pairs differ somewhat in their emphasis and are therefore likely to be used in different discourse situations. Nevertheless, they are semantically similar. *The vet examined Esmeralda* and *Esmeralda was examined by the vet* express the same proposition. In both cases the action of *examining* is predicated to the arguments *the vet* and *Esmeralda*, with *the vet* as the agent and *Esmeralda* as the patient.

As [11.28b] indicates, the agent may be left out in a passive. We can say:

[11.29]   Esmeralda was examined.

Such a construction is called an **agentless passive**.

Study the Welsh data below.

[11.30]   **a.**   Rhybuddiodd   y dyn   y merched.
              warned   the man   the girls
              'The man warned the girls.'

        b.   Rhybuddiwyd   y merched   gan y dyn
              warned   the girls   by   the man
              'The girls were warned by the man.'

a.   What is the order of the subject, object and verb in active voice sentences in Welsh?
b.   How is the passive formed?

In Welsh the order of the major constituents in active voice sentences is

Verb-Subject-Object. In the passive, the subject and object exchange places. Furthermore, the subject is realised as an oblique NP governed by the preposition *gan* 'by'. Morphologically, the verb suffix changes from *-iodd* to *-iwyd*:

[11.31]   Active: *V-iodd*   Subject NP   Object NP
          Passive: *V-iwyd*   Object NP   *gan* Subject NP

This is typical. Rules that disrupt the canonical match between GFs and θ-roles have very important morphological consequences.

11.5.2 Antipassive

The **antipassive** is the process used in ergative languages to turn a transitive verb into an intransitive verb. It causes the object NP to be realised as an oblique NP, or to be deleted. The antipassive is quite widespread. It is found Basque; it occurs in some native American languages like Eskimo; and it is very common in Australian languages (cf. (10.4.3).

In ergative languages, the unmarked argument of the transitive verb is the object. The object is the only argument of a transitive verb that is always obligatorily present. It is in the absolutive case, like the subject of the intransitive verb (see section 10.4.3)). Furthermore, the absolutive case is usually morphologically unmarked in a very literal sense: it is signalled by zero inflection. The antipassive rule takes the object noun in the absolutive case and makes it an oblique NP (or deletes it).

The effect of the antipassive is comparable to that of the passive. Just as the passive demotes the original subject to an oblique NP in a nominative-accusative language (see [11.28] above), the antipassive demotes the original object of a transitive sentence to an oblique NP and the underlying agent NP argument which should otherwise be in the ergative is put in the absolutive.

This can be seen in the example below from Dyirbal (which is borrowed from Dixon (1980: 444,446):

[11.32]   a.   bala   yugu   baŋgul yara-ŋgu gunba-n baŋgu   barri-ŋgu
                 it-ABS tree-ABS he-ERG man-ERG cut-PAST it-INST   axe-INST
                 'the man cut the tree with an axe.'

          b.   *Antipassive*
                 *bayi   yara   gunba-l-*ŋa-nyu   bagu   yugu-gu baŋgu
                 he-ABS man-ABS cut-ANTIPASS-PAST   it-DAT tree-DAT it-INST

                 barri-ŋgu
                 axe-INST
                 'The man was cutting the tree with an axe.'

Notes: ABS = absolute; ANTIPASS = antipassive; ERG = ergative; INST = instrumental; DAT = dative (roughly equivalent to the benefactive case – see [11.2]).

As Dixon (1980: 447) points out, the antipassive has no effect on the peripheral constituents. Hence, 'the axe', which has the θ-role of instrument, remains unchanged when an antipassive sentence is formed in [11.32b]. What the antipassive changes are the GFs of the agent and patient NPs which are centrally involved in transitivity. In [11.32a], the ergative inflection of 'the man', the agent NP, is replaced by absolutive inflection. The 'tree', the patient NP, which is in the absolutive in [11.32a] becomes an oblique NP in the dative case in [11.32b].

Study the following examples and comment on the effect of the antipassive in Greenlandic Eskimo (data from Woodbury, 1977: 322–3):

[11.33] **a.** Miirqa-t   paar   -ai
                 child-(ABS)pl take care of   -IND:3sg,3pl (i.e. 3sg <u>Subj</u>, 3pl
                                                                         <u>Obj</u>)

                 'She takes care of the children.'

           b. miirqu-nik paara-ši   -vuq
                 -INSTpl   -APAS-IND:3sg
                 'She takes care of the children.'

Notes: APAS = antipassive, 3sg = 3rd person singular, INST = instrumental, pl = plural, ABS = absolute, IND = indicative.

The situation in Greenlandic Eskimo is similar to that in Dyirbal. The object, 'children', which is in the absolute case in the unmarked sentence in [11.33a] is realised in [11.33b] as an oblique NP in the instrumental case. The verb receives the suffix *-i-* which renders it intransitive so that it can occur without an object. Notice also that the verb agrees with the object if there is one.

11.5.3 Applicative

The **applicative** (or applied) is another common GF-changing rule with significant morphological consequences. Baker (1987: 9) characterises it using this schema:

[11.34]

$$\left.\begin{array}{c} \text{oblique} \\ \text{indirect object} \\ \text{null} \end{array}\right\} \rightarrow \text{object; object} \rightarrow \text{'2nd object'}$$

(or oblique)

The applicative covers a wide range of GF-changing processes. The most important of these are listed below.

(i) **Benefactive**: a NP in the benefactive case that has the GF of second object (roughly 'indirect object') can be realised as a direct object when the applicative rule applies. This can be seen in [11.35]:

[11.35] a. Andrew gave   the flowers   to Helen
            agent   theme   benefactive   (θ-roles)
            Subject   Object   Oblique NP   (GFs)
         b. Andrew gave   Helen   the flowers
            agent   benefactive   theme   (θ-roles)
            Subject   Object   2nd Object   (GFs)

The usual name for this process in English is **dative shift** or **indirect object movement**. Dative shift has no inflectional ramifications in English verbal morphology.

Describe the differences between the two Chamorro sentences in [11.36] (data from Baker, 1987).

[11.36] a. Hu tugi'   i   kätta   pära   i   che'lu-hu
            1sS-write   the   letter   to   the   sibling-my
            'I wrote the letter to my brother.'
         b. Hu tugi' -i   i   che'lu-hu   ni   kätta
            1sS-write-APPL   the   sibling-my   OBL   letter
            'I wrote my brother a letter.'

Note: 1sS = 1st person singular subject.

In Chamorro, as in English, the unmarked structure is one where the benefactive appears as an oblique NP (preceded by the preposition *pära* 'to' in second object position). The difference is that, unlike its English equivalent, the Chamorro dative shift rule triggers morphological marking in the verb (*-i*).

(ii) **Locative**: in many languages the applicative can be used with a locative meaning which is expressed in English using prepositions like *in*, *on*, *at*, and so on.

Contrast the following sentences from Kinyarwanda. The sentence in [11.37a] is basic while the one in [11.37b] contains the applicative used with a locative meaning (data from Kimenyi, 1980; Baker, 1987).

[11.37]   **a.**   Umwaana   y-a-taa-ye   igitabo   mu maazi.
child   SP-past-throw-ASP   book   in water
'The child has thrown the book into the water.'

b.   Umwaana   y-a-taa-ye-mo   amaazi   igitabo
child   SP-past-throw-ASP-APPL (in)   water   book
'The child has thrown the book into the water.'

Notes: SP = subject pronoun, APPL = applicative, ASP = aspect (perfective).

Describe in detail the changes caused by the applicative in Kinyarwanda. (Ignore the alternation between *amaazi* and *maazi* which here depends on whether or not a noun appears after the preposition *mu* 'in'.)

As Kimenyi and Baker both point out, the applicative causes a considerable amount of syntactic and morphological marking. In [11.37a], the object *igitabo* 'book' appears immediately after the verb and the locative NP *amaazi* 'water' following it is governed by the preposition *mu* and hence is an oblique NP. But in [11.37b] *amaazi* appears in object position immediately after the verb and the original object, *igitabo*, is now a second object. In addition to these syntactic changes, there is a morphological change: the verb *y-a-taa-ye* is suffixed with the applicative suffix *-mo*.

(iii)   **Possessor raising** is the final GF-changing rule involving the applicative that we will consider. When possessor raising takes place, an NP which functions as the 'possessor' modifying the head of a possessive noun phrase is turned into the object of the verb. The original object is shunted into a new slot and becomes the second object. (Refer back to pp. 240–1 in section (10.4.3) and [11.26] above.)

Possessor raising is shown schematically in [11.38]:

[11.38]   **a.**   NP$_{\text{possessor in possessive NP}}$ → object of Verb
b.   object of verb → 2nd object

Now study the following Luganda data.
a.   Describe the way in which possessor raising is signalled.
b.   What is the semantic restrictions on possessor raising?

[11.39] **a.** a-li-menya   okugulu kw-a Kapere
            S/he-fut-break   leg   of   Kapere
            'S/he will break Kapere's leg.'

         b. A-li-menya   Kapere   okugulu
            S/he-fut-break   Kapere   leg
            'S/he will break Kapere's leg.'

[11.40] **a.** a-li-menya   omuggo   gw-a Kapere
            S/he-fut-break   stick   of   Kapere
            'S/he will break Kapere's stick.'

         b. *a-li-menya   Kapere   omuti
            S/he-fut-break   Kapere   tree
            'S/he will break Kapere's tree.'

[11.41] **a.** a-li-mu-menya   okugulu
            S/he-fut-him-break   leg
            'S/he will break his leg.'

         b. *a-li-mu-menya   omuti
            S/he-fut-him-break   tree
            'S/he will break his tree.'

The sentence in [11.39a] illustrates what is generally known as the **associative construction** in Bantu. The possessor is realised by an oblique NP governed by a preposition-like linker -*a* that agrees in noun class with the possessed noun in the associative construction.

Compare [11.39a] with [11.39b], which expresses the same proposition without using the associative construction. In the latter sentence, the possessor NP is promoted to object and is placed immediately after the verb. In this particular case, there is no morphological marking to reflect the GF changes. Word order change and the omission of the associative marker are sufficient. As we saw on p. 271 in the examples in [11.35] illustrating dative shift in English, GF-changing rules need not necessarily leave morphological changes in their wake.

In Luganda, possessor raising is subject to this semantic restriction: the possessor NP can be raised to object only if it represents an inalienably possessed noun, e.g. a part of the body. Hence, while *okugulu* 'leg' can be turned into an object, *omuggo* 'stick' cannot since it is not an integral part of Kapere's body. (Cf. *She slapped his face / She slapped him in the face* vs *She hit his tree / *She hit him in the tree*.)

Finally, in [11.41a] we see that the pronominal object marker -*mu*-referring to the possessor noun can be incorporated in the verb if the possessed NP is inalienably possessed (e.g. *okugulu*). However, this is disallowed in [11.41b] where the possessed noun is not inalienably possessed (e.g. *omuti*).

11.5.4 Causative

The changes in grammatical function caused by the **causative** GF process can be stated in this way:

[11.42] a. null → subject
         b. subject → object
         c. object → 2nd object

Causative increases the valency of a verb, allowing it to take a fresh NP with a new θ-role as its argument (cf. section (11.5)). We have already seen in section (10.2.2) that the causative can be expressed lexically, syntactically or morphologically. Our concern here is with morphological causatives. Consider the following example from Luganda:

[11.43] a. Abalenzi   ba-li-fumb-a   lumonde
             boys   sp-future-cook-bvs   potatoes
             'The boys will cook potatoes.'

         b. Kaperæ-li-fumb-is-a   abalenzi   lumonde
             boys   sp-future-cook-bvs boys   potatoes
             'Kapere will make the boys cook potatoes.'

Note: sp = subject prefix; bvs = basic verbal suffix.

As you see when you compare [11.43a] with [11.43b], the causative introduces a new agentive NP as subject in [11.43b]. The original subject becomes the object and the original object becomes a second object. Equally important, the verb receives the causative suffix *-is-*.

Now describe in detail the changes brought about by causativisation in the Swahili example below:

[11.44] a. Wanafunzi   wa-ta-imb-a
             pupils   sp-future-sing-bvs
             'The pupils will sing.'

         b. Mwalimu   a-ta-wa-imb-ish-a   wanafunzi
             teacher   sp-future-om-sing-cause-bvs   pupils
             'The teacher will make the pupils sing.'

Here we see that the causative can also function as a transitivising affix. Causativisation turns the intransitive, one-predicate verb *-imba* 'sing' into a transitive verb in [11.44b] and a new subject, *mwalimu* 'teacher' is introduced. The NP *wanafunzi* 'pupils', which is the subject of 'sing' in [11.44a], becomes its object in [11.44b]. At the same time, the form of the verb undergoes three morphological changes: (i) a new subject marker prefix *a-* agreeing with *mwalimu* is introduced; (ii) an object marker prefix *-wa-* is added before the verb root where there was none in [11.44a]; and, crucially, (iii) a causative suffix *-ish-* is attached to the verb.

Let me summarise. Normally, GF-changing processes are morphologically marked: when the mapping of semantic functions on to GFs is disrupted by GF-changing rules, morphological changes take place. Significantly, the verb is the main locus of these changes. The morphology of verbs with arguments whose functions are obscured by GF-changing rules tends to be more complex than that of corresponding verbs whose arguments are related to the verbs in a straightforward way.

11.6 THE MIRROR PRINCIPLE

The nature of the relationship between the morphological and the syntactic dimension of GF-changing rules is an issue that has increasingly exercised the minds of linguists in the last few years. The standard view in earlier generative grammar was essentially functional. It was assumed that since GF-changing rules obscure the function of an NP, morphological marking is useful (or even necessary) because it leaves behind an overt mark which enables the hearer to unravel the real underlying syntactic function of an NP (and very indirectly its semantic role).

Recently, generative grammarians have questioned this earlier position. For instance, Baker (1987: 13), while not dismissing functional expressions out of hand, is sceptical about their merits for two reasons. First, he asks, if leaving behind a morphological marker to indicate the disruption of grammatical structure mapping in a sentence is functionally important, why do rules like relative clause formation and Wh-question formation, which cause as much disruption of the canonical surface pattern of a sentence as the GF-changing rules, not require similar morphological marking to that of GF-changing rules? You can see below, for example, that no marking of the verb is needed in [11.45a] when a Wh-question shifts the NP *some students* from the embedded sentence in [11.45a] to the matrix sentence in [11.45b]:

[11.45]   **a.**   Peter thinks that the man who came yesterday asked you about some students.

b. Which students does Peter think that the man who came yesterday asked you about?

Secondly, even if we accepted that the functional explanation was sound, we would still not be able to explain why the morphological marking of GF-changing processes is on the verb rather than on its NP arguments. In principle, the chances of forming the passive by affixing a passive morpheme to a noun (as in the putative example in [11.46b]) should be as great as the chances of attaching a passive morpheme to a verb (as in [11.47b]).

[11.46] **a.** The vet examined Esmeralda (Active)
 b. *Esmeralda-PASS examined by the vet (Passive)

[11.47] **a.** The vet examined Esmeralda (Active)
 b. Esmeralda examin-PASS by the vet (Passive)
 (Esmeralda was examin-*ed* by the vet.)

Whereas passives on the pattern of [11.46b] are theoretically possible, the only passives encountered in language are on the pattern of [11.47b]. For an adequate account of the principles behind the interaction between syntax and morphology that is manifested by GF-changing rules we need to look beyond the functional explanations that have been proposed.

The essential property of morphology is that it is concerned with the structure of words; the essential property of syntax is that it is concerned with the structure of sentences. However, the two components of the grammar are intimately related and the demarcation lines between them are sometimes fuzzy.

This is particularly clear when we examine situations where more than one GF-changing rule applies in a sentence. What strikes us is the fact that, in language after language, the morphological changes are, as a rule, in step with the syntactic operations that they are associated with. This generalisation follows from the principle of universal grammar known as the **mirror principle** (Baker, 1985, 1987):

[11.48] The Mirror Principle (Baker, 1985)

Morphological derivations must directly reflect syntactic derivations (and vice versa).

The Mirror Principle can be seen at work in the Luganda data below where both the causative and the passive apply:

[11.49] **a.** Subj.   Verb
              Abaana   basoma
              children   read
              'The children read.'

      b. Causative
              Subj.   Verb   Obj.
              Nnaaki   asom-es-a   abaana
              Nnaaki   read-CAUS-BVS   children
              'Nnaaki makes the children read.'

      c. Causative–Passive
              Subj.   Verb   Obj.
              Abaana   basom-es-ebw-a   Nnaaki
              children   read-CAUS-PASS-BVS   (by) Nnaaki
              'The children are made to read by Nnaaki.'

As [11.49b] shows, the subject *abaana* 'children' becomes the object of the verb 'read' when the causative applies. Once the original subject has been demoted to object, the original object NP *abaana* can be promoted to subject in [11.49c] by the passive rule.

This shows that, syntactically, causativisation must apply before passivisation. Given that causative creates a transitive verb from an intransitive verb and only transitive verbs can passivise, causative must apply before passive (cf. also [11.44] on p. 274 above). The morphological consequence of this is that the causative suffix is attached first, and is closer to the verb root than the passive suffix. The syntactic derivation is isomorphic with the morphological derivation. This follows from the mirror principle.

Revise the section on applicatives (11.5.3) and examine the following sentences from Chichewa, a Bantu language of Malawi.

Using the Mirror Principle, explain why [11.50e] is ill-formed while the rest of the sentences are grammatical.

[11.50] **a.** Mbidzi zi-na-perek-a   mpiringidzo kwa mtsikana
              zebras   SP-PAST-hand-ASP   crowbar   to girl
              'The zebras handed the crowbar to the girl.'

      b. Mbidzi zi-na-perek-er-a   mtsikana mpiringidzo
              zebras   SP-PAST-hand-APPL-ASP   girl   crowbar
              'The zebras handed the girl the crowbar.'

      c. Mpiringidzo u-na-perek-edw-a   kwa mtsikana ndi mbidzi
              crowbar   SP-PAST-hand-PASS-ASP   to girl   by   zebras
              'The crowbar was handed to the girl by the zebras.'

d. Mtsikana a-na-perek-er-edw-a   mpiringidzo ndi mbidzi
   girl   SP-PAST-hand-APPL-PASS-ASP   crowbar   by   zebras
   'The girl was handed the crowbar by the zebras.'

e. *Mtsikana a-na-perek-edw-er-a   mpiringidzo ndi mbidzi
   girl   SP-PAST-hand-PASS-APPL-ASP crowbar   by   zebras
   'The girl was handed the crowbar by the zebras.'

(from Baker, 1988: 14)

In [11.50a] grammatical θ-roles are mapped onto GFs in a simple and unmarked manner; [11.50b] is a paraphrase of [11.50a] in which the applicative has applied while [11.50c] is another paraphrase of [11.50a], this time involving the passive. In [11.50d], both the applicative and the passive apply. As you can see, the applicative suffix *-er-* is closer to the verb root than the passive suffix *-edw-* in [11.50d].

This morphological sequencing reflects the syntactic ordering of operations: the applicative feeds the passive (cf. (6.2.1) on 'feeding'). *Mtsikana* 'girl', having been turned into an object by the applicative, is available for promotion to subject by the passive rule. This is precisely what happens in [11.50d]. The ungrammaticality of [11.50e] stems from the fact that morphological sequencing is out of kilter with syntactic sequencing: the passive suffix *-edw-* incorrectly precedes the applicative *er*, but these operations apply in reverse order in the syntax.

Next we will consider **reciprocals**. Reciprocalisation derives an intransitive verb from an underlyingly transitive verb. Before reciprocalisation, there are two sentences with transitive verbs that have subjects and objects in agent and patient roles who do something to each other, e.g. *Bill punched Paul/Paul punched Bill*. After reciprocalisation, the two sentences are conflated and the subject of the verb refers to two (or more) participants and the object function is eliminated, rendering the derived verb intransitive. In English the reciprocal involves the *each other* construction, e.g. *Bill and Paul punch each other*.

Look closely at the following Luganda data:

[11.51]   **a.**   ba-   li-   sal-   agan-   ir-   a
               they-   FUT-   cut-   RECIPR-   APPL-   BVS
               'they will cut-each-other for/at'
               (e.g., they will cut each other for a reason/at a place)

b. ba-   li-   sal-   ir-   agan-   a
they-   FUT-   cut-   APPL-   RECIPR-   BVS
'they will cut for a reason/at a place each other'

Comment on the relationship between the order of the suffixes and the
semantic interpretation of these words.

As the glosses indicate, the linear order of the applicative and reciprocal
suffixes which follows the mirror principle reflects differences in semantic
scope. The scope of these two suffixes in [11.51a] and [11.51b] is shown by
the bracketing in [11.52a] and [11.52b] respectively:

[11.52]   **a.**   [[they-cut-each other] for/at]
       b.   [[they-cut-for/at] each other]

Baker's mirror principle comes free with universal grammar. It applies by
default in unmarked situations. It handles in a non-ad hoc manner the fact
that the order of morphemes, like the applicative and the reciprocal in a
Bantu verb, can vary (just as the order of words in a sentence can vary) in
ways that reflect differences in semantic scope. In this respect morphology
does resemble syntax.

However, the mirror principle is not absolutely inviolable. It works well
where affixation is cyclic such that each syntactic process triggers a round
of affixation starting near the root and going outwards, as we have seen in
the Chichewa data above. But it fails in languages that have **non-cyclic
affixation**. For instance, the morphology of many Native American languages, e.g. Classical Nahuatl, stipulates **position class**. Morphemes occurring in a word of a certain class must always occupy a certain pre-ordained
position in a sequence. In this kind of language, the order of morphemes is
not dictated by the mirror principle.

Suárez (1983: 60–1) reports that in Classical Nahuatl (dating from the
sixteenth and seventeenth centuries), there was a very rich array of position classes in the verb. Morphemes in the verb had to occur in the
sequence shown in [11.53].

[11.53]   Tense/Mode-Person of subject-Person of object-Plural of object-
Directional-Indefinite subject-Reflexive-Indefinite object-STEM-
Transitive-Causative-Indirective-Voice-Connector-Auxiliary   1-
Tense/Aspect-Auxiliary 2- Plural of subject

No single verb would exhibit all the positions in [11.53]. The compatibility
of positions in any one instance would depend on the particular verb stem.
Suárez cites these examples:

[11.54] **a.** ni-miṣ-teˑ -tḷa-makiˑ -1ti -s
I-you-him-it-give-causative-future.
'I shall persuade somebody to give it to you.'

   b. ni-te-tla-kow-i-a
I-him-it-buy-indirective-present
'I buy it for someone.'

With the verb in [11.54a] causative is obligatory, as is indirective with the verb in [11.54b] (contrast *ni-te-tla-kow-i-a* 'I buy it for someone' with *ni-tla-kow-a* 'I buy it'). In a system of this kind where position classes stipulated by morphology are the determinant of morpheme order, the mirror principle seems to be inapplicable, or at best marginalised.

We earlier observed the mirror principle at work in Chichewa, a typical Bantu language. But even in Bantu the mirror principle is sometimes defied. Hyman and Katamba (1992) show that there are two types of factors that may result in violations of the mirror principle:

(i)   Affixation may be morphologically determined, with morphemes occurring in words in fixed, pre-determined **morphological positions**, regardless of considerations of the ordering of syntactic operations and questions of semantic scope. (This is somewhat similar to Classical Nahuatl.)
(ii) Phonological considerations rather than the mirror principle may be the sole determinant of the order of morphemes in a word.

To illustrate, let us first consider the example in [11.55] from Kinande. We see here pre-determined morphological positions dictating the sequencing of morphemes:

[11.55] **a.** -imb-ir-an-a   'sing for each other'
sing-APPL-RECIPROCAL-BVS

   b. hum-ir-an-a   'hit each other for/at'
hit-APPL-RECIPROCAL-BVS   (e.g. hit each other for a reason/at a
place)

Kinande morphology requires the applicative to precede the reciprocal morpheme – notwithstanding the mirror principle and issues about semantic scope. The right interpretation of scope in [11.55] depends on the discourse context or the meaning of the verb.

This shows that because morphology is a separate level of analysis, with properties of its own, the order of morphemes expected on the basis of the sequencing of syntactic operation may be different from the order stipulated on the basis of pre-determined morphological positions.

As mentioned, in some instances, it is phonological rather than morphological considerations that are paramount. So in Bantu, any monophone verbal suffix, simply because it is monophone (i.e. one that is realised by a single sound e.g. ɪ y u or w), tends to appear after all other suffixes – except the final vowel representing the basic verbal suffix (Meusen, 1959). This is a purely phonological determinant. It applies regardless of the function of the monophone suffix. This requirement also normally supersedes any other factor, including the mirror principle.

[11.56] a. -tsap-   'get wet' (intr.)
             -tsap-ɪ-   'wet something' (= cause to get wet)
             -tsap-an-ɪ-   'wet each other'

         b. *-song-   ———
             -song-ɪ-   'gather' (tr.)
             -song-an-ɪ   'gather each other'

In [11.56a] we see that reciprocal -an- precedes causative ɪ in -tsap-an-ɪ-. However, this morpheme order does not reflect the order of syntactic operations responsible for the suffixation. From a syntactic perspective, causative -ɪ- must be added first in order to transitivise -tsap- since the reciprocal suffix -an- cannot be added to intransitive verbs.

Let us now turn to our attention to -song-ɪ in [11.56b]. The bare root -song- never occurs in isolation in the language. It must be assumed that -song-ɪ was lexicalised with the frozen causative suffix -ɪ. This means that -song-ɪ must be entered as -song-ɪ in the lexicon.

If we assumed a derivation of [11.56a] that reflects the semantic scope of the causative and reciprocal suffixes and faithfully obeys the mirror principle, -ɪ would be suffixed before -an- and we would have the derivation in [11.57a]. But this turns out to be incorrect. Instead, what we find is the derivation in [11.57b] where -an- is inserted before -ɪ.

[11.57] a. -tsap- →   -tsap-ɪ →   *tsap-ɪ-an →   [[[tsap-]ɪ]an]
             'get wet'   'wet' (tr.) 'wet each other'

         b. -tsap- →   -tsap-ɪ →   tsap- an-ɪ-
             'get wet'   'wet' (tr.) 'wet each other'

Observe at work here the phonological requirement that -V- suffixes must occur last, following all other verbal suffixes (except the final vowel) of the verb. This phonological requirement overrides the mirror principle.

Let me summarise. We have established in this section that the mirror principle predicts that the order of morphological operations reflects the order of syntactic operations. Affixes attached by morphological rules that are a consequence of the application of earlier syntactic processes are nearer the root than those that attach affixes representing morphemes introduced as a morphological side-effect of later syntactic rules. The mirror principle applies in unmarked situations when GF-changing rules are used.

Baker (1985) has further argued that the ordering of some GF-changing processes relative to each other is fixed. Thus, for example, the passive rule applies *after* the applicative in Chichewa. Baker claims that this is not an idiosyncratic property of Chichewa (and other Bantu languages): passivisation follows applicativisation in *all* languages. He concludes that there must be something in the nature of these processes that requires them to be combined in this specific order.

We have ended the section by indicating that although the mirror principle seems to be essentially correct, it is not an inviolable universal. Languages may show some variation in the way in which they set this parameter. The use of the mirror principle is the default case. It applies if neither morphological positioning nor phonological factors dictate a particular order of morphemes.

11.7 INCORPORATION

Baker claims that what GF-changing processes have in common is the movement of a **lexical category** (i.e. word) from one position in a sentence to a new position, which happens to be inside another word. The process whereby one semantically independent word is moved by syntactic rules to a new position and comes to be found 'inside' another word is called **incorporation**.

According to Baker, a key property of incorporation is the fact that it alters the *government* relations between predicates and arguments. This results in a change in grammatical relations. The syntax of GF-changing rules can be accounted for using general syntactic principles that regulate movement rules and government relations. No specific properties need to be seen as unique to a specific GF-changing rule like passive. All that is needed is a range of **movement rules** that shift constituents around in sentences. Sometimes the shifted constituents end up as elements within phrases, and sometimes they end up as elements inside words (Baker, 1988). If the latter happens, significant morphological changes take place. When the syntax requires the placing of one word inside another the result is a complex word containing two words. This is a kind of compounding. Words produced by incorporation are very similar to compounds produced by derivational morphology.

The rest of this chapter surveys noun, verb and preposition incorporation phenomena.

11.7.1 Noun Incorporation

Noun incorporation is the term used for a compounding process whereby a noun that has a θ-role; of patient, benefactive, theme, instrument or location and the **grammatical function** of object combines with the verb to form a compound verb. Noun incorporation is perhaps the most syntax-like of all morphological processes (Mithun, 1984).

Study the following data:

[11.58] *Siberian Koryak* (data from Mithun, 1984)

    a. Tınmékın qoyáwġe   (without noun incorporation)
       'I-slaughter reindeer'

    b. Tıqoyanmátekın   (with noun incorporation)
       'I-reindeer-slaughter'

Describe the effect of the noun incorporation rule.

As seen, whereas the object NP *qoya-* 'reindeer' is a separate word in [11.58a], it is moved and inserted inside the verb to form a kind of compound verb in [11.58b]. Noun incorporation is a productive syntactic rule that builds a special type of compound verb.

Mithun has observed that if a language has words formed by incorporating the object of the verb into the verb itself, it will also almost invariably have paraphrases that express the same propositional content with the object noun standing as an independent constituent of the sentence.

To the English-speaking reader, noun incorporation might look like an exotic phenomenon. But it should not. It occurs even in English. Find at least three English examples of nominal object incorporation.

Many familiar English verbal compounds are formed using noun incorporation:

[11.59]   *Verb*$_{ing}$ *Nobj*   *N*$_{obj}$ *-Verb*$_{ing}$ *Example*
       hunting (for) souvenirs   souvenir-hunting to go souvenir-hunting
       picking   potatoes   potato-picking   to go potato-picking
       rattling   sabres   sabre rattling   they are sabre-rattling

11.7.2 Verb Incorporation

The kind of analysis applied to noun incorporation is extended by Baker (1987: 21) to grammatical function changing processes like passivisation and causativisation. He shows that, for instance, morphological causatives in Chichewa can be re-analysed in terms of incorporation:

[11.60]   **a.**   Mtsikana   a-na-chit-its-a   kuti   mtsuko   u-gw-*e*
            girl   she-past-do-cause-asp   that   waterpot   it-*fall*-asp
            'The girl made the waterpot fall.'

       b.   Mtsikana   a-na-*gw-ets*-a   mtsuko.
            girl   she-past-*fall-cause*-asp   waterpot
            'The girl made the waterpot fall.'

Note: the causative has two allomorphs due to vowel height harmony. Where the last vowel in the verb root is one of /a i u/ allomorph -*its*- is selected; elsewhere -*ets*- is used.

Not only are the sentences in [11.60] paraphrases of each other, they also do contain the same morphemes. Furthermore, those morphemes express the same grammatical relations in both sentences. The crucial difference is that the morphemes -*its*- 'cause' and -*gw*- 'fall' are separate lexical roots, found in separate words in [11.60a], but they appear in [11.60b] as one single complex predicate, namely -*gw-ets*- 'fall-cause'. The verb -*gw*- is moved from its original position to a new position inside another word where it is combined with -*its*- in a manner reminiscent of noun incorporation.

Describe the incorporation phenomenon in the following data from Samoan (from Chung, 1978):

[11.61]   **a.**   E   tausi-pepe   'oia
            UNS   care-baby   he
            'He takes care of babies.'

       b.   E   tausi   e   ia   pepe
            UNS   care   ERG   he   babies
            'He takes care of the babies.'

Note: UNS = unspecified tense-aspect-mood; ERG = ergative.

In her description of the process of verb incorporation in the above data, Chung (1978) observes that an object NP in a single clause is bereft of all articles and becomes incorporated in the verb itself under certain circumstances. The circumstances being: (i) the clause has frequentative or habitual aspect, (ii) the NP object is a nonspecific common noun like 'babies' (not a definite or specific NP like 'the babies' or 'his babies').

The incorporation creates a new complex word. Once it is incorporated, the object NP becomes an integral part of the verb and loses its separate identity. Consequently, processes like the cliticisation of pronominal copy =*ai* that affect simple verbs also affect verbs with an incorporated NP. So, just as =*ai* cliticises to the right edge of a simple verb in [11.61a], it also cliticises to the right edge of the verb, including the incorporated object in [11.61b]:

[11.62] **a.** Po 'o   afea   e   tausi=ai   e   ia   tama?
              Q PRED   when?   UNS   care=PRO   ERG   he   child
              'When does he take care of the babies?'

            b. Po 'o   afea   e   tausi-tama=ai   'oia?
              Q PRED   when?   UNS   care-child=PRO   he
              'When does he babysit?'

Note: Q = question; PROC = pronominal copy; PRED = predicate.

11.7.3 Preposition Incorporation

In many Austronesian languages (e.g. Chamorro) and Bantu languages (e.g. Chichewa) there are applicative sentences with direct objects which can be paraphrased by other sentences in which the same NP occurs as an oblique NP governed by a preposition.

Again, Baker (1988) contends that no special GF-changing rule is required to account for these facts. They can be adequately dealt with if we assume that such languages have a movement rule that takes a preposition out of a prepositional phrase and plants it inside a verb. This phenomenon is illustrated by the following Chichewa sentences:

[11.63] **a.** Mfumu   a-na-tumiz-a   mbuzi *kwa*   mlimi
              chief   SP-PAST send-ASP   goat   to   farmer
              'The chief sent a goat to the farmer.'

            b. Mfumu   a-na-tumiz-*ir*-a   mlimi   mbuzi
              chief   SP-PAST-send-APPL-ASP   farmer goat
              'The chief sent the farmer a goat.'

Note: SP = Subject Pronoun (equivalent to what we called SM, i.e. subject marker elsewhere).

In [11.63a] the preposition *kwa* 'to' occurs as an independent constituent of the prepositional phrase *kwa mlimi*. But, in [11.63b], virtually the same morpheme is found incorporated in the verb *a-na-tŭmiz-ir-a*. Although in the latter case the complex of morphosyntactic properties realised by *kwa* in [11.63a] is realised by the bound morph *-ir-*, there is no substantive semantic or syntactic difference between the two sentences. Admittedly, this case is somewhat less convincing since there are two different forms *-ir-* and *kwa* that are involved.

11.8 CONCLUSION

We started the chapter by establishing the nature of the distinction between θ-roles and grammatical functions and showing how the former are mapped on to the latter.

We went on to examine the mirror principle, a hypothesis based on the claim that the relationship between morphology and syntax is a very close one. We suggested that, perhaps, this is to be expected, since much of the morphological marking found in languages serves the function of encoding the θ-roles and grammatical functions of verb arguments. However, in order to fulfil various discourse requirements, languages have rules that mask the relationship between grammatical functions and θ-roles. These GF-changing rules target the verb as the pivotal element. Normally they create morphologically marked forms of the verb. (In addition, they may also modify the form of its arguments.)

We have noted that, in unmarked situations, the mirror principle ensures that the order of morphemes reflects the sequence of the syntactic operations. But in marked cases morphological position classes or phonological considerations may frustrate the mirror principle.

In the last section, we have considered incorporation. The grammars of many languages provide alternative ways of expressing the same kind of propositional meaning. There may be a choice between using independent words, or incorporating words such as object NPs, verbs (e.g. 'cause') or prepositions into verbs to form complex words. It is claimed by Baker that incorporation is effected by rules that are essentially the same as the syntactic movement rules that shift around constituents of sentences. Incorporation puts one word inside another word rather than shift it to another place in a phrase or sentence.

Implanting a word within another word produces a complex word containing two words. In this regard, incorporation is similar to compounding, which is the subject of the next chapter. Both processes result in the creation of words which contain 'words' as their constituents.

Obvioulsy, incorporation obscures the boundary between syntax and morphology. The existence of compounds formed by incorporation also casts doubt on the common assumption that all derivational word-building takes place in the lexicon and morphology while sentence building is done in the syntax. Nevertheless, morphology maintains its distinct identity. This is particularly clear where morphological material introduced by GF-changing rules is subject to specific morphologically determined morpheme ordering constraints of the kind we observed in Kinande, and even more so in Classical Nahuatl.

Furthermore, incorporation blurs the boundary between inflection and derivation. The standard assumption is that inflection is syntactically determined word-formation that does not create new lexemes but derivation is word-formation that does create new lexemes (cf. section 10.2). Incorporation does not fit in this picture since it is a syntactically motivated word-formation process that results in the creation of a new lexical item, albeit one that is not necessarily institutionalised and listed in the dictionary.

Finally, in view of the nature of the problems which defy the mirror principle, it is worth considering an alternative that might fare better. Whereas the mirror principle assumes a symbiotic relationship between morphology and the lexicon, a different approach to the issue of the morphology syntax interaction might be more enlightening. Sadock (1985, 1986) has proposed a theory of **autolexical syntax** which envisages both morphology and syntax being autonomous although they are held together by universal principles.

Rather than viewing morphology as applying cyclically to the output of syntactic rules, as Baker does, Sadock proposed that the questions of syntactic well-formedness are separate from questions of morphological well-formedness. Syntactic well-formedness does not guarantee morphological well-formedness, and vice versa. According to this view, there is no mirror principle. Sadock's evidence comes mainly from noun-incorporation in Eskimo. The discussion of position classes in this chapter also lends support to this approach. Much more research needs to be done in this area to improve our understanding of the interaction between morphology and syntax.

EXERCISES

1. List three verbs whose lexical entries would include the subcategorisation template below:

   #(your verb)   SUBJ   OBJ   OBLIQ
                     ↑   ↑   ↑
               AGENT THEME LOCATIVE

2. Suggest subcategorisation templates for the different senses of the verbs *fly*, *sing* and *walk* in the following:

   (i)   Birds fly.
          Peter is flying a kite.

   (ii)   The dog walked.
          Barbara walked the dog.

   (iii)   Peter sang.
          Peter sang a lullaby.
          Peter sang a lullaby for the baby.
          Peter sang the baby a lullaby.

3. (a) Identify the θ-role and grammatical function of each argument in the sentences below.
   (b) Use your analysis in (3a) to justify the distinction between grammatical relations and theta roles.
   (c) How could the semantic relationship between the verbs *lend* and *borrow* on the one hand, and *buy* and *sell* on the other, be accounted for using the localist hypothesis and θ-rules? (See p. 243.)

   (i)   Peter lent the car to Lesley.
         Lesley borrowed the car from Peter.

   (ii) Janet bought the car from Helen.
         Helen sold the car to Janet.

4. Study the Malayalam data below (from Mohanan, 1983) and answer the questions that follow.

   Kuṭṭi   aanaye   ṇuḷḷ-i
   child-NOM   elephant-ACC pinched
   'The child pinched the elephant.'

   Kuṭṭiyaal   aana   ṇuḷḷappeṭṭu
   child-INSTR elephant-NOM pinch-PASS-PAST
   'The elephant was pinched by the child.'

Amma   kuṭṭi-ye-koṇṭə   annaye   ṇuḷḷ-icc-uccu
mother child-ACC with   elephant-ACC pinch-cause-past
'Mother made the child pinch the elephant.'

Ammayaal   aana   ṇuḷḷ-ikk-appeṭṭ-u
mother-INSTR elephant-NOM pinch-CAUSE-PASSIVE-PAST
'The elephant was caused to be pinched by mother.'

*Ammayaal   kuṭṭi   annaye   ṇiḷḷ-ikk-appeṭṭ-u.
mother-INSTR child-NOM   elephant-ACC pinch-CAUSE-PASSIVE-PAST
'The child was caused to pinch the elephant by the mother.'

Notes: NOM = nominative, INSTR = instrumental, ACC = accusative.

(a) What GF-changing processes can be identified in the data? What
    is their syntactic and morphological manifestation?
(b) State the interaction between the passive and the causative.
(c) What bearing do these data have on the mirror principle?

5. Study the following Kinyarwanda data from (Kimenyi, 1980) and
   answer the questions that follow:

Set A
(i) Umugabo   y-a-boon-ye   amáaso y'úmugóre
    man   he-past-see-asp   eyes   of woman
    'The man saw the eyes of the woman.'

    Umugabo   y-a-boon-ye   umugóre amáaso
    man   he-past-see-asp   woman   eyes
    'The man saw the woman's eyes.'

(ii) Umugóre   y-a-vun-nye   ukúboko k'úmwáana
     woman   she-past-break-asp arm   of child
     'The woman broke the arm of the child.'

     Umugóre   y-a-vun-nye   umwaana ukuboko.
     woman   she-past-break-asp child   arm
     'The woman broke the child's arm.'

Set B
(i) umuhuûngu y-a-twaa-ye   igitabo ey'úmukoôbwa
    boy   he-past-take-asp   book   of girl
    'The boy took the book of the girl.'

    umuhuûngu y-a-twaa-ye   umukoôbwa igitabo
    boy   he-past-take-asp   girl   book
    'The boy took the girl's book.'

(ii)   Umujuura   y-iib-ye   amafaraanga   y'úmunyéeshuûri.
       thief   he-steal-asp   money   of student
       'The thief stole the money of the student.'

       Umujuura   y-iib-ye   umunyéeshuûri   amafaraanga.
       thief   he-steal-asp   student   money
       'The thief stole the student's money.'

Set C
(i)   Abáana   ba-rá-kubit-a   ímbwa   y'úmugabo
       children   they-pres-beat-asp   dog   of man
       'The children are beating the dog of the man.'

       Abáana   ba-rá-kubit-ir-a   umugabo   ímbwa
       children   they-pres-beat-ben-asp   man   dog
       'The children are beating the dog of the man.'

(ii)   Umuhuûngu   a-ra-som-a   igitabo   ey'umukoôbwa
       boy   he-pres-read-asp   book   of girl
       'The boy is reading the book of the girl.'

       Umuhuûngu   a-ra-som-er-a   umukoôbwa   igitabo
       boy   he-pres-read-ben-asp   girl   book
       'The boy is reading the girl's book.'

Notes: pres = present; asp = aspect. Due to vowel height harmony,
   the applicative has two allomorphs: *-er-* after mid vowels and *-ir-*
   elsewhere (i.e. after /i a u/).

Questions
(i)   Write down, complete with glosses, the expected form of the
       sentence translated as '*The student is reading the boy the book.*'
(ii)   Identify the GF-changing rule exemplified by each set of
       examples in A, B and C.
(iii) State as explicitly as you can the syntactic and morphological
       properties of the GF-changing rule exemplified in each set of
       examples in A, B and C. Where generalisations can be made
       across sets of examples, make them.
(iv)   Use the data provided to argue either in favour or against the
       claim that Kinyarwanda has incorporation.

12 Idioms and Compounds: The Interpretation of the Lexicon, Morphology and Syntax

12.1 INTRODUCTION: THE INTERFACE BETWEEN MODULES

In this chapter we will continue our exploration of the nature of the relationship between morphology and syntax. We will examine the similarities and differences between words, the objects of morphological investigations, on the one hand and phrases (and sentences) the objects of syntactic investigation on the other.

The nature of **idioms** will form an important part of our investigations. The reason for this is that idioms raise interesting questions about the interaction between syntax and morphology. Idioms (e.g., *eat humble pie*, i.e. 'submit to humiliation') are lexical entities and function very much like a single word although they contain several words and are comparable to syntactic phrases or clauses (e.g., *[eat Swiss chocolate]$_{VP}$*). The question that has to be asked is this: how are idioms to be distinguished from phrases? To put it differently, how do lexical objects (i.e. words) differ from syntactic objects? Are they generated using rules that are essentially different? (See section (12.3.2) below.)

Most of our attention, however, will be devoted to compounds rather than idioms not only because of the greater importance of compounding as a method of word-building in English, but also because compounds raise perhaps even more acutely the issue of the interpenetration of morphology, the lexicon and syntax.

You will recall (from section (3.4)) that a prototypical **compound** is a word made up of at least two bases which can occur elsewhere as independent words, for instance, the compound *greenhouse* contains the bases *green* and *house* which can occur as words in their own right (e.g. in the noun phrase *the green house*, i.e. the house that is green). In the latter case we have a syntactic unit rather than a single lexical item. If compounds (which are lexical items) and phrases (which are syntactic units) contain words, how are they to be distinguished? In what respects are the rules that generate compound words and those that generate syntactic phrases similar, and in what respects are they different?

291

12.2 PHONOLOGICAL FACTORS IN COMPOUNDING

Although we are primarily interested in the syntactic aspects of compounds, we will not overlook altogether the role of phonology in compounding. So I will begin by briefly outlining some phonological considerations that may play a role in the formation of compound words.

The formation of some English compounds is at least in part motivated by phonology. There is a tendency to form compounds by joining together pre-existing words that rhyme, such as:

[12.1] Black-Jack   claptrap   night-light

Naturally enough such words are called **rhyming compounds**. Often rhyming compounds are made up of identical words:

[12.2] goody-goody   'good in a sentimental and naïve manner'
        pretty-pretty   'prettiness that goes over the top'
        preachy-preachy 'boringly moralising'

There is a tendency for words formed using this pattern to share the semantic trait of being pejorative, as you will have noticed in [12.2].

However, in some rhyming compounds neither of the bases is a word in its own right. Words such as the following are very marginal members of the class of compounds:

[12.3] nitwit   titbit   skitwit
        helter-skelter hobnob   namby-pamby

The reason for treating the forms in [12.3] as compounds, although their bases are not words that occur in isolation, is that they each contain two word-forming units, neither of which is an affix.

The situation here is analogous to that dealt with by Bauer (1983: 213–14), who proposes that non-rhyming words like *biocrat*, *electrophile*, *galvanoscope* and *homophile* which contain Neo-classical bases (borrowed from Latin or Greek) that only occur as bound morphemes should be treated as compounds. The merit of this analysis is that it enables us to avoid the embarrassing and contradictory alternative of regarding them as words that contain a prefix *bio-*, *electr(o)-*, *hom(o)* and a suffix (*-crat*, *-phile*, *-scope*) but no root morpheme (cf. section (12.5.2)).

In other words, although normally the bases that are combined to form a compound are autonomous words, the possibility of occurring as independent words is not a pre-requisite that all bases in compounds must satisfy. Very marginal members of the class of compounds are recognised which only contain bound bases.

Rhyme is not the only phonological motivation behind compounding. Determine the phonological factor that plays a role in the following:

[12.4]   zigzag   tittle-tattle
          dilly-dally   tick-tock
          sing-song   wishy-washy

The above compounds are motivated by **ablaut**. At the skeletal tier, the two bases have identical consonants in the syllable onsets and codas, but the vowels are different. The derivation of a form like *zigzag* proceeds in a manner reminiscent of some of the reduplication phenomena discussed in Chapter 9:

a.   A root (e.g., *zig*) is taken as the input.
b.   A syllable template is suffixed to the root. The C slots of the template are linked to the C slots of the input root, but the V-slots are left unlinked.
c.   A non-rhyming vowel is inserted and linked to the V slot.

The derivation of *zigzag* can be seen in [12.5]:

[12.5]   a.   b.   c.

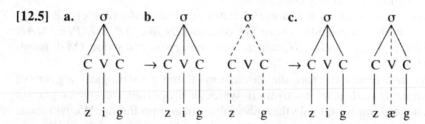

Such marginal cases, where apparent compounds contain bases that do not exist as independent words, are problematic. We will ignore them until section (12.5). Between now and then, we will only deal with clear cases of compounding where at least two bases that are potentially independent words are combined to form a new word.

12.3 ARE COMPOUNDS DIFFERENT FROM SYNTACTIC PHRASES?

It is not always obvious when we have a compound. Orthographic conventions offer limited help in distinguishing compounds from phrases. Some very well established compounds are written as one word, with or without a hyphen (e.g. *breakfast* and *ice-cream*). However, many other compounds are not conventionally identified as such by the orthography. Thus, a compound like *free trade* sometimes appears as two separate words separated by a space, and sometimes as *free-trade*, a single hyphenated word. There is a considerable degree of inconsistency in the orthographic representation of compounds – even among linguists. (We find writers on morphology producing articles and books with titles like *Some Problems of Wordformation* (Rohrer, 1974), *Word Formation in Generative Grammar* (Aronoff, 1976) and *English Word-formation* (Bauer, 1983) .)

Clearly, orthographic conventions are a poor guide to compounding. It has been suggested that phonology is a more reliable indicator. According to Bloomfield (1933: 228), **accent subordination** is the hallmark of compounds. One word accent dominates the rest in a compound, but not in a comparable syntactic phrase. (In the examples capital letters represent main stress in a word and lower case letters represent secondary stress.) So, Bloomfield argues, *ICEcream*, with main stress on *ICE* and reduced stress on *cream* is a compound, but *ICE CREAM*, with equal stress on the two words, is a syntactic phrase. This analysis would also apply to the American President's residence, the *WHITEhouse*, which is a compound, as opposed to *WHITE HOUSE* (i.e. any house that is white), or the market gardener's *GREENhouse*, as opposed to just any *GREEN HOUSE* (i.e. any house that is green).

Accent subordination is a useful criterion. But, unfortunately, it is not found in all compounds. There are compounds like *APPLE PIE*, *MAN MADE* and *EASY-GOING* which show no accent reduction (Marchand, 1969: 20).

We will now cut short the discussion of the phonological aspects of compounds and move on to their syntactic properties for syntax plays a much more important role than phonology in compounding. We will focus on prototypical compounds (e.g. *moonlight*, *birthday*, *eyelid*) which contain more than one potentially independent word.

12.3.1 The Notion 'Word' Revisited

In order to recognise compound words, we need to have a clear theoretical idea of what words in general are in the first place. This is essential if our analysis of compounds is to make any headway. So, in this section we are going to further clarify what we mean by the term **word**. The discussion will draw on Williams (1981b) and Di Sciullo and Williams (1987).

Recall that in Chapter 2 we distinguished the following three senses of the term word:

(i)   the **lexeme** (i.e. vocabulary item) (cf. 2.1.1),

(ii)   the **word-form** (i.e. a particular physical realisation of that lexeme in speech or writing) (cf. 2.1.2),

(iii) and the **grammatical word** (i.e. a lexeme that is associated with certain morphosyntactic properties) (cf. 2.1.3).

Our concern will be the nature of the word in the senses of lexeme and grammatical word. First, in section (12.3.2) we will compare the representation of lexemes in the lexicon with the representation of phrases and sentences in syntax. Later on, we will compare the properties of grammatical words with those of syntactic phrases (cf. 12.4). This discussion will take place against the backdrop of the wider issue of the nature of the similarities and differences between morphology and syntax.

12.3.2 Listemes

The lexicon contains a list of **lexical items** (e.g. nouns, adjectives, verbs, adverbs). Di Sciullo and Williams (1987) refer to the items listed in the lexicon as **listemes**. Most listemes are single vocabulary items such as *mediatrix*. The use of the term listeme is meant to highlight the fact that words in this sense must be *listed* in the lexicon because they have idiosyncratic properties (not governed by general principles) that speakers must simply memorise. By contrast, syntactic phrases are generated by general rules and are analysable in terms of those general rules. So they do not need to be listed in the lexicon. The idiosyncratic properties of listemes typically include:

(a)   morphological properties: *mediatrix* is borrowed from Old French; it takes the suffix *-ices* for plural;

(b)   semantic properties: *mediatrix* means 'a go-between'; *mediatrix* is human and female and the male equivalent is *mediator*;

(c)   phonological   properties: indicating   pronunciation   (e.g. /miːdɪətrɪks/);

(d)   syntactic properties: *mediatrix* is a noun, countable, feminine, etc.

Why are the phrases in [12.6] problematic in the light of the claim that phrases need not be listed in the lexicon?

[12.6]   take it out on   eat humble pie
         pass the buck   be in the red

Normally the meaning of a phrase can be worked out if we know the meaning of the words it contains and the ways in which they are syntactically related to each other. For instance, if we know what *pass* means and what *salt* means, and if we know that *salt* is the object of the verb *pass*, with the θ-role of theme, we can work out what *pass the salt* means, i.e. *pass the salt* is semantically **compositional**: its meaning can be inferred from the meaning of its parts.

In contrast, the meaning of phrases like those in [12.6] is *not* compositional. Such phrases must be listed in the dictionary with their meanings and memorised, just as the meaning of a single vocabulary item like *mediatrix* must be listed, because the meanings of those phrases cannot be worked out on the basis of the meanings of the words which they contain. Such phrases are referred to as **listed syntactic objects** by Di Sciullo and Williams. More commonly, they are called **idioms**. Idioms thus are listemes which represent a counter-example to the idea that no phrases should be listed in the lexicon.

12.3.3 Unlisted Morphological Objects

We have seen above that lexical items (e.g. the word *mediatrix*) are listemes. Does listedness *per se* necessarily distinguish lexical items (including idioms) from syntactic phrases? Unfortunately, the answer is no. True, sentences and phrases that are generated by the syntax need not be listed. But, equally, languages do also contain **unlisted morphological objects**, i.e. unlisted fresh words created routinely by morphological rules. Many words need not be listed in any dictionary because they are produced in a predictable manner by word-formation rules. For instance, a word like *re-draw* need not be listed in the lexicon and memorised. A speaker who knows the meanings of *re-* and *draw* as well as the relevant morphological rule of prefixation, can easily prefix *re-* to *draw* to form *re-draw*.

Many **nonce words** (i.e. words expressly created by speakers) are built using standard rules. Recently, I encountered *Prime ministerable* and *uncomplicatedness* in these contexts:

[12.7]   a.   Mr Kinnock is now *Prime ministerable*.
                    (= has what it takes to be made into a Prime Minister)

          b.   the *uncomplicatedness* of language
                    (= the fact that language is not complicated)

Like me, you will have experienced no difficulty in working out the meanings of the italicised words, although none of them is listed in any dictionary.

In principle there is no limit to the number of new words that can be created. It is true that generally speakers tend to memorise a very large number of the words – but not the sentences – of their language.

Nonetheless, both morphology and syntax – since they are creatively rule-governed – are open-ended (see Chapter 4).

This is especially clear in an agglutinating language where a word may have thousands of different forms. It is implausible to imagine that speakers memorise all the forms of all the words which they use. For instance, a verb root in Luganda can be preceded by up to six prefixes and be followed by up to three suffixes:

[12.8]   te- ba- li-   gi- tu- mu- fumb- ir-   iz-   a
            not they future it us her cook for cause basic verbal suffix
            'they will not make us cook it for her'

Speakers do not memorise such words. They use rules to manufacture them as necessary. Just coping with the task of memorising all forms of even a few hundred common verbs would impose an intolerable load on the memory; memorising all the words one uses is simply out of the question. Nevertheless, the claim that words tend to be listed more frequently than phrases and sentences is valid. Why should that be so? Di Sciullo and Williams observe that a listeme usually offers a concise way of encoding a specific and arbitrary notion (see section (10.2.2) above). The less compositional the nature of a form the greater the likelihood of its being listed. An item composed of other items is normally built using general principles of composition. Its meaning can be composed and decomposed using these general principles. The items lower down the linguistic hierarchy are less compositional in nature and hence more likely to be listed. Di Sciullo and Williams's observations are summarised in [12.9]:

[12.9]	*morpheme >*	*word >*	*compound >*	*phrase >*	*sentence*
	All morphemes must be listed since they are minimal, non-decomposable semantic entities.	Most complex words containing several morphemes are listed. (But they are not listed if their meanings are compositional.)	Many compound words made up of two or more words are less liable to be listed and memorised.	Some syntactic phrases (i.e. idioms) are idiosyncratic. They must be listed.	The vast majority of sentences need not be listed since they are composed of units lower down the hierarchy, using general rules.

12.3.4 Syntactic Objects and Syntactic Atoms

As we have seen, the syntactic objects whose meaning is not compositional are idioms. What is meant by saying that idioms are syntactic objects is that:

(i)   they always fall into one of the recognised syntactic units of a
       language such as:

[12.10]   **a.**   *S'* : (i.e. subordinate clause) when the chips are down
                   VP : rule the roost
                   NP : (post-modified by *PP*) the man on the Clapham omnibus
                   PP : round-the-clock

          b.   *S'* : (i.e. subordinate clause) when the chaps are here
                   VP : rule the country
                   NP : (post-modified by PP) the man on the boat
                   PP : round the house

(ii)   they have the internal structure of normal syntactic units, and
(iii) they behave just like other syntactic units of the same type in the
       syntax.

A group of words that does not form a syntactic unit could not be an
idiom. So a string like *week was a* in the sentence *The entire week was a
success* could not be an idiom, contributing a non-compositional meaning
to the whole sentence, since it is not a syntactic unit.

Note, however, that many idiomatic expressions contain VPs that do not
look like coherent syntactic units. Di Sciullo and Williams cite VPs such as
take (_____NP) in hand; *push (_____NP) too far*, etc., which have a *blank*
NP following the V. Superficially the VPs seem to contain a transitive verb
followed by a prepositional phrase like *in hand* or an adverbial phrase like
too far despite the fact that neither of these constructions is sanctioned by
the PS rules of English.

Draw phrase structure trees to show the structure to *take (——————NP) in hand*
and *push (_____NP) too far.*

Your trees should look like these:

[12.11]   **a.**   **b.**

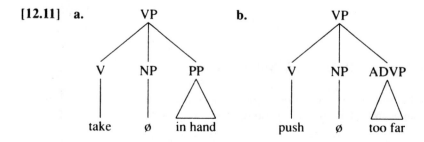

In spite of the fact that these idiomatic VPs have a blank NP position, they function like normal VPs with a transitive verb. For instance, we can say *push (somebody)$_{NP}$ too far* and we can derive the passive *(somebody)$_{NP}$ was pushed too far*. Similarly, with the idiom *keep tabs on* we can derive the passive *Tabs are being kept on the new students by the teacher* from *The teacher is keeping tabs on the new students*. Clearly, syntactic rules can see the internal structure of these idioms and apply to them in the normal way.

It is not syntactic awkwardness but lack of compositionality that makes the listing of idioms in the lexicon necessary. Listed items may be idioms formed by regular syntactic rules or words formed by morphological rules. In either case the meaning of the listed item cannot be worked out by rule. If (lack of) compositionality applies equally to listed lexical items, be they idioms (syntactic objects) or words (morphological objects), is there no significant difference between the two? As we shall see in the next paragraph, the answer is that there is a major difference.

Syntactic rules applying at the level of the phrase do not take into account the internal structure of words. This position is known as the **lexicalist hypothesis** (Chomsky, 1970; S. R. Anderson, 1988; Williams, 1978). The essential claim of the lexicalist hypothesis is the following:

[12.12]   Syntactic rules apply to words regardless of their internal struc-
           ture. The morphological and syntactic characteristics of a word
           are independent of each other. So, in the syntax, morphologi-
           cally complex words do not behave differently from words con-
           taining underived simple roots.

The key difference between words – in particular compound words – and syntactic phrases lies in the fact that, whatever internal structure a compound has, that structure is inaccessible to the rules of syntax. That is why Di Sciullo and Williams argue that all words are **syntactic atoms**. Although, as we shall see in (12.4), the internal structure of compound words is created using the syntactic rules of the language that generate syntactic phrases, syntax cannot access that internal structure. By contrast, syntax can see the internal structure of phrases, including idioms.

So a syntactic rule like Wh-movement, which moves to the front of an item about which information is being requested and places a Wh-word before it, cannot extract a part of a compound and front it, and ask for information about it, as in [12.13a] any more than a Wh-question could be used to seek information about elements inside an affixed word as in [12.13b]:

[12.13]   a.   She saw the greengrocer.
                 Which greengrocer did she see?
                 *Which green grocer did she see?
                 (i.e. which green one did she see?)

b. He is unrepentant.
  How unrepentant is he?
  *How un repentant is he?
  (i.e. is he very *un-* or not so *un-*)

Compounds, like affixed words, are treated as indivisible units by syntactic rules. Syntactic rules can manipulate the elements inside phrases but they cannot manipulate the elements inside words, no matter whether they are affixed words or compounds. In this respect compounds differ from idioms which, as we saw above, are syntactic units subject to syntactic operations.

This is not to suggest that syntactic rules are totally insensitive to any idiosyncratic properties that words may have. Certain words have very specific requirements of the syntactic contexts where they can be inserted (see p. 219).

Suggest an explanation for the ill-formedness of the sentences in [12.14b]:

[12.14] **a.** I blinked my eyes.   **b.** *We blinked his eyes.
            We blinked our eyes.   *They blinked his eyes.
            You blinked your eyes.   *She blinked your eyes.
            She blinked her eyes.   *She blinked his eyes.
            He blinked his eyes.   *I blinked her eyes.

         c. I held my tongue.   **d.** *We held his tongue.
            We held our tongues.   *They held his tongue.
            You held your tongue.   *She held your tongue.
            She held her tongue.   *She held his tongue.
            He held his tongue.   *I held her tongue.

Note: Here *hold X's tongue* is an idiom meaning 'say nothing'.

The above examples illustrate a peculiarity of the verb *blink*. The X position after *blink* in *blink X's eyes* must be filled by a pronoun that refers back to the subject as in [12.14a]. All the putative sentences in [12.14b] where the X position is occupied by a pronoun that does not refer to the subject are ill-formed.

Interestingly, although idioms are syntactic objects, sometimes they too are subject to lexical restrictions. Thus, the genitive noun that fills the X position in *hold X's tongue* must be co-referential with the subject. All the correct sentences in [12.14c] have a genitive pronoun that refers to the subject; the incorrect ones in [12.14d] do not. When *hold X's something* appears outside the idiom, no such restriction applies. We can say *we held his books*, *she held my bag*, etc.

The lexicalist hypothesis was initially a reaction against **generative semantics**. Generative semantics was an approach to semantics and syntax inthe 1970s which allowed syntactic rules free access to the internal structure of words. In this theory, syntactic rules could manipulate the innermost structure not only of idioms, but also of simple words. Indeed, it was claimed that many words, including monomorphemic ones (e.g. *drop*, *kill*), were not listed in the lexicon but were created as a result of syntactic manipulations (Morgan, 1969; McCawley, 1968, 1971; Lakoff, 1971; Postal, 1970).

A classic example was the treatment of the verb *kill* by McCawley (1968). He argued that *kill* can be decomposed into at least four underlying predicates *CAUSE-BECOME-NOT-ALIVE*. So the lexical entry for *kill* would be:

[12.15]

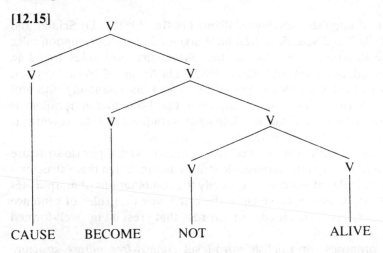

Various transformational rules would apply to that abstract underlying structure which would in due course be replaced by the phonological representation /kɪl/.

This analysis was used to explain the fact that a sentence containing *kill* and an adverb like *almost*, e.g. *Jack almost killed Jill* is ambiguous. The ambiguity was attributed to the alleged fact that rules of the syntax could place *almost* in different positions so that it could modify one or other of the underlying predicates. In other words, syntactic rules could apply *inside* the lexical item before it was eventually constituted as the surface verb *kill*. This would yield the two interpretations *Jack almost caused Jill to die* or *Jack caused Jill almost to die*.

We shall reject the speculations of the generative semanticists that syntax can access the internal structure of words and maintain the lexicalist position. In the foregoing paragraphs (see especially [12.13]) we have already shown the inaccessibility of word-internal structure to syntax.

Furthermore, even if we were eager to accept the generative semanticists claims, we would have to reject the specific analysis of *kill* in [12.15]. Given the fact that four predicates underlying *kill* are posited, the sentence *Jack almost killed Jill* should be four-ways ambiguous, since in principle *almost* can modify any one of the four predicates. But this is not the case. The theory predicts many more ambiguities than the two that actually occur. This casts doubt on the existence of all those underlying predicates (cf. Chomsky, 1972a).

12.4 THE CHARACTER OF WORD-FORMATION RULES

A number of linguists, notably Williams (1981a, 1981b). Di Sciullo and Williams (1987) and Selkirk (1982), have argued that word-formation rules are phrase-structure rules akin to the phrase-structure rules found in syntax. Indeed, as the title of Selkirk (1982), *The Syntax of Words*, implies, morphological rules are not regarded by Selkirk as essentially different from syntactic rules. She explicitly argues in that book that morphology is the study of 'the syntax of words' while what is traditionally called syntax is the study of 'the syntax of sentences'.

In the study of both 'word syntax' and 'sentence syntax' phrase structure rules are used to generate permissible strings and to assign them structural descriptions. Rules of word syntax specify the combinations of morphemes that words are made up of much in the same way that rules of sentence syntax specify concatenations of words that result in well-formed sentences.

Selkirk proposes for English word-level *context-free phrase structure rules* (i.e. phrase-structure rules of the type A →B that are not restricted to a specific environment) like the following:

[12.16]   Phrase-structure rule   Example
          stem → affix stem   expel
          stem → stem affix   fraternal
          word → word affix   book-ing
          word → affix word   re-wind
          word → word word   footpath
          (Here 'stem' stands for any bound non-affix morpheme.)

A more explicit account of the use of phrase-structure rules in morphology is provided below. But before we introduce that account, we need to explain another important concept, that of *headedness*.

The discussion that follows is not intended to provide a thorough survey of English compounding. Others have already done that. See the work of Bauer (1983) for a textbook treatment, and Jespersen (1954) and, above all, Adams (1973), Marchand (1969), Selkirk (1982). Rather, my aim is to examine English compounds as a starting point of our continuing exploration of the relationship between morphology and other components of the grammar, in particular syntax.

12.4.1 Headedness of Compounds

The notion of head plays a key role in the work of generative morphologists like Williams (1981a, 1981b), Di Scullio and Williams (1987) and Selkirk (1982). Using the theory of **X-bar syntax** they highlight the fact that just as phrases in syntax have **heads**; words also have heads.

In this theory, the representation of a syntactic category has two elements. The first is the **category type**, e.g. NP, VP, N, etc., and the second is the **level**, which is represented by the number of bars above the category, e.g. $\bar{\bar{N}}$, $\bar{N}$, N (Chomsky, 1970; Jackendoff, 1977; Selkirk, 1982: 4–9).

A head has the following properties:

(i) it assigns its category features to the constituent of which it is the head, e.g. the head of a NP is a noun, the head of a VP is a verb, etc. So, because the noun *books* is the head of the phrase *new books*, the entire phrase $[new]_{ADJ}[books_N]_{NP}$ is a NP.

(ii) it is one level lower in the X-bar hierarchy than the constituent of which it is the head. As you can see in the tree in [12.17b], where the NP in [12.17a] is presented using X-bar notation, N is one level lower than NP etc.; a lexical category like the noun *book* has zero bar (i.e. no bar).

[12.17] **a.**   **b.**

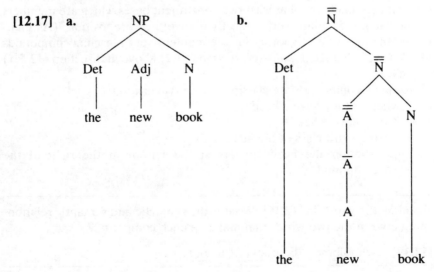

The well-formedness condition known as **percolation** guarantees that the head is specified with features identical to those of the entire constituent, and vice versa. This entails that the head of a noun phrase is a noun, the head of a verb phrase is a verb, the head of a preposition phrase is a preposition, and so forth.

Selkirk (1982) proposes an X-bar analysis of noun compounds parallel to the syntactic analysis of NPs in [12.17] as shown below in [2.18]:

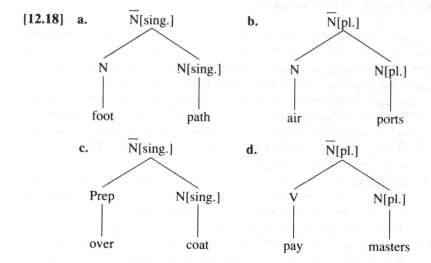

[**12.18**]   a. $\overline{\text{N}}$[sing.]   b. $\overline{\text{N}}$[pl.]

              N   N[sing.]   N   N[pl.]

              foot   path   air   ports

   c. $\overline{\text{N}}$[sing.]   d. $\overline{\text{N}}$[pl.]

         Prep   N[sing.]   V   N[pl.]

         over   coat   pay   masters

(For the time being we are ignoring issues concerning the representation of inflection in compounds (cf. sections (12.4.2) and (12.4.3) below.)

Compounds in English belong to the word-classes noun, verb or adjective, and are made up of at least two constituent bases which are members of the categories noun, verb, adjective, adverb or preposition. (As mentioned before, we are ignoring for the moment very marginal compounds like *hob-nob* and *Anglophile* (cf. section (12.2) above and section (12.5.1) below).

Normally compounds are classified using two criteria:
(i)   whether they have a head
(ii)   if they have a head,
    a.   the word-class of the head
    b.   whether the head appears at the left or at the right of the compound

---

Examine the words in [12.19]. What is the syntactic and semantic relationship between the two words that make up each compound?

[**12.19**]   school<u>boy</u>   bed<u>room</u>   tea<u>pot</u>

The compounds in [12.19] are typical examples of **endocentric compounds** (i.e. headed compounds). It has been recognised for a long time that most English compounds are endocentric, with the head normally on the **right** (cf. Bloomfield, 1933; Marchand, 1969; Williams, 1981; Selkirk, 1982).

Syntactically the head is the dominant constituent of the entire compound word (cf. [2.17] and [2.18]). Normally inflectional properties of the head (e.g. tense, number and so on) percolate to the entire compound.

Semantically an endocentric compound indicates a sub-grouping within the class of entities that the head denotes. The head is underlined in the examples in [12.19]. Thus, a *schoolboy* is a kind of *boy*, a *bedroom* is a kind of *room*, and a *teapot* is a kind of *pot*. The first word in each case functions as a modifier of the head which specifies the meaning of the head more precisely.

Let us take a closer look at English endocentric compounds, starting with noun compounds. As seen in [12.20], a compound noun may contain a noun followed by another noun, an adjective followed by a noun or a preposition followed by a noun:

[12.20] **a.** N N **b.** A N **c.** Prep N
water-lily   hothouse   undergraduate
bookcase   sour-dough   near-sightedness
motor-car   greenfly   outskirts
skyline   high-court   underdog
India-rubber   wet-suit   oversight

(For the time being, I am ignoring the very common type of verbal compound represented by *moneylender* and *roadsweeper*. We shall consider such compounds in [12.27] below.)

Write a (context-free) phrase structure rule that generates the noun compounds in [12.20].

The phrase structure rule required is given in [12.21]:

[12.21]

$$N \rightarrow \left\{ \begin{array}{c} N \\ V \\ A \\ Prep \end{array} \right\} \quad N$$

Thus the kind of context-free phrase-structure rewrite rule that is used to account for phrase-structure in syntax is also used to generate compound words.

As we saw in section (4.1), compounding in English is recursive in principle and so there is no limit to the size of compound words. For example, noun-noun compounds like *lakeside* and *grammar school* can be combined to form a compound noun like *Lakeside Grammar School* with the structure:

$$_N[_N[_N[lake]_N \; _N[side]_N]_N \; _N[_N[grammar]_N \; _N[school]_N]_N]_N$$

The context-free phrase-structure rule N → N N in [12.21] is sufficient to generates such compounds.

English also has many compound adjectives. Some examples are listed in [12.22]:

[12.22]   a.   *N A*   b.   *A A*   c.   *P A*
          world-wide   short-lived   overwhelming
          user-friendly   hard-hearted   under-mentioned
          seaworthy   good-natured   outspoken
          foolproof   long-winded   near-sighted

    a. Add two fresh examples to each column.
    b. Write a phrase-structure rule to generate the adjectives in [12.22].

       (Refer back to the discussion of past-participles and adjectives in section (10.2.1.3). Note that the forms *lived*, *hearted*, *natured* and *winded* in [12.22b] are bound adjectival bases which do not occur in isolation as adjectives (**a very lived-/natured/hearted/winded person*).

These adjectives contain a noun followed by an adjective in [12.22a]. An adjective is followed by an adjective (derived from the past participle form of a verb) in [12.22b], and a preposition is followed by an adjective (derived from the present or past-participle form of a verb) in [12.22c]. So the phrase structure rule required is the following:

[12.23]

$$A \rightarrow \left\{ \begin{array}{c} N \\ A \\ Prep \end{array} \right\} A$$

Lastly, we will survey compound verbs. One fairly common type consists of a prepositional or adverbial particle (P) followed by a verb.

[12.24]   P V

undersell	overrate	upstage
outstay	offload	upset

But by far the commonest type of compound verb in English is the **phrasal verb**, which contains a verb plus a prepositional or adverbial particle as shown in [12.25]. Many of these verbs have an idiomatic meaning. I propose that for our present purposes we consider the examples in [12.25a] in their literal meanings and those in [12.25b] in their idiomatic meanings:

[12.25]   **a.**   VP

look through	look into	look up	look out
take away	take in	take up	take out
put through	put down	put up	put out

     b.

turn off	hand out	let out	take over
show off	blow over	catch up	put away
make off	go off	kick off	do in

Formulate a phrase structure rule that generates the compound verbs in [12.24] and [12.25].

Those compound verbs are generated by the rule below:

[12.26]
$$V \rightarrow \left\{ \begin{array}{cc} P & V \\ \\ V & P \end{array} \right\}$$

Let me summarise: so far we have restricted our investigations to a class of endocentric compounds with these properties:

(i)   They contain a constituent which functions as the syntactic head.

(ii) The syntactic properties of the head percolate to the entire compound word.

(iii) The head is on the right hand (phrasal verbs form a clear and large class of exceptions to this generalisation (cf. section (12.4.3)).

(iv) There is a tendency for the semantic relation between the head and nonhead to be one of modification (e.g. a *bedroom* is a kind of *room*) but there are a number of significant exceptions as we will see below.

Idiomatic phrasal verbs are a blatant exception to the last claim. For instance, *go off* in the idiomatic sense of 'turn sour' is not 'a kind of going' nor is *take in*, meaning 'deceive' a kind of taking. As is the case with all idioms, the meaning of idiomatic phrasal verbs cannot be computed from the meaning of the verb and the preposition or particle.

Even in non-idiomatic endocentric compounds frequently there is no one semantic relationship tht systematically holds between a head and its modifier. Thus a *pincushion* is a little cushion for sticking pins in, *seaweed* is a kind of weed that grows in the sea, *fleabite* is a bite given by a flea etc. Knowing that we have a noun compound does not enable us to work out automatically the semantic relation between its constituents.

One class of endocentric compounds, which is referred to technically as **verbal compounds**, stands out from the rest in that it exhibits quite consistent semantic readings that match the syntactic characteristics of the compounds (see Selkirk, 1982; and also Downing, 1977; Allen, 1978; Dowty, 1979). Verbal compounds have these characteristics:

(i)   a complex head adjective or noun, which is *derived from a verb*;
(ii)   the nonhead constituent is interpreted as a **syntactic argument** of the deverbal noun or adjective head;
(iii)   the θ-role of the nonhead is that of agent, patient, etc.;
(iv)   the meaning of the compound is transparent.

In the remainder of this section we shall highlight the important role played by the head in the semantic interpretation of verbal compounds.

First, drawing on Adams (1973) and Selkirk (1982), I list below a range of verbal compounds:

[12.27]   <u>NOUNS</u>   <u>ADJECTIVES</u>

a. [*Noun-verb-er*]$_N$ b. [*Noun-verb-en*]$_A$
    moneylender   hand-written
    gamekeeper   computer-matched
    shoemaker   hand-sewn
    bookseller   time-worn
    anteater   guilt-ridden

c. [*Noun-verbing*]$_N$ d. [*Noun-verb-ing*]$_A$
    bear-baiting   God-fearing
    hay-making   awe-inspiring
    brick-laying   self-seeking
    sheep-shearing   eye-catching

There is a striking similarity in argument-structure between deverbal compounds and syntactic phrases containing the same words, and having the same argument-structure. The semantic relationship is predictable, general and systematic. For instance, in the compound noun *moneylender*, *money* is the theme and the deverbal agentive noun *lender* is the agent. This is comparable to the phrase *lend$_V$ money$_{THEME}$* in which *money* functions as the object of the verb. So, we can paraphrase *moneylender* as *lender of money*. All the other compounds in [12.27a] can be analysed in the same way.

Is there any generalisation to be made about the thematic roles in [*Noun-verb-en*]$_A$ in [12.27b], [*Noun-verb-ing*]$_N$ in [12.27c] and [*Noun-verb-ing*]$_A$ in [12.27d]?

In the compound adjectives in [12.27b] the noun and verb show the same relationship to each other as they do in the corresponding phrase. For example, in both the compound *hand-written* and in the phrase *written by hand*, the *hand* is the *instrument* with which the writing is done.

In the compound nounds in [12.27c] the noun in each compound functions as the patient argument of the deverbal noun on its right. In [12.27c], in a compound noun like *bear-baiting*, the noun is a *patient* and undergoes whatever action is indicated by the V-ing that follows it. Thus, the noun *bear-baiting* is analogous in its argument-structure to the verb phrase *baiting$_{VERB}$ bears$_{PATIENT}$*. The noun *brick-laying* is analogous to *laying$_{VERB}$ bricks$_{PATIENT}$*, and so on.

The compound adjectives in [12.27d] require the same treatment. An adjective like *God-fearing* can be paraphrased using the verb phrase *fearing$_{VERB}$ God$_{PATIENT}$*; *eye-catching* can be paraphrased as (metaphorically) *catching$_{VERB}$ eyes$_{PATIENT}$*, and so forth.

What is the argument-structure of the following compounds?

[12.28]		Nouns		Adjectives
	a.	[*Noun-verb-er*]$_N$	c.	[*Noun-verb-en*]$_A$
		chain smoker		capacity filled (= filled to capacity)
		party drinker		guilt-laden
		daydreamer		alcohol related (= related to alcohol)
	b.	[*Noun-verb-ing*]$_N$	d.	[*Adjective-verb-ing*]$_A$
		Sunday closing		low-flying
		church-going		hard-working
		spring-cleaning		hard-hitting

I expect that you saw the catch. Unlike the compounds in [12.27], the compounds in [12.28] are not *verbal compounds* in the special sense defined on p. 308. The noun or adjective constituent is not an argument of the deverbal element that follows it, with a θ-role of agent, patient, theme, instrument etc. Rather, the function of the noun or adjective constituent is to specify place, time or manner in which the process, action, etc., indicated by the verb takes place. In [12.28a] the noun *chain*, for example, is not a patient of the verb *smoke*: a *chain smoker* does not smoke chains! Similarly, in [12.28b], *Sunday* is not the patient that undergoes the action of *closing*. Rather, *Sunday closing* has the temporal meaning of 'shutting down (of a commercial establishment) on Sundays'. In [12.28c] a noun like *capacity* is not an argument. Its role is that of an adjunct indicating the extent or circumstances in which an event, action or state takes place. Likewise, in [12.28d], the adjective preceding *V-ing* does not function as an argument of the verb. Rather, it is an adjunct that indicates manner. In *low-flying*, for example, *low* is not the agent, instrument or patient of *flying* but a manner adjunct.

Assuming the validity of the assumption that in verbal compounds a noun satisfies the argument-structure of a deverbal noun, show why the words in [12.29] should not be treated as verbal compounds.

[12.29]	think tank	copycat	punch-line
	call girl	playboy	showman

Unlike the earlier examples of verbal compounds in [12.27] and [12.28], with a deverbal head that appears *on the right*, the compounds in [12.29] have a verb that appears on the left and which does not function as the head of the compound (cf. section (12.4.2) below). Furthermore, in each case the verb is followed by a noun. But the noun does not function as an argument of the verb. We would normally expect a noun following a verb to have the grammatical function of object and the θ-role of patient or theme. However, here the noun following the verb has none of these properties.

   No general rule can provide a phrasal source, complete with the right argument structure for compounds such as these. So they are semantically unpredictable. We cannot predict their meaning beyond stating rather vaguely that 'N is in some sense associated with V'. The precise nature of the semantic interpretation is established virtually afresh for each individual example (cf. Adams, 1973: 61; Selkirk, 1982: 24; Jespersen, 1954).

   To sum up, following Selkirk (1982), I reserve the label **verbal compound** for those compounds in which the nonhead is an argument of the deverbal head as in [12.27]. I will exclude words that are superficially similar to the verbal compounds in [12.27] but crucially lack a noun that functions as an argument of the verb.

Recall that we saw in (section (11.3)) that in order for a syntactic phrase to be well-formed, the argument-structure of a predicate (normally a verb) which is the head of that syntactic phrase must be satisfied by the appropriate arguments. What we have just seen shows that morphology resembles syntax in this regard. If a compound has a deverbal predicate as head (as in [12.27]), the argument structure of the head must be satisfied by the nonhead noun argument. For example, a transitive verb like *shear*, occurring as the head of either a verb phrase, or as a deverbal noun in a compound word, must have a patient as one of its arguments.

However, where the verbal noun constituent is not a head, it is not necessary to have its argument-structure satisfied. Thus, though *copy* is a transitive verb that normally requires a patient, it is allowed to occur without a patient in the compound *copycat*: the noun *cat* is not the object of *copy* in this word (cf. Roeper and Siegel, 1978; Lieber, 1983). In cases such as these, the satisfaction of the argument-structure of the verb, which is mandatory in syntax, is disregarded by morphology. What is required in syntax is not necessarily also required in morphology.

12.4.2 The Right-hand Head Rule (RHR)

Most compounds in English are endocentric, i.e. they have a head. In such compounds the head element (normally) appears as the right-handmost constituent of the word. It is clear from the phrase structure rules in [12.21], [12.23] and [12.26] above and from the examples in [12.20] above and [12.30] below that the right-hand constituent is the one whose syntactic category like (noun, verb, adjective) percolates to the entire compound word. (In other words, the head determines the category of the entire compound.)

[12.30] $[\text{bird}_N \text{ watch }_V]_V$ $\qquad$ $[\text{over}_P \text{ react}_V]_V$
$\qquad$ $[\text{sugar}_N \text{ daddy}_N]_N$ $\qquad$ $[\text{blue}_A \text{ book}_N]_N$
$\qquad$ $[\text{blue}_A \text{ black}_A]_A$ $\qquad$ $[\text{wind}_N \text{ screen}_N]_N$

The generalisation about the data above is captured by Williams's Right-hand Head Rule:

[12.31] Right-hand Head Rule (RHR)
In morphology we define the head of a morphologically complex word to be the right-hand member of that word. (Williams, 1981a: 248)

This correctly predicts that, in each instance in [12.30], the rightmost member of the word is the head of the word.

The question we are now going to address concerns headedness and the analysis of inflected compounds. In the literature two possible analyses of inflected compounds have been discussed (cf. Selkirk, 1982: 55). Both of them are shown in [12.32]:

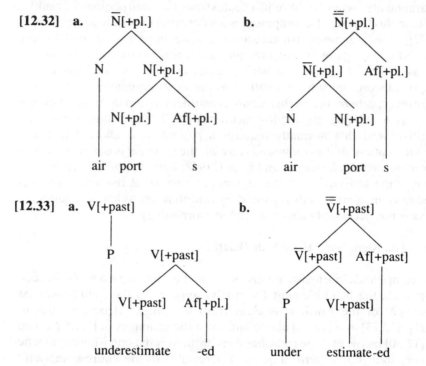

In the examples in [12.32b] and [12.33b] the inflection marks the compound as a whole, which implies that compounding is done first and then affixation takes place later. But in [12.32a] and [12.33a] the reverse order is assumed. Affixes are first attached to the head of the compound and then subsequently compounding occurs. Since either way we would get the same semantic reading for these compounds, is there a principled way of choosing between the two analyses?

The fact that compounding can feed inflection has been used by Williams (1981a) to make a case for treating affixes as heads of their words. It is claimed that the RHR applies to affixes: the rightmost suffix in a word assigns its properties by percolation to the entire word. As Di Sciullo and Williams put it (1987: 25):

> For an affix to determine the properties of its word, it must appear in the 'ultimate' head position (the head of the head of the head . . .).

This is a revision of the traditional position presented in Chapter 3, where it was assumed that stems and bases select their affixes: verbs select verbal affixes (e.g. *-ed*), adjectives select adjectival suffixes (e.g. *-er*, *-est*) and so on (cf. [3.13]).

Williams's proposal turns the argument on its head: a suffix determines the category of the word because, it is claimed, like words, affixes belong to a lexical category like N, V, Adj. Affixes differ from lexical nouns, verbs, adjectives, etc., in that they can only occur as bound morphemes. So the lexical entries for the derivational suffixes *-er*$_N$ and *-ion*$_N$ would include the category N. If a derivational suffix like *-er* or *-ion* comes last in a word, e.g. *work-er*, *educat(e)-ion*, then the entire word has to be a noun. Inflectional affixes would also be marked in the same way, e.g. *-s*$_N$ for the noun plural suffix and *-ed*$_V$ for the past tense verbal suffix. In each case the category of the right-hand element (i.e. the head) would determine the category of the entire word. (Di Sciullo and Williams, 1987: 25, recognise the unfortunate blurring of the distinction between inflection and derivation inherent in this proposal since they both are said to have the potential of determining word category.)

If right-headedness in morphology essentially boils down to appearing in the last position in the word, there is nothing surprising about suffixes being the heads of words, including compounds. Once a compound like [[*wet*$_A$] [*suit*$_N$]]$_N$ has been formed, it can receive a plural *-s* suffix (to form [[*wetsuit*]$_N$]*s*$_N$). The plural suffix is attached to the word on the right which was the head before the inflectional suffix was added (see the (b) examples in [12.32] and [12.33]). If, instead, the word on the left is inflected for plural, the resulting word is ill-formed (*[[[*wet*] *s*]$_N$[*suit*]]). (We return to this in [12.42] below.)

Against this it should be noted that where the word on the right in a compound has irregular inflection, as in *firemen, field mice, underwrote, outdid, oversaw*, there is a good case for adopting the analysis in [12.32a] and [12.33a] since there is no separate inflectional affix and the inflection is assigned directly to the second element. The rightmost word in the compound percolates its properties, including irregular inflection, to the entire compound:

[12.34]   **a.** $\overline{N}$[+pl.]   **b.** $\overline{V}$[+past]

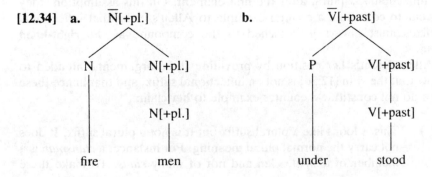

Now extend the above analysis to the data in [12.35]. Show how the right-hand head rule (RHR) throws light on the plural formation shown here. (Refer back to section (7.2.5.2).)

[12.35]	Column A	Column B	Column C
	*scissor	scissors	scissor-handles
	*trouser	trousers	trouser-hangers
	*pant	pants	pant-liners
	*binocular	binoculars	binocular-cases

The somewhat more complex facts in [12.35] can be accounted for if we assume that compounding feeds affixation. A compound word like *scissor-handle* must be formed first before plural suffixation can occur. That is the case even where an uninflected form of the nonhead like *scissor does not exist in isolation. Afterwards, when the plural suffix is added, it must go after the head *-handle*. The requirement to place the suffix after the head at the end of words is paramount. Some have gone so far as to argue that there is a universal law that bans inflectional affixes from appearing inside compounds (Allen, 1978: 112).

Show how the above account (which assumed (i) that the head of a compound is the rightmost element and (ii) that the head is the element to which it is inflected in cases of irregular inflection) can be extended to cover the following.

[12.36]	a.	craftsman	craftsmen	b.	groundsman	groundsmen
		swordsman	swordsmen		clansman	clansmen

The forms in [12.36] are problematic because they appears to contain an inflectional *-s* plural after the first element. On this assumption, they appear to constitute a counter-example to Allen's claim that inflectional suffixes must always be attached to the compound as the **right-hand head**.

Allen defends her position by providing three arguments intended to show that the *-s* in [12.36] is not an inflectional suffix, and that hence these data do not constitute a counter-example to her claim:

(i)   This *-s* looks like a plural suffix but it is not a plural suffix. It does not carry the normal plural meaning. For instance, a *clansman* is a member of a single clan and not of many *clans*. To make these words plural the rightmost word takes the plural form *men*, i.e. *craftsmen, swordsmen, groundsmen, clansmen*.

(ii) The compound-internal *-s* can be attached to words that are plural in meaning which elsewhere always appear without any plural inflection, e.g. *kins* as in *kinsman* (but *kin* (plural) in isolation (cf. *kith and kin*); *deer* as in *deersman* (but *deer* (plural) in isolation).

(iii) Interestingly, the second member of these compounds with an internal *-s* is *-man*. Allen argues that there are two *-man* morphs which represent different morphemes. These morphs can be distinguished phonologically. In pseudo-compounds like those in [12.33b] *-man* is pronounced [mən], with a reduced vowel. By contrast, in true compounds, like those in [12.33a], *-man* is pronounced [mæn], similar to the way it is pronounced in isolation.

Allen argues that *-man* ([mən]), is a derivational linking suffix and it is distinct from the *-man* [mæn] that appears in authentic compounds. As seen in [12.37], a linking derivational *-s* can appear before [mən] but not before [mæn]:

[12.37] **a.** door [mæn] *doors [mæn] **b.** doors [mən] *doors [mən]
  oar [mæn] *oars [mæn] oars [mən] *oars [mən]

The problem with Allen's analysis (at least in British English) is that in both *oarman* and *doorman* reduction of *-man* [-mæn] to [mən] is possible. So it is difficult to justify the claim that there are two distinct forms of *-man* in [12.37]. The third argument against treating *-s* as an inflectional suffix is doubtful but the other two appear to be sound.

12.4.3 Left-headed Compounds

But even if we were persuaded by Allen's arguments, not all apparent cases of non-right-hand headed compounds would be so easily disposed of as we shall see in this section.

Show how the universality of RHR is put into question by the data below from Italian.

[12.38] pesce[Nsg., masc.] cane [Nsg., masc.] → pesci[Npl., masc.]   cane[Nsg., masc.]
fish   dog (= 'shark')   fish (pl.)   dog ('sharks')
                                                              ↓
                                        [pesci cane][Npl., masc.]

capo[Nsg., masc.] famiglia[Nsg., fem.] capi[Npl., masc.]   famiglia[Nsg., fem.]
head   family   heads   family
'head-of-family'   'heads-of-family'
                                                    ↓
                                        [capi famiglia][Npl., masc.]

The right-hand head rule is certainly not universal. While in a language like English the head of a compound is (normally) on the right, there is another type of language, of which Italian is an example, where the head of a compound is normally on the left (Scalise, 1984: 125).

Italian is not unique in having compound nouns with heads on the left. French too has left-headed compound nouns some of which are joined by *de*, as in [12.39a], and others are not linked by anything, as in [12.39b].

[12.39]

	singular		plural
a.	un chef d'atelier	'foreman'	des chefs d'atelier
	une chemise de nuit	'nightdress'	des chemises de nuit
	un bureau de change	'foreign exchange office'	des bureaux de change
	un billet de banque	'banknote'	des billets de banque
b.	un timbre-poste	'postage stamp'	des timbres-poste

The plural inflection is attached to the first noun in the compound in each case. The RHR does not apply here.

Let us now return to English. Although the RHR normally applies, even here there is a small minority of endocentric compounds with left-hand heads. They include nouns which form their plural by adding the plural morpheme to the noun in first position such as the following which are listed in Quirk and Greenbaum (1973: 84):

[12.40]

singular	plural	
passer-by	passers-by	(*passer-bys)
notary public	notaries public	(*notary publics)
grant-in-aid	grants-in-aid	(*grant-in-aids)
coat of mail	coats of mail	(*coat of mails)
mother-in-law	mothers-in-law	(*mother-in-laws)

But such is the pressure exerted by the RHR to have the inflectional ending after the right-hand member of the compound that in [12.41] there is a growing tendency to put the plural suffix after the right-hand nonhead element in these compounds which formerly only had their plural inflection on the left-hand element:

[12.41]	attorney general	attorneys general	or	attorney generals
	mouth-ful	mouths-ful	or	mouth-fuls
	spoon-ful	spoons-ful	or	spoon-fuls

Selkirk (1982: 52) also deals with a number of compounds whose first member carries the plural affix. Part of her list is reproduced in [12.42].

a. Identify the head of each compound.
b. What is odd about the plural marking in compounds of this kind?

[12.42]	overseas investor	sales receipt
	parks commissioner	parts distributor
	arms merchant	arms race
	buildings inspector	weapons analysis

These nouns are endocentric compounds, with a right-hand head. Semantically, we know that *investor, commissioner, receipt*, etc., are the heads. An *overseas investor* is a kind of investor, a *parks commissioner* is a kind of commissioner, an *arms merchant* is a kind of merchant, etc. The fact that there is a plural marking on the left-hand, nonhead element does not indicate plurality of the entire compound.

This follows from the percolation principle. As a rule, it is the head which drips its morphosyntactic features onto the rest of the compound. Assuming that the rightmost affix (which surfaces as part of the rightmost word in a compound) is the head, it is to be expected that it will determine the properties of the whole compound.

Thus, in [12.42], the plural suffix in *overseas, parks, arms* etc. does not mark the plurality of the whole compound but rather the plurality of the nonhead constituent. In order to pluralise the entire compound noun, we have to attach the plural suffix to the head constituent which is on the right to yield *overseas investors, parks commissioners, arms merchants*, etc.

Let us now leave noun compounds and turn once more to **particle verbs** made up of a verb followed by a preposition or adverb, e.g. *phone in, drive out, lock out, lay off, melt down*. As a group, particle verbs constitute the majority of examples of English compounds whose head is on the left. The verb, which is the head of the compound comes first and the particle, e.g. *in, out*, as in *phone in* and *lock out*, specifies more narrowly the meanings of the verb. Being the head, the verb itself gets all the inflections of the compound as in *drove out*, *phoned in* and *locks out* (not **drive outed*, **phone ined* and **lock outs*).

---

a. What is the word-class of the compounds below?
b. How are they formed?
c. Represent the structure of each compound using a tree diagram.

[12.43]   see-through (curtains)   live-in (nanny)

These compounds are adjectives derived from particle verbs by conversion. The input verbs are represented by the trees in [12.44a] and [12.45a]. Without any overt change, a new Adj node is added to dominate the V node as shown in [12.44b] and [12.45b]. Compounding feeds conversion:

[12.44]   **a.**   **b.**

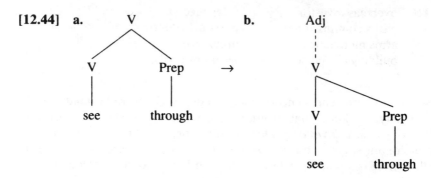

[12.45]   **a.**   **b.**

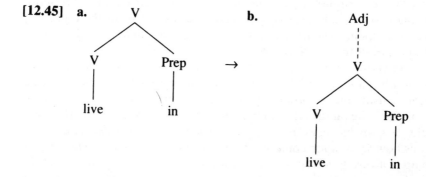

We have established that the RHR is not a universal principle that regulates compounding in all languages. Left-hand heads are not rare in Romance languages like French and Italian. Furthermore, although right-hand heads are the norm in English, nonetheless English has a minority of nonright-hand headed compounds. The genuine generalisation which we should attempt to capture is that inflectional affixes are (normally) attached to the head, regardless of whether it is on the left or on the right of a compound.

12.4.4 Headless Compounds

There is an important difference between syntax and morphology in the way in which headedness is treated. All syntactic phrasal constituents generated by X-bar phrase structure rules must have a **head**. But in morphology compound words generated by X-bar phrase structure rules need not always contain a head.

12.4.4.1 Exocentric compounds
Let us begin by considering headless compounds which do not contain an element that functions as the semantic head which is modified by the nonhead element. Such compounds are called **exocentric compounds**. They are also often referred to by their Sanskrit name of **bahuvrihi** compounds.

Analyse the following compounds and contrast their syntactic properties with their semantic properties:

[12.46] **a.** dark-room   football   plain-chant
            b. greenhouse   lazy-bones   blue-nose

Both sets of examples contain an adjective followed by a noun. The words in [12.46a] are endocentric noun compounds. A *dark-room* is a kind of *room*, *plain-chant* is a kind of *chant* and a *football* is a kind of *ball*.

By contrast, from a semantic perspective, [12.46b] contains *headless* compounds made up of an adjective and a noun. The constituents in [12.46b] do not have a head-modifier semantic relationship. A *greenhouse* is not a house that is green. Similarly, *lazy* does not specify more narrowly a set of bones in *lazy-bones*. *Lazy-bones* is a lazy person. Similarly, *blue-nose* is not a *nose* at all but a purplish variety of potato grown in Nova Scotia (hence the use of 'blue-nose' as a pejorative nickname for a Nova Scotian).

From a syntactic point of view, however, these compounds are not headless. They are generated by the standard rule in [12.21] N → A N that generates regular endocentric compounds since they have the structures *[[green$_A$][house$_N$]]$_N$*, *[[lazy$_A$] [bones$_N$]]$_N$*, *[[blue$_A$][nose$_N$]]$_N$*. They obey the RHR. The rightmost word determines the category of the compound. And, to inflect these compounds for plural, we attach the inflection in the standard fashion at the very end, e.g. *greenhouses* (cf. also the adjective noun compounds *loudmouths*, *dimwits*, *high fliers*).

Other exocentric compounds can be treated in the same way. Standard, context free phrase-structure rules like N → V N will generate compounds like *spoil-sport* and *dare-devil* where a verb is followed by a noun. But here the noun is not taken in a literal sense as an argument (specifically the theme or patient) of the verb. For the purposes of semantic interpretation, a special stipulation is required to state that a *spoil-sport* 'is a grumpy person who frustrates others' efforts to have fun'. A *dare-devil* is 'a reckless person' – not one who literally dares the devil. Plural inflection is again attached to the second word, according to the RHR, to give *spoil-sports* and *dare-devils*.

Now describe both the syntactic and the semantic relationships between the words that constitute the following compounds.

[12.47]   butterfingers   blockhead   turncoat

These too are exocentric compounds. There is no element that functions as the *semantic* head of the compound which is modified by the nonhead element. So the word does not have the meaning of 'x is a kind of y'. We know that *butterfingers* (*butter$_N$* + *fingers$_N$*) is neither a kind of *fingers* nor a kind of *butter* but rather a person who is apparently incapable of holding things without dropping them (especially a person who fails to hold a catch at cricket). The noun *blockhead* is analysable as (*block$_N$* + *head$_N$*). But a *blockhead* is neither a kind of *block* nor a kind of *head* but rather an idiot. A *turncoat* (*turn$_V$* + *coat$_N$*) is not a kind of *coat* but a renegade.

Obviously, the meaning of an exocentric compound is opaque. It is impossible to work out what an exocentric compound means from the sum of the meanings of its constituents. For this reason exocentric compounding tends to be used much less frequently than endocentric compounding in the creation of new words.

How should the grammar deal with exocentric compounds? As mentioned, from a semantic point of view, exocentric compounds are opaque like idioms, i.e. they are not subject to compositionality. So there is a case for listing their meanings in the lexicon as we do for idioms.

As for their syntax, Selkirk (1982: 25) proposes that exocentric compounds are generated by the kind of context-free phrase-structure rules given in [12.21], [12.23] and [12.26] which generate endocentric compounds. What is special about exocentric compounds is that they each have idiosyncratic rules of semantic interpretation. For instance, the exocentric compound *butterfingers* ($[butter_N + fingers_N]_N$) is produced by the phrase structure rule N → N N which also produces the endocentric compound *schoolboys*. But a special rule of semantic interpretation applies to *butterfingers* to yield the meaning 'inept catcher'.

I am sceptical about the use of rules of semantic interpretation to account for the idiosyncratic meaning of exocentric compounds. I would prefer a simple listing of the meanings. Just as idioms are syntactic objects and listemes at the same time, exocentric compounds are both unlisted morphological objects with regard to their syntax, which is regular word syntax, and listemes with regard to their semantics, which is opaque.

We will now leave exocentric compounds and turn to another class of headless compound.

12.4.4.2 Copulative compounds

These compounds are called **copulative compounds**, because they have two words which are coupled (or conjoined). They are also often referred to by their Sanskrit name of **dvandva** compounds. They have the structure shown in [12.48]:

[12.48]

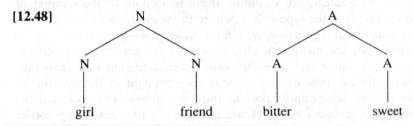

From a syntactic point of view, however, copulative compounds are headed. The rightmost noun is the head. These compounds obey the RHR. Affixation processes attach affixes to the element on the right as in *player–managers*, *worker–priests* and *boyfriends*.

From a semantic point of view, the coupled elements are of equal status, with neither element being regarded as the head that dominates the entire word. Copulative compounds are not semantically opaque. Rather, each element characterises a separate aspect of the meaning of the entire word as you can see:

[12.49] boyfriend north–west
Urbana–Champaign player–manager
Harper–Collins worker–priest

In a word like *north–west* the northerly and westerly directions are equally important. In the name of a corporation like Harper–Collins the names of

the two companies that merged to form a new publishing company are
semantically of equal status. The cities of Urbana and Champaign that
form the twin city of Urbana–Champaign in Illinois enjoy parity, and so
on.

12.5 COMPOUNDING AND DERIVATION

We are revisiting the problem of word-building elements that seem to be
bases rather than affixes which never occur as independent words (see
section (12.2)). When such elements are combined in a word, should that
word be regarded as a compound?

12.5.1 Cranberry Words

It has been demonstrated above that compounds are created by phrase-
structure rules like those in [12.21], [12.23] and [12.26]. They are not
listable in the lexicon. No finite list can be drawn up containing all the
compounds of English. For example, there is no limit to the number of
compound nouns of the type N N_{er} (where the second is a deverbal noun)
such as *moneylender, gamekeeper, schoolteacher, time-keeper*, etc.

But this does not mean that all compounds are generated by phrase-
structure rules. There are some compounds which, though analysable into
their syntactic constituents, are permanently resident in the lexicon. A
number of such compounds contains the form *-berry*. The most famous
member of this set is *cranberry*. Hence they are all called **cranberry words**:

[12.50]   a.   N → N N   b.   N → A N
              cranberry   blueberry
              huckleberry   blackberry
              elderberry
              strawberry
              gooseberry

The compounds words in [12.50b] are unproblematic. The colour adjective
found in the compound is a word that occurs elsewhere with a meaning that
is relateable to that found in the compound.

The compounds in [12.50a] also seem to be analysable as containing the
word *berry* as head, preceded by a modifier. In [12.50a] the element that
precedes *berry* is a noun. In *elderberry*, the nound *elder* refers to the elder
tree that produces *elderberries*; *elder* can be an independent word. So
elderberry is a compound.

But in the other NN compounds, the situation is confused. It is not clear whether the form that occurs with *berry* in the compound is the related to the form found elsewhere. Does the form *goose* in *gooseberry* belong to the same lexeme as *goose* in the word that refers to a species of bird? Although they have the same form, there is no reason to assume that these elements are semantically related. Likewise, there is no reason to regard *straw* in *strawberry* as a form related to *straw* that is found in a stable. There is no clear motivation for regarding *strawberry* as a compound.

Huckle- and *cran-* in *huckleberry* and *cranberry* are even more intriguing. There is no 'huckle' or 'cran' shrub. Neither *huckle* nor *cran* occurs in any other word with this meaning. These morphs only appear in these words.

While syntactically it might make good sense to analyse these noun compounds as containing NN, given their semantic opacity it is extremely unlikely that speakers construct them from scratch each time they use them. It is more likely that they are simply listed as compounds in the mental lexicon (cf. section (12.4.4.1)).

S. R. Anderson (1990) argues persuasively that we need not presume that *cran-* is listed separately in the dictionary. All that is needed with regard to semantics and syntax, is a lexical entry for *cranberry*, which gives the appropriate sense for the whole and which shows that this word contains two nominal bases, the first one of which is bound (cf. section (3.1.1)):

[12.51] $[_N [_N$ cran- $[_N$ berry]

Anderson suggests that, in addition to the very general syntactic-type phrase-structure rules, there are **analogical rules** of compounding. The existence of the phrase structure rule N → N N, which licenses compound nouns like $_N [_N$ *school* $[_N$ *boy*], provides the basis of **analogical formations** like $_N [_N$ *cran* $[_N$ *berry*] within the lexicon.

To sum up, *schoolboy* is a canonical example of a compound word generated by phrase-structure rules. It contains two bases that occur as independent words. *Cranberry* is a non-canonical example of a compound. It contains *berry* and the **bound base** *cran-* which is uniquely found in this word. Non-canonical examples of compounds are problematic because it is unclear whether one of the putative bases is indeed a base and not an affix.

12.5.2 Neo-classical Compounds

In English there are other words (besides the cranberry ones) which appear to straddle the borderline between compounding and affixation. In most of these words part of the word is a form borrowed from Greek or Latin. Hence they are called **neo-classical compounds** in the literature (Adams, 1973; Bauer, 1983).

For instance, it is not clear whether the latinate form *multi-* which occurs in words like *multi-media*, *multi-lateral* etc. is a prefix or a base and hence whether the words which contain it are compounds or not.

A similar problem is raised by Greek borrowings like the following:

[12.52]   *hydro-* 'water'   *theo-* 'god'
          hydrology   theology
          hydrometer   theocracy
          hydrolysis   apotheosis
          hydrogeology   theosophy

In each case a base (*hydro-* and *theo-*) has something attached to it. Whether or not speakers have a clear idea as to the status of these forms may depend more on their knowledge of the classical languages where they come from than on their knowledge of English. Speakers who do not know Greek might regard these as unanalysable words, or (if they are aware of the distributional patterns) they might perhaps regard them as words with bound bases of indeterminate meaning similar to the cranberry words we saw above. But those with some knowledge of Greek would readily classify them as compounds for they know, for instance, that *hydro-meter* is a compound with two bases: *hydro-* 'water' and *meter* 'measure'; and that *theo-cracy* also a compound (combining *theo* 'god' with *-cracy* 'rule' (i.e. rule over a nation by god, or god's representatives, e.g. clerics).

Consider these other neo-classical compounds which contain the following elements: *phone*, *pseudo*, *graph*, *tele*, *hyper*, *narco*, *neo*, *chem*, *amphi*, *crat* and *ology*.
a. Find two examples of words that contain each one of these forms.
b. Should these forms be treated as affixes or as bases? On what basis can a decision be made?

A case could be made for treating such forms as affixes. Many of them can be added to a stem just like the prototypical affixes like *-er* or *-ly*. Compare the last two columns in [12.53]. The forms *neo-* and *hyper-* behave in a similar way to the prefixes *pre-* and *in-*.

[12.53]   **a.** colonial   pre-colonial   *neo*-colonial
          b. active   in-active   *hyper*-active

Should all so-called bases in neo-classical compounds be regarded as affixes? That would be unwise. As Bauer (1983: 213) observes, treating such forms as affixes would force us to take the bizarre position that there are English words containing neo-classical elements that are made up of prefixes and suffixes only. But we know that words like *un-ish-ness* that consist solely of affixes do not exist. In the pair *phonic* and *narcotic,-ic* is obviously a suffix, as we already know. So, *phon-* and *narco-* must be bases. In *theo-cracy* and *hydr-ology*, at least one of the elements must be a bound base or else these words will be said to contain only affixes. If we identify forms like *theo* and *cracy* as bases in some contexts, when they come together in a word like *theocracy* we have good cause to treat it as a compound.

12.6 CONCLUSION

After being eclipsed by syntax and phonology (in the sense of being exhaustively divided between the two) in the early years of generative linguistics, morphology is no longer considered marginal. For some time now it has been widely recognised that word-formation must be given an important place in the theory of language.

Although morphology has been presented from the standpoint of generative linguistics, proselytising has not been my objective. So I have tried to avoid being partisan and polemical. The pages of this book bear testimony not only to the insights that have come from generative theoreticians in recent years, but also to the findings of generations of theoretical and descriptive linguists who have investigated a multitude of languages in other frameworks.

While broadly indicating satisfaction with the way in which the generative model of morphology addresses the gamut of issues concerning word-structure, I have not hesitated to point out problems for which totally satisfactory solutions still elude us. For instance, although I have adopted the hierarchical model of lexical morphology, I have shown the reservations that remain, e.g. with regard to finding reliable criteria for recognising lexical strata and resolving bracketing paradoxes.

It has been my aim to present a coherent generative model of morphology and the lexicon. This has meant being selective, and omitting, or not emphasising, some alternatives to the approaches that I have focused on. After outlining traditional and structuralist morphological concepts in Part I, I presented in Part II the interaction between phonology, morphology and the lexicon using the theory of level-ordered lexical morphology and prosodic morphology. The latter was shown to be particularly suitable for the analysis of nonconcatenative morphological systems. In the discussion viable non-level-ordered lexical morphology alternatives were only briefly dealt with. I concluded Part II by pointing the reader in the direction of non-hierarchically ordered theories of the lexicon like that of Rubach (1984). Earlier I had also shown the merits of Word-and-Paradigm

(WP), another non-cyclic theory, which treats complex systems of fusional morphology insightfully. I expect future research will find a way of synthesising the findings of these approaches in a new coherent model.

Part III dealt with the relationship between syntax and morphology. After clarifying the nature of the distinction between inflection and derivation and surveying a broad range of inflectional processes, the central issue that we tackled was the morphological mapping of grammatical relations. We saw that Baker's mirror principle, to which I am sympathetic, seems to work well in cases where morphological operations are isomorphic with syntactic operations, but fails us where they are not. We briefly noted that Sadock's autolexical syntax may be more successful where morphological and syntactic operations do not coincide.

In the final chapter of the book we have examined the interaction between the lexicon, morphology and syntax through an investigation of idioms and compounds. These investigations have provided us with a window on the nature of lexical items (i.e. words (including compounds) and idioms) and syntactic phrases. The question we have attempted to answer is this: what is the difference between lexical items such as compound words and idioms on the one hand and syntactic phrases on the other? Obviously, if there is no significant difference, much of the justification for having a separate morphological module in our grammar evaporates. We have shown that there are both similarities and differences between lexical items and syntactic units. The differences justify the existence of morphology as a separate module.

We have seen that the structure of complex lexical items (compounds and idioms) is constructed by the same phrase-structure rules that are used to construct phrases. Compounds and syntactic phrases are subject to the same syntactic principles (e.g. headedness and the right-hand head rule). However, while idioms are syntactic constituents subject to syntactic rules, just like any other syntactic units, compounds are not. Syntactic operations cannot apply to the constituents of a compound.

We have also demonstrated that lexical items differ from phrases and sentences in the way their meanings are accessed. Typically morphemes and simple words have idiosyncratic semantic properties and so must be listed in the dictionary. By contrast, syntactic phrases and sentences are subject to general rules and their meanings can be computed from the meanings of their constituents. So they need not be listed in the lexicon.

In between there lie some awkward items: (i) complex words whose meaning is compositional, like the meaning of sentences, and hence need not be listed in the lexicon; (ii) idioms and compounds (especially exocentric ones) which have regular syntactic structure but idiosyncratic noncompositional meanings. The ordinariness and regularity of their syntax means that the syntactic dimensions of idioms and compounds need not be listed in the lexicon. But their semantic unpredictability makes it necessary to list their meanings in the lexicon.

No doubt, refining the answer to the question 'what is the difference between a lexical item and a phrase?' is a task that is likely to engage for many years morphologists interested in the nature and extent of the interpenetration of the lexicon, morphology and syntax. There are problems such as the nature of cranberry words and the dividing line between compounding and derivational affixation, the line between recursive compounds and syntactic phrases, differences between syntactic heads and morphological heads, etc. that are not yet fully resolved. But the most intriguing problems are probably posed by idioms.

Expounding the standard generative theory of word-formation has not been the sole purpose of this book. Theories come and go because answers to linguistics questions, like all answers to scientific questions, are provisional. All our solutions can, at least in principle, be improved on as our knowledge and understanding of the issues grows. Mastery of the subtleties of a particular theory only represents temporary triumph. Some of the theoretical positions outlined in this book will no doubt be refuted in the coming years. But the analytical skills acquired through this practical course in morphological theory and analysis are likely to last.

EXERCISES

1. (a) Explain whether the compounds below are headed.
   (b) Comment on the plural marking in each compound.

   menservants   women priests   gentlemen farmers

2. Write rules to account for the formation of the following words:

   rub-a-dub   jig-a-jig   crackerjack   chock-a-block

3. (a) Draw a phrase structure tree showing the internal structure of each of the compounds below.
   (b) What similarities do these compounds have with syntactic phrases and sentences?

will-o'-the-wisp   merry-go-round   fly-by-wire
up-to-the-minute   devil-may-care   stay-at-home
stick-in-the-mud   kiss-me-quick   commander-in-chief
what's-his-name   kiss-and-tell   lady-in-waiting
what-d'ye-call-'em   touch-and-go   man-of-war

4. Study the data below and answer the questions that follow.

    a. self-deception   b. money changer
        child-molestation   camp follower
        crime-detection   road sweeper

    c. examining magistrate   d. tuning fork
        managing director   hiking boots
        flying doctor   gardening gloves

    e. flying saucer   f. factory packed
        winding stair   farm-bred
        rocking horse   home made

    g. blue-collar (worker)   h. earthquake-resistant
        green belt   moisture-repellent
        red carpet   fire retardant

    (a) Using labelled bracketing, show the structure of these com-
         pounds. In each case, indicate the word-class of every constituent
         of the compound as well as the word-class of the entire com-
         pound.
    (b) State the argument structure, if any, displayed by each compound.
    (c) What generalisations can be made about the meanings of each set of
         compounds?

5. Study the following words:

    get up   lock up   push on   do up   touch down
    lock out   turn over   buy out   lift-off   teach in
    mix-up   touch up   stand-by   look-out   touch off

    (a) Determine the word-class of each compound.

Note: Compounds can be subject to conversion. So, some of these
words may belong to more than one class.)
(b) Determine which compounds are headed and which ones are not.
(c) Underline the head of each compound and indicate whether it
obeys the right-hand head rule.
(d) Bearing in mind the fact that compounds may undergo conversion,
draw a tree representing the structure of each compound and
reflecting its derivational history.

6. Which of the following are compounds and which are syntactic
phrases? What is your evidence?

coal merchant	old age pensioner
extended warranty	European currency unit
mother hennish	parking meter attendant
caravan theft	night shift worker
teacher training college	schoolboy humour
capital punishment	wooden spoon
heavyweight champion	new hospital building
capital city	tabloid newspaper
life insurance salesman	analytical skills
Big Apple	car crime prevention year

bull's-eye	Parkinson's disease
farmer's wife	Mothers' Day
ladies' man	parson's nose
lamb's wool	Aaron's rod
dog's life	Achilles' heel
smokers' cough	Ben's birthday

Glossary

The explanations presented below are not formal, technical definitions. They are only practical guides to the use of linguistic terminology.

Ablaut A change in a vowel in the root of a word that signals a change in grammatical function as is *ride ~ rode*.

Accusative A case inflection that marks the object of a transitive verb, e.g. *him* in *he sees him*.

Adjectival phrase A phrase whose head is an adjective, e.g. *far too long* is an adjectival phrase.

Adverbial phrase A phrase whose head is an adverb, e.g. *very soon* is an adverbial phrase.

Affricate A stop consonant that is released gradually, e.g. [tʃ] as in *cheap*.

Agreement A grammatical constraint requiring that if one word has a particular form, other words appearing in the same construction must take the appropriate corresponding form, e.g. *I jump*, *he jumps* (not **I jumps* and **she jump*).

Alveolar A speech sound produced using the tip or blade of the tongue and the alveolar ridge (i.e. teeth ridge above the upper front teeth), e.g. [t] in *tar*.

Animacy The linguistically relevant distinction between living and non-living things. It also encompasses the further distinction between human and non-human living things.

Approximant A speech sound produced with a gap between the articulators that is sufficiently wide for air to escape without causing turbulence. In Standard British English [r, l, w, j] as in *yes*, *lead*, *reed*, and *wall* are approximants.

Articulator An organ used in the production of speech.

Assimilation The modification of a sound in order to make it more similar to some neighbouring sound, e.g. *good girl* (gud gɜ:l] being realised in casual speech as [gug gɜ:l] in anticipation of the [g] of *good*.

Aspect A grammatical category that characterises the action or state denoted by a verb as complete or incomplete.

Asterisk(*) A diacritic showing a disallowed linguistic form.

Attributive adjective An attributive adjective modifies the head of a noun phrase, e.g. *new* in *a new idea*. (See **predicate adjective**.)

Auxiliary verb An auxiliary verb is used together with a lexical verb to convey grammatical meaning or to signal grammatical contrasts, e.g. *can, might, will*, as in *She can swim, She might swim, She will swim, Will she swim*.

Bilabial   A speech sound produced using both lips, e.g. [b] in *big*.

Borrowing   The importation of linguistic forms (e.g. words, morphemes) from another language, e.g. *pizza* was borrowed by English from Italian.

Case   A marking on a noun, pronoun or adjective that shows its grammatical function, e.g. the difference in the form of the pronoun *she* vs *her* shows a difference in case.

Class/classifier   A morpheme that shows the grammatical sub-category of a noun. Usually classifiers have some residual semantic basis. They may group together nouns referring to humans vs non humans, long objects, round objects etc. (See **gender**.)

Clause   A syntactic unit containing a verb but smaller than a sentence, e.g. *(It is obvious)* <u>*that our team will win*</u>; *(I know the girl)* <u>*who plays the flute*</u>.

Complement   A constituent of a verb phrase that completes the meaning of the verb. For example, *a good teacher* is a noun phrase complement in *She is a good teacher*. *Rich* is an adjectival phrase complement in *They became rich*.

Complementary distribution   If two units are in complementary distribution, they occur in mutually exclusive environments. Hence, they do not contrast.

Constituent   A unit that functions as a building block of a larger construction.

Contrast   A difference in linguistic form that distinguishes meaning or grammatical function.

Coreference   If elements are coreferential, they are bound to each other such that one element is only interpretable by referring to the other, e.g., in *Mary enjoyed herself* it is understood that both *Mary* and *herself* refer to the same individual.

Dental   A speech sound produced using the tip of the tongue and the upper front teeth, e.g. [t] in French *thé* 'tea'.

Determiner   A non-adjectival form that occurs in a noun phrase as a modifier of the head, e.g. *the*, *a* as in *the boy* and *a boy*.

Diachronic   Relating to the historical, evolutionary changes in a language or languages in general.

Diacritic   A mark used to modify the value of a symbol, e.g. [ʷ] in [tʷ] indicates a [t] made with rounded lips.

Distribution   The sum of the environments where a linguistic form occurs is its distribution.

Ditransitive   A verb that takes two objects is said to be ditransitive, e.g. *I gave* <u>*Mary a cheque*</u>.

Epenthesis   The insertion of a sound inside a word.

Finite verb   A form of a verb capable of occurring on its own in a sentence. Such a verb must be marked for tense.

Foot   A foot is a prosodic unit, normally consisting of two syllables one of which is stressed and the other unstressed (e.g. <u>un</u>der<u>ta</u>king has two feet) (stressed syllables are underlined).

Fricative   A sound produced with a considerable degree of turbulence as air squeezes through a narrow gap between the articulators, e.g. [ʃ] in *sheep*.

Functional   A functional approach to linguistics highlights what language is used for. Some functional linguists claim that the structure of language is shaped by the uses to which language is put.

Geminate   A 'true' geminate vowel or consonant consists of two adjacent identical segments in the same morpheme, e.g. Luganda *dda* 'long ago'. (Contrast this with a 'fake' geminate which has adjacent identical segments belonging to different morphemes like the two [p]s in *top people*.)

Gender   Grammatical classification of nouns as *masculine*, *feminine* and *neuter*. The degree to which this classification reflects semantic sex-gender is highly variable. (See **class**.)

Glide   A sound produced with the articulators moving fast towards (or away from) another articulation, e.g. [j] and [w].

Glottal   A sound made with a narrowed or closed glottis.

Glottis   The space between the vocal cords in the larynx.

Head   The pivotal constituent of a construction on which the rest of the elements in the construction depend.

Historical linguistics   See **diachronic** linguistics.

Homorganic consonants   Adjacent consonants that share the same place of articulation, e.g. the alveolar consonants [t] and [l] (as in *kettle*).

Inalienable possession   Refers to non-accidental (and hopefully lasting) possession. For example, branches are inalienably possessed by a tree and body-parts like legs and eyes are inalienably possessed by animals. But cars and shoes are not inalienable possessions.

Indo-European   The hypothetical reconstructed parent language from which most European languages (e.g. English, Latin, French, German, Russian) and some Asian languages (e.g. Sanskrit, Hindi, Persian) are ultimately descended.

Intransitive   A verb that does not take a direct object is intransitive, e.g. *snore* as in *he snored* (not *he snored <u>something</u>*).

Labial   An articulation involving at least one of the lips.

Lexical item   A lexical item is a 'content' word listed in the dictionary which has an identifiable meaning and is capable of occurring independently, e.g. *sky, scream, under*, etc.

Marked   If two units contrast, often one of them is **unmarked** (i.e. neutral) and the other is **marked** (i.e. non-neutral). Often the non-neutral form carries an extra marking (hence the term 'marked'), e.g. the infinite of the verb is realised by the bare root, as in *jump*, but the past tense form is marked by attaching *-ed*, as in *jumped*.

Mood Grammatical expression of attitudes (like doubt, certainty, possibility, permission, necessity, obligation), e.g. *she may come*, *she must come*, etc.

Mora The mora is the minimal rhythmic unit. It is smaller than the syllable. It is the unit taken into account in assigning tone or stress in some languages.

Nasal A sound like [m] [n] [ŋ] which is made with the soft palate lowered to allow air to escape through the nose.

Noun phrase A phrase whose head is a noun (or pronoun), e.g. *The little monkeys*; *them*.

Obligatory possession See **inalienable possession**.

Obstruent A consonant made with a drastic obstruction of the airstream. Obstruents are stops, fricatives or affricates.

Perfective aspect A form of the verb indicating completed action.

Person A grammatical form distinguishing speaker roles in a conversation. First person (*I*) is the speaker, second person (*you*) is the addressee and third person (*he/she/it*) is neither speaker nor addressee.

Phoneme A sound that distinguishes word meaning in a particular language, e.g. /p/ and /f/ are phonemes of English. They distinguish *pin* /pin/ from *fin* /fin/.

Phonology The study of the properties, patterns, functions and representations of speech sounds in a particular language and in language in general.

Phonotactics The dimension of phonology that deals with constraints on the combination of speech sounds.

Phrase A syntactic constituent whose head is a lexical category, i.e. a noun, adjective, verb, adverb or preposition. (See **noun phrase**, **adjectival phrase**, etc.)

Place of articulation The position in the mouth or throat where the obstruction involved in the producing of a consonant takes place. (See **bilabial**, **velar**, etc.)

Predicative adjective A predicative adjective functions as a complement of an intransitive verb in a verb phrase, e.g. *That student is new*; *it seems very wrong*; *she became angry*. (See **attributive adjective**.)

Prepositional phrase A phrase whose head is a preposition, e.g. *in the house*.

Proposition A meaning asserted to be true by a statement, e.g. *Life is never boring*.

Reciprocal A grammatical form indicating mutual action, e.g. *Margaret and Jane admire each other*.

Recursive A rule which can apply again and again is said to be recursive. Thus the rule that introduces prepositional phrases in this sentence is recursive: *I know the girl standing near the stout man with a scrawny dog with a bald patch on its back*.

Reduced vowel A centralised vowel found in unstressed positions in words. Compare the reduced vowel [ə] of ['mɪlkmən] (*milkman*) (where stress is on *milk*), with full [æ] of ['mæn] (*man*).

Reflexive A construction where both the subject and object refer to the same individual, e.g. *I hurt myself*.

Schwa The name of the mid-central vowel [ə].

Sonorant A voiced sound with a high propensity to voicing. Sonorants are non-obstruents, e.g. [r l m n w j].

Stop A sound whose production involves a total obstruction and closure of the path of the airstream at the place where the articulators meet. In a stop air is not allowed to escape through the centre of the mouth during the duration of the closure. Examples of stops include [p], [d], [g].

Stress The relative auditory prominence of a syllable due to factors like its pitch, duration or loudness. Articulatorily, stress involves using increased respiratory energy.

Subject The part of the sentence which the rest of the sentence says something about and which agrees with the verb, e.g. *The teacher is writing* vs *The pupils are playing*.

Synchronic linguistics This is the study of language focusing on one period in its history. (See **diachronic**.)

Syncope The deletion of sounds from the middle of a word.

Tense This is reference to the time when some action, event or state takes place (e.g. as 'past', 'present' or 'future') in relation to the moment of speaking. Tense is normally marked by attaching affixes to the verb.

Tone A pitch difference that conveys part of a contrast in word meaning or grammatical function.

Transitive verb A verb that takes a direct object, e.g. *kick* in *I kicked the ball*.

Typology The study of significant structural similarities and differences among languages.

Unmarked See **marked**.

Velar A consonant produced using the back of the tongue and the soft palate, e.g. [g] in [gɑːd] *guard* and [ŋ] in [lɒŋgə] *longer*.

Verb phrase A phrase whose head is a verb, e.g. *They should have left*; *The farmer bought some heifers*.

Voiced A voiced sound is produced with the vocal cords vibrating. In English all the vowels as well as some consonants like [l m d z] are voiced.

Voiceless A voiceless sound is produced without vibration of the vocal cords. Consonants like [f s t h ʔ] are voiceless.

Word-class This refers to a set of words that occur in the same syntactic environments, e.g. determiners, nouns, verbs, adjectives, adverbs and prepositions.

References

ABBREVIATIONS

BLS	*Berkeley Linguistic Society*
CLS	*Chicago Linguistic Society*
IULC	Indiana University Linguistic Club
LI	*Linguistic Inquiry*
Lg	*Language*
NLLT	*Natural Language and Linguistic Theory*
YM	*Yearbook of Morphology*

Adams, V. (1973) *An Introduction to Modern English Word-formation* (London: Longman).

Allen, M. (1978) 'Morphological Investigations', doctoral dissertation, University of Connecticut.

Andersen, T. (1987) 'An Outline of Lulubo Phonology', *Studies in African Linguistics*, **18** pp. 38–65.

Anderson, J. M. (1971) *The Grammar of Case* (Cambridge: CUP).

Anderson, J. M. (1973) 'Maximi Planudis in Memoriam', in Kiefer, F. and Ruwet, N. (eds) (1973) *Generative Grammar in Europe* (Dordrech: Reidel).

Anderson, J. M. (1977) *A Localist Theory of Case Grammar* (Cambridge: CUP).

Anderson, S. R. (1977) 'On the Formal Description of Inflection', *CLS*, **13**, pp. 15–44.

Anderson, S. R. (1982) 'Where is Morphology', *LI*, **13**, pp. 571–612.

Anderson, S. R. (1984) 'Kwakwala Syntax and the Government-binding Theory', in Cook, E. D. and Gerdts, D. (eds) *Syntax and Semantics*, vol. 16: *The Syntax of Native American Languages* (New York: Academic Press).

Anderson, S. R. (1985) 'Typological Distinctions in Word Formation', in Shopen (1985).

Anderson, S. R. (1988a) 'Morphological Theory', in Newmeyer (1988).

Anderson, S. R. (1988b) 'Inflection', in Hammond and Noonan (1988).

Anderson, S. R. (1990) *A-morphous Morphology*, MS, Cognitive Science Center (Baltimore: The Johns Hopkins University).

Anderson, S. R. and Kiparsky, P. (eds) (1973) *A Festschrift for Morris Halle* (New York: Holt, Rinehart and Winston).

Andrews, A. (1988) 'Lexical Structure', in Newmeyer (1988).

Archangeli, D. (1983) 'The Root CV-template as a Property of the Affix: Evidence from Yawelmani', *NLLT*, **1**, pp. 347–84.

Archangeli, D. (1988) 'Aspects of Underspecification Theory', *Phonology*, **5**, pp. 183–207.

Arnott, D. W. (1970) *The Nominal and Verbal Systems of Fula* (Oxford: Clarendon Press).

Aronoff, M. (1976) *Word Formation in Generative Grammar* (Cambridge, Mass.: MIT Press).

Aronoff, M. and Oehrle, R. (1984) *Language Sound Structure: Studies in Phonology Presented to Morris Halle by his Teacher and Students* (Cambridge, Mass.: MIT Press).

Ashton, E. (1944) *Swahili Grammar* (London: Longman).

Baker, M. (1985) 'The Mirror Principle and Morphosyntactic Explanation', *LI*, **16**, pp. 373–416.

Baker, M. (1988) *Incorporation: A Theory of Grammatical Function Changing* (Chicago: University of Chicago Press).

Bauer, L. (1983) *English Word-formation* (Cambridge: CUP).

Bauer, L. (1988) *Introducing Linguistic Morphology* (Edinburgh: Edinburgh University Press).

Bloomfield, L. (1933) *Language* (New York: Holt).

Bloomfield, L. (1939) 'Menomini Morphophonemics', *Travaux du Cercle Linguistique de Prague*, **8**, pp. 105–15.

Boas, F. (1947) 'Kwakiutl Grammar, with a Glossary of Suffixes', in Boas, H. Y. (ed.) *Transactions of the American Philosophical Society*, vol. 37, part 3 (Philadelphia: American Philosophical Society).

Booij, G. (1985) 'The Interaction of Phonology and Morphology in Prosodic Phonology', in Gussmann (1985).

Bopp, F. (1816) *Über das Conjugationssystem der Sanskritsprache in Vergleichung mit jenem der griechischen, lateinischen, persischen und germanischen Sprachen* (a partial English translation of this work prepared in 1820 was published in Amsterdam in 1974 by John Benjamins Publishing Co.).

Bresnan, J. (1982a) 'The Passive in Lexical Theory', in Bresnan (1982b).

Bresnan, J. (ed.) (1982b) *The Mental Representation of Grammatical Relations* (Cambridge, Mass.: MIT Press).

Bresnan, J. and Mchombo, S. (1987) 'Topic Pronoun and Agreement in Chichewa', *Lg*, **63**, pp. 741–82.

Bromley, H. M. (1981) 'A Grammar of Lower Grand Valley Dani', *Pacific Linguistics*, C63.

Broselow, E. (1983) 'Salish Double Reduplication: Subjacency in Morphology', *NLLT*, **1**, pp. 347–84.

Broselow, E. and McCarthy, J. (1983) 'A Theory of Internal Reduplication', *The Linguistic Review*, **3**, 25–88.

Brown, K. and Miller, J. E. (1980) *Syntax: An Introduction to Sentence Structure* (Hutchinson: London).

Bybee, J. (1985) *Morphology: A Study of the Relation between Meaning and Form* (Amsterdam: John Benjamins).

Carrier-Duncan, J. (1984) 'Some Problems with Prosodic Accounts of Reduplication', in Aronoff and Oehrle (1984).

Carstairs, A. (1987) *Allomorphy in Inflection* (Beckenham: Croom Helm).

Carstairs, A. (1988) 'Some Implications of Phonologically Conditioned Suppletion', *YM*, **1**, pp. 67–94.

Chomsky, N. (1957) *Syntactic Structures* (The Hague: Mouton).

Chomsky, N. (1965) *Aspects of the Theory of Syntax* (Cambridge, Mass.: MIT Press).

Chomsky, N. (1970) 'Remarks on Nominalization', in Jacobs and Rosenbaum (eds) (1970) *Readings in English Transformational Grammar* (Waltham, Mass.:

Blaisdell).

Chomsky, N. (1972a) 'Some Empirical Issues in the Theory of Generative Grammar', in Chomsky (1972b).

Chomsky, N. (1972b) *Studies on Semantics in Generative Grammar* (The Hague: Mouton).

Chomsky, N. (1973) 'Conditions on Transformations', in Anderson, S. R. and Kiparsky, P. (eds) *Festschrift for Morris Halle* (New York: Holt, Rinehart and Winston).

Chomsky, N. (1975) *Reflections on Language* (New York: Pantheon).

Chomsky, N. (1980) *Rules and Representations* (Oxford: Basil Blackwell).

Chomsky, N. (1981) *Lectures on Government and Binding* (Foris: Dordrecht).

Chomsky, N. (1986) *Knowledge of Language: Its Nature Origin and Use* (New York: Praeger).

Chomsky, N. and Halle, M. (1968) *The Sound Pattern of English* (New York: Harper and Row).

Chung, S. (1978) *Case Marking and Grammatical Relations in Polynesian* (Austin: University of Texas Press).

Clark, E. V. and Clark, H. H. (1979) 'When Nouns Surface as Verbs', *Lg*, **55**, pp. 767–811.

Clements, G. N. (1985) 'Compensatory Lengthening and Consonant Gemination in Luganda', in Wetzels, L. and Sezer, E. (eds) (1985) *Studies in Compensatory Lengthening* (Dordrecht: Foris).

Clements, G. N. and Ford, K. C. (1979) 'Kikuyu Tone Shift and its Synchronic Consequences' *LI*, **10**, pp. 179–210.

Clements, G. N. and Goldsmith, J. (1984) 'Autosegmental Studies in Bantu Tone: Introduction', in Clements, G. N. and Goldsmith, J. (eds) (1984) *Autosegmetal Studies in Bantu Tone* (Dordrecht: Foris).

Clements, G. N. and Keyser, S. J. (1983) *CV Phonology* (Cambridge, Mass.: MIT Press).

Cole, P. and Sadock, J. (1977) *Syntax and Semantics* vol. 8: *Grammatical Relations* (New York: Academic Press).

Comrie, B. (1976) *Aspect* (Cambridge: CUP).

Comrie, B. (1985) *Tense* (Cambridge: CUP).

Comrie, B. (1981) *Language Universals and Linguistic Typology* (Oxford: Basil Blackwell).

Conklin, H. (1959) 'Linguistic Play in its Cultural Context', *Lg*, **35**, pp. 631–6.

Cook, E. D. and Gerdts, D. (eds) (1984) *Syntax and Semantics* vol. 16: *The Syntax of Native American Languages* (New York: Academic Press).

Corbett, G. (1990) *Gender* (Cambridge: CUP).

Crystal, D. (1980) *A First Dictionary of Linguistics and Phonetics* (London: André Deutsch).

Davies, W. D. (1984) 'Choctaw Switch-reference', in Cook and Gerdts (1984).

Di Sciullo, A. M. and Williams, E. (1987) *On Defining the Word* (Cambridge, Mass.: MIT Press).

Dixon, R. (1972) *The Dyirbal Language of West Queensland* (Cambridge: CUP).

Dixon, R. (1977) *A Grammar of Yidiɲ* (Cambridge: CUP).

Dixon, R. (1979) 'Ergativity', *Lg*, **55**, pp. 59–138.

Dixon, R. (1980) *The Languages of Australia* (Cambridge: CUP).

Dobson, E. J. (1957) *English Pronunciation 1500–1700* (London: Oxford University Press).

Donaldson, T. (1980) *Ngiyambaa: Language of the Wangaaybuwan* (Cambridge: CUP).

Downing, P. (1977) 'On the Creation and Use of English Compound Nouns', *Lg*, **53**, pp. 810–42.

Dowty, D. (1979) *Word Meaning and Montague Grammar* (Amsterdam: North-Holland).

Dressler, W. (1985) *Morphology* (Ann Arbor: Karoma).

Everaert, M., Evers, A., Huybregts, R. and Trommelen, M. (eds) (1988) *Morphology and Modularity* (Dordrecht: Foris).

Fillmore, C. J. (1968) 'The Case for Case', in Bach, E. and Harms, R. T. (eds) *Universals in Linguistic Theory* (New York: Holt, Rinehart and Winston).

Firth, J. R. (1948) 'Sounds and Prosodies', *Transactions of the Philological Society*, pp. 127–52; reprinted in Firth, J. R. (1957) *Papers in Linguistics 1934–1951* (Oxford: OUP) and in Palmer (1970).

Foley, W. (1986) *The Papuan Languages of New Guinea* (Cambridge: CUP).

Fortescue, M. (1984) *West Greenlandic* (London: Croom Helm).

Gleason, H. A. (1955) *Workbook in Descriptive Linguistics* (New York: Holt, Rinehart and Winston).

Gleason, H. A. (1961) *An Introduction to Descriptive Linguistics* (New York: Holt, Rinehart and Winston).

Goldsmith, J. (1976) 'Autosegmental Phonology', doctoral dissertation, MIT, Cambridge. (Published by Garland: New York, 1979).

Goldsmith, J. (1990) *Autosegmental and Metrical Phonology* (Oxford: Basil Blackwell).

Greenberg, J. H. (1954) 'A Quantitative Approach to the Morphological Typlogy of Languages', in Spencer, R. (ed.), *Method and Perspective in Anthropology* (Minneapolis: University of Minnesota Press).

Greenberg, J. H. (1963) 'Some Universals of Grammar with Particular Reference to the Order of Meaningful Elements', in Greenberg (1966).

Greenberg, J. H. (ed.) (1966) *Universals of Language* (Cambridge, Mass.: MIT Press).

Grimm, J. (1819–37/1970–98) *Deutsche Grammatik* (Göttingen: Dietrich; Berlin: Dümmler; Güttersloh: Bertelsmann).

Gruber, J. S. (1965) 'Studies in Lexical Relations', doctoral dissertation, MIT.

Gruber, J. S. (1976) *Studies in Lexical Relations* (Amsterdam: North Holland).

Gussmann, E. (1985) *Phonomorphology* (Lubin: Redakcja Wydawnictw Katolickiego Uniwersytetu, Lubelskiego).

Haegeman, L. (1991) *Introduction to Government and Binding Theory* (Oxford: Blackwell).

Halle, M. (1973) 'Prolegomena to a Theory of Word-formation', *LI*, **4**, 3–16.

Halle, N. and Clements, G. N. (1983) *Problem Book in Phonology* (Cambridge, Mass.: MIT Press).

Halle, M. and Mohanan, K. P. (1985) 'Segmental Phonology of Modern English', *LI*, **16**, pp. 57–116.

Halle, M. and Vergnaud, J-R. (1988) *An Essay on Stress* (Cambridge, Mass.: MIT Press).

Hammond, M. (1988)'Templaic Transfer in Arabic Broken Plurals', *NLLT*, **6**, pp. 247–70.

Hammond, M. and Noonan, M. (eds) (1988) *Theoretical Morphology: Approaches in Modern Linguistics* (Orlando: Academic Press).

Harris, Z. S. (1942) 'Morpheme Alternants in Linguistic Analysis', *Lg*, **18**, pp. 169–80; reprinted in Joos (1957) pp. 109–15.

Harris, Z. S. (1944) 'Simultaneous Components in Phonology', *Lg*, **20**, pp. 181–205; reprinted in Joos (1957) pp. 124–38.

Harris, Z. S. (1946) 'From Morpheme to Utterance', *Lg*, **22**, pp. 161–83; reprinted in Joos (1957) pp. 142–53.

Harris, Z. S. (1951) *Methods in Structural Linguistics* (Chicago: University of Chicago Press); reprinted as *Structural linguistics* (1961).

Harrison, S. P. (1973) 'Reduplication in Micronesian Languages', *Oceanic Linguistics*, **12**, pp. 407–54.

Hayes, B. (1982) 'Extrametricality and English Stress', *LI*, **13**, pp. 227–76.

Hilders, J. H. and Lawrence, J. C. D. (1957) *An Introduction to the Ateso Language* (Kampala: The Eagle Press, East African Literature Bureau).

Hockett, C. (1947) 'Componential Analysis of Sierra Popoluca', *International Journal of American Linguistics*, **13**, pp. 259–67.

Hockett, C. (1952) 'A Formal Statement of Morphemic Analysis', *Studies in Linguistics*, **10**, pp. 27–39.

Hockett, C. (1954) 'Two Models of Grammatical Description', *Word*, **10**, pp. 210–31; reprinted in Joos (1957) pp. 386–99.

Hockett, C. (1955) *A Manual of Phonology* (Baltimore: Waverley Press).

Hockett, C. (1958) *A Course in Modern Linguistics* (New York: Macmillan).

Hogg, R. and McCully, C. (1987) *Metrical Phonology: A Coursebook* (Cambridge: CUP).

Hooper, J. (1976) *Introduction to Natural Generative Phonology* (New York: Academic Press).

Hooper, P. and Thompson, S. (1982) *Syntax and Semantics*, vol. 15: *Studies in Transitivity* (New York: Academic Press).

Horrocks, G. (1987) *Generative Syntax* (London: Longman).

van der Hulst, H. and Smith, N. (1982a) *The Structure of Phonological Representations, Part I* (Dordrecht: Foris).

van der Hulst, H. and Smith, N. (1982b) *The Structure of Phonological Representations, Part II* (Dordrecht: Foris).

van der Hulst, H. and Smith, N. (1982c) 'Introduction', in van der Hulst, H. and Smith, N. (1982a).

van der Hulst, H. and Smith, N. (1985) *Advances in Nonlinear Phonology* (Dordrecht: Foris).

Hurford, J. (1981) Review of Anderson, J. M. (1977) *A Localist Theory of Case Grammar* (Cambridge: CUP) in *The Journal of Linguistics*, **17**, pp. 374–8.

Hyman, L. M. (1975) *Phonology: Theory and Analysis* (New York: Holt, Rinehart and Winston).

Hyman, L. M. (1985) *A Theory of Syllable Weight* (Dordrecht: Foris).

Hyman, L. M. and Katamba, F. (1990) 'Final Vowel Shortening in Luganda', *Studies in African Linguistics*, **21**, pp. 1–45.

Hyman, L. M. and Katamba, F. (1992) 'Cyclicity in the Bantu Verb Stem',

Berkeley Linguistic Circle, **17**, pp. 134–44.

Hymes, D. and Fought, J. (1981) *American Structuralism* (The Hague: Mouton).

Inkelas, S. (1989) 'The Representation of Invisibility' (unpublished MS; Los Angeles: UCLA).

Jackendoff, R. (1975) 'Morphological and Semantic Regularities in the Lexicon', *Lg*, **51**, pp. 639–71.

Jackendoff, R. (1976) 'Towards an Explanatory Semantic Representation', *LI*, **7**, pp. 89–150.

Jackendoff, R. (1977) *X̄ Syntax: A Study of Phrase Structure* (Cambridge, Mass.: MIT Press).

Jensen, J. (1990) *Morphology* (Amsterdam: John Benjamins).

Jespersen, P. (1909) *A Modern English Grammar on Historical Principles*, vol. I: *Sounds and Spellings* (Heidelberg: Carl Winter).

Jespersen, P. (1954) *A Modern English Grammar on Historical Principles* (London: George Allen and Unwin).

Jones, W. (1786) 'The Third Anniversary Discourse, on the Hindus, Delivered 2nd February 1786', in Jones (1799) **I**, pp. 19–34.

Jones, W. (1799) *The Works of Sir William Jones*, 6 vols (London: Robinson and Evans).

Joos, M. (ed.) (1957) *Readings in Linguistics* (Chicago: University of Chicago Press).

Kaisse, E. (1985) *Connected Speech: The Interaction of Syntax and Phonology* (Orlando: Academic Press).

Kaisse, E. M. and Shaw, P. (1985) 'On the Theory of Lexical Phonology', *Phonology Yearbook*, **2**, pp. 1–30.

Katamba, F. (1985) 'A Nonlinear Account of the Syllable in Luganda', in D. L. Goyvaerts (ed.) *African Linguistics: Essays in Memory of M. W. K. Semikenke*, Studies in the Science of Language Series 6 (Amsterdam: John Benjamins).

Katamba, F. (1989) *An Introduction to Phonology* (London: Longman).

Keenan, E. L. and Comrie, B. (1977) 'Noun Phrase Accessibility and Universal Grammar', *LI*, **8**, pp. 63–99.

Keenan, E. L. and Comrie, B. (1979) 'Data on the Noun Phrase Accessibility', *Lg*, **55**, pp. 333–51.

Kennedy, B. H. (1948) *The Revised Latin Primer* (London: Longmans).

Kenstowicz, M. and Kisseberth, C. (1977) *Topics in Phonological Theory* (New York: Academic Press).

Kenstowicz, M. and Kisseberth, C. (1979) *Generative Phonology: Description and Theory* (New York: Academic Press).

Kimenyi, A. (1980) *A Relational Grammar of Kinyarwanda* (Berkeley and Los Angeles: University of California Press).

Kiparsky, P. (1968) *How Abstract is Phonology?* (Bloomington: IULC).

Kiparsky, P. (1973) ' "Elsewhere" in Phonology', in Anderson and Kiparsky (1973).

Kiparsky, P. (1982a) 'From Cyclic Phonology to Lexical Phonology', in van der Hulst and Smith (1982a).

Kiparsky, P. (1982b) 'Lexical Morphology and Phonology', in Yang, I. S. (ed.) *Linguistics in the Morning Calm* (Seoul: Hanshin).

Kiparsky, P. (1983) 'Word Formation and the Lexicon', in Ingemann, F. (ed.)

Proceedings of the 1982 Mid-America Linguistics Conference (University of Kansas).

Kiparsky, P. (1985) 'Some Consequences of Lexical Phonology', *Phonology Yearbook*, **2**, pp. 85–138.

Klavans, J. (1982) 'Some Problems in a Theory of Clitics' (Bloomington: IULC).

Klavans, J. (1985) 'The Independence of Syntax and Phonology in Cliticization', *Lg*, **61**, pp. 95–120.

Koopman, H. (1984) *The Syntax of Verbs: From Verb Movement Rules in the Kru Languages to Universal Grammar* (Dordrecht: Foris).

Koutsoudas, A., Sanders, G. and Noll, C. (1974) 'The Application of Phonological Rules', *Lg*, **50**, pp. 1–28.

Krupa, V. (1966) *Morpheme and Word in Maori* (The Hague: Mouton).

Kuno, S. (1978) 'Japanese: a Characteristic OV Language', in Lehmann, W. P. (ed.) *Syntactic Typology* (Sussex: Harvester Press).

Lakoff, G. (1971) 'On Generative Semantics', in Steinberg, D. and Jakobovits, L. (eds) (1971) *Semantics* (Cambridge: CUP).

Langacker, R. W. (1972) *Fundamentals of Linguistic Analysis* (New York: Harcourt, Brace, Jovanovich).

Lass, R. (1976) *English Phonology and Phonological Theory* (Cambridge: CUP).

Lass, R. (1984) *Phonology* (Cambridge: CUP).

Leben, W. (1973) 'Suprasegmental Phonology', doctoral dissertation, MIT.

Leben, W. (1978) 'The Representation of Tone', in Fromkin, V. (ed.) *Tone: A Linguistic Survey* (New York: Academic Press).

Lees, R. (1960) *The Grammar of English Nominalizations* (The Hague: Mouton).

Levin, B. (1983) 'On the Nature of Ergativity', doctoral dissertation, MIT.

Levin, B. (1985) 'Lexical Semantics in Review: an Introduction', MS (Cambridge, Mass.: MIT).

Lewis, G. (1967) *Turkish Grammar* (Oxford: OUP).

Li, C. N. and Thompson, S. A. (1978) 'An exploration of Mandarin Chinese', in Lehmann, W. P. (ed.) *Syntactic Typology* (Sussex: Harvester Press).

Lieber, R. (1980) 'The Organisation of the Lexicon', doctoral dissertation, MIT (distributed by IULC).

Lieber, R. (1983) 'Argument linking and compounding', *LI*, **14**, pp. 251–86.

Lightner, T. (1975) 'The Role of Derivational Morphology in Generative Grammar', *Lg*, **51**, pp. 617–38.

Lipka, L. (1990) *An Outline of English Lexicology* (Tübingen: Max Niemeyer).

Lyons, J. (1968) *Introduction to Theoretical Linguistics* (Cambridge: CUP).

Lyons, J. (1977) *Semantics*, vol. 2 (Cambridge: CUP).

McCarthy, J. (1979) *Formal Problems in Semitic Phonology and Morphology*, doctoral dissertation, MIT (distributed by IULC; republished New York: Garland, 1982b).

McCarthy, J. (1981) 'A Prosodic Theory of Nonconatenative Morphology', *LI*, **12**, pp. 373–418.

McCarthy, J. (1982a) 'Prosodic Templates, Morphemic Templates, and Morphemic Tiers', in van der Hulst and Smith (1982a).

McCarthy, J. (1986) 'OCP Effects: Gemination and Antigemination', *LI*, **17**, pp. 207–64.

McCarthy, J. (1988) 'Feature Geometry and Dependency: a Review', *Phonetica*,

43, pp. 84–108.

McCarthy, J. and Prince, A. S. (1990) 'Foot and Word in Prosodic Morphology', *NLLT*, **8**, pp. 209–83.

McCawley, J. (1968) 'Lexical Insertion in a Generative Grammar Without Deep Structure', *CLS*, **4**, pp. 71–80; reprinted in McCawley (1976).

McCawley, J. (1971) 'Interpretive Semantics meets Frankenstein', *Foundations of Language*, **7**, pp. 285–96. Reprinted in McCawley (1976).

McCawley, J. (1976) *Grammar and Meaning* (New York: Academic Press).

Marantz, A. (1982) 'Re reduplication', *LI*, **13**, pp. 483–545.

Marantz, A. (1984) *On the Nature of Grammatical Relations* (Cambridge, Mass.: MIT Press).

Marchand, H. (1969) *The Categories and Types of Present-day English Word-formation* (Munich: C. H. Beck Verlagsbuchhandlung).

Martin, S. (1975) *A Reference Grammar of Japanese* (New Haven: Yale University Press).

Mascaró, J. (1976) 'Catalan Phonology and the Phonological Cycle', doctoral dissertation, MIT (distributed by IULC).

Matthews, P. (1972) *Inflectional Morphology* (Cambridge: CUP).

Matthews, P. (1974) *Morphology* (Cambridge: CUP).

Mayerthaler, W. (1988) *Naturalness in Morphology* (Ann Arbor: Karoma).

Melville, H. (1851/1961) *Moby Dick* (New York: Signet Classics).

Meussen, A. E. (1959) *Essai de grammaire rundi* (Tervuren: Musée Royale de l'Afrique Centrale).

Miller, J. E. (1985) *Semantics and Syntax* (Cambridge: CUP).

Mithun, N. (1984) 'The Evolution of Noun Incorporation', *Lg*, **60**, pp. 847–94.

Mithun, N. (1986) 'On the Nature of Noun Incorporation', *Lg*, **62**, pp. 32–7.

Mohanan, K. (1982) *The Theory of Lexical Phonology*, doctoral dissertation, MIT (published in 1986, Dordrecht: Reidel).

Mohanan, K. P. (1983) 'Move NP or Lexical Rules? Evidence from Malayalam Causativisation', in Levin, L., Rappaport, M. and Zaenen, A. (eds) (1983) *Papers in Lexical-functional Grammar* (distributed by IULC).

Mohanan, K. and Mohanan, T. (1984) 'Lexical Phonology of the Consonant System in Malayalam', *LI*, **15**, pp. 575–602.

Morgan, J. (1969) 'On Arguing about Semantics', *Papers in Linguistics*, **1**, pp. 49–70.

Moortgat, M., van der Hulst, H. and Hoekstra, T. (eds) (1981) *The Scope of Lexical Rules* (Dordrecht: Foris).

Moravcsik, E. A. (1978) 'Reduplicative Constructions', in Greenberg, J. H. (ed.) *Universals of Human Language*, vol. 3: *Word structure* (Stanford: Stanford University Press) pp. 297–334.

Mutaka, N. and Hyman, L. M. (1990) 'Syllables and Morpheme Integrity in Kinande Reduplication', *Phonology*, **7**, pp. 73–119.

Müller, F. M. (1899) *Three Lectures on the Science of Language and its Place in General Education* (Reprinted 1961, Benares: Indological Book House).

Nash, D. (1980) *Topics in Walpiri Grammar*, doctoral dissertation, MIT (published New York: Garland).

Nespor, M. and Vogel, I. (1986) *Prosodic Phonology* (Dordrecht: Foris).

Newmeyer, F. (ed.) (1988) *Linguistics: The Cambridge Survey*, vol. 1: *Linguistic*

Theory: Foundations (Cambridge: CUP).

Nida, E. (1949) *Morphology: the Descriptive Analysis of Words* (Ann Arbor: University of Michigan Press).

Odden, D. (1986) 'On the Obligatory Contour Principle', *Lg*, **62**, pp. 353–83.

Odden, D. (1988) 'Anti-gemination and the OCP', *LI*, **19**, pp. 451–75.

Odden, D. and Odden, M. (1988) 'Ordered Reduplication in Kihehe', *LI*, **16**, pp. 497–503.

Oomen-van Schendel, A. J. G. (1977) *Aspects of the Rendille Language*, unpublished MA dissertation, University of Nairobi.

Palmer, F. R. (1970) *Prosodic Analysis* (Oxford: OUP).

Perlmutter, D. (ed.) (1983) *Studies in Relational Grammar I* (Chicago: University of Chicago Press).

Perlmutter, D. (1988) 'The Split-morphology Hypothesis: Evidence from Yiddish', in Hammond and Noonan (1988).

Perlmutter, D. and Postal, P. (1977) 'Toward a Universal Characterisation of Passivisation', *BLS*, **4**, pp. 394–417.

Perlmutter, D. and Rosen, C. (1984) *Studies in Relational Grammar*, vol. 2 (Chicago: Chicago University Press).

Pesetsky, D. (1979) 'Russian Morphology and Lexical Theory', unpublished paper, MIT.

Pesetsky, D. (1985) 'Morphology and Logical Form', *LI*, **16**, pp. 193–246.

Pike, K. (1948) *Tone Languages* (Ann Arbor: University of Michigan Press).

Plank, F. (1979) *Ergativity* (New York: Academic Press).

Postal, P. (1970) 'On the Surface Verb Remind', *LI*, **1**, pp. 37–120.

Pulleyblank, D. (1986) *Tone in Lexical Phonology* (Dordrecht: Reidel).

Pulleyblank, D. (1988) 'Tone and the Morpheme Tier Hypothesis', in Hammond and Noonan (1988).

Pullum, G. K. and Zwicky, A. M. (1988) 'The Syntax-phonology Interface', in Newmeyer (1988).

Quirk, R. and Greenbaum, S. (1973) *A University Grammar of English* (London: Longman).

Radford, A. (1988) *Transformational Grammar* (Cambridge: CUP).

Rice, K. (1985) 'On the Placement of Inflection', *LI*, **16**, pp. 155–61.

van Riemsdijk, H. and Williams, E. (1986) *Introduction to the Theory of Grammar* (Cambridge, Mass.: MIT Press).

Ringen, C. O. (1972) 'Arguments for Rule Ordering', *Foundations of Language*, **8**, pp. 266–73.

Robins, R. H. (1958) *The Yurok Language: Grammar, Text, Lexicon* (Berkeley and Los Angeles: University of California Press).

Robins, R. (1959) 'In Defence of WP', *Transactions of the Philological Society*; reprinted in Robins, R. (1970) *Diversions of Bloomsbury* (Amsterdam: North Holland).

Roeper, T. and Siegel, D. (1978) 'A Lexical Transformation for Verbal Compounds', *LI*, **9**, pp. 199–260.

Rohrer, C. (1974) 'Some Problems of Wordformation', in Rohrer, C. and Ruwet, N. (eds) *Actes du colloque Franco-allemand de grammaire transformationelle II* (Tübingen: Niemeyer).

Rosen, C. (1984) 'The Interface Between Semantic Roles and Initial Grammatical

Relations', in Perlmutter and Rosen (1984).

Rubach, J. (1984) *Cyclic and Lexical Phonology* (Dordrecht: Foris).

Sadock, J. (1980) 'Noun Incorporation in Greenlandic', *Lg*, **56**, pp. 300–19.

Sadock, J. (1985) 'Autolexical Syntax: A Proposal for the Treatment of Noun Incorporation and Similar Phenomena', *NLLT*, **3**, pp. 379–440.

Sadock, J. (1986) 'Some Notes on Noun Incorporation', *Lg*, **62**, pp. 19–31.

Sagey, E. (1988) 'On the Ill-formedness of Crossing Association Lines', *LI*, **19**, pp.109–18.

Sapir, E. (1911) 'The Problem of Noun Incorporation in American Languages', *American Anthropologist*, **13**, pp. 250–82.

Sapir, E. (1921) *Language* (New York: Harcourt, Brace and World).

Sapir, E. (1925) 'Sound Patterns in Language', *Lg*, **1**, pp. 37–51; reprinted in Joos (1957) pp. 19–25.

Scalise, S. (1984) *Generative Morphology* (Dordrecht: Foris).

Selkirk, E. O. (1981) 'English Compounding and the Theory of Word Structure', in Moortgat et al. (1981).

Selkirk, E. O. (1982) *The Syntax of Words* (Cambridge, Mass.: MIT Press).

Selkirk, E. O. (1984) *Phonology and Syntax: The Relation Between Sound and Structure* (Cambridge, Mass.: MIT Press).

Sells, P. (1985) *Lectures on Contemporary Syntactic Theories* (California: Stanford University, Center for the Study of Language and Information).

Shibtani, M. (ed.) (1976) *Syntax and Semantics*, vol. 6: *The Syntax of Causative Constructions* (New York: Academic Press).

Shopen, D. (ed.) (1985) *Language Typology and Grammatical Description*, vol. 3: *Grammatical Categories and the Lexicon* (Cambridge: CUP).

Siegel, D. (1971) 'Some Lexical Transderivational Constraints in English', unpublished paper, MIT.

Siegel, D. (1974) 'Topics in English Morphology', doctoral dissertation, MIT. (Published New York: Garland, 1979.)

Simpson, J. A. and Weiner, E. S. C. (compilers) (1989) *The Oxford English Dictionary* (second edition) (Oxford: Clarendon Press).

Sloat, C. and Taylor, S. (1978) *The Structure of English Words* (Eugene, Oregon: Pacific Language Associates).

Spencer, A. (1991) *Morphological Theory* (Oxford: Blackwell).

Spiegel, F. (1987) *In-words and Out-words* (London: Elm Tree Books).

Sproat, R. (1988) 'Bracketing Paradoxes, Cliticization and Other Topics: The Mapping Between Phonological and Syntactic Structure', in Everaert et al. (1988).

Staalsen, P. (1972) 'Clause Relationships in Iatmul', *Pacific Linguistics*, **A7**, pp. 69–76.

Steriade, D. (1987) 'Redundant Values', in Bosch, A., Need, B. and Schiller, E. (eds) (1987) *Twenty-third Annual Regional Meeting of the Chicago Linguistic Society. Part Two: Parassession on Autosegmental and Metrical Phonology* (Chicago Linguistic Society).

Steriade, D. (1988) 'Reduplication and Syllable Transfer in Sanskrit and Elsewhere', *Phonology*, **5**, pp. 73–155.

Stowell, T. (1981) 'Origins of Phrase Structure', doctoral dissertation, MIT.

Strauss, S. (1982a) *Lexicalist Phonology of English and German* (Dordrecht: Foris).

Strauss, S. (1982b) 'On "Relatedness Paradoxes" and Related Paradoxes', *LI*, **19**, pp. 271–314.

Suárez, J. A. (1983) *The Mesoamerican Indian Languages* (Cambridge: CUP).

Swadesh, M. (1934) 'The Phonemic Principle', *Lg*, **10**, pp. 117–29; reprinted in Joos (1957) pp. 32–7.

Twaddell, W. F. (1935) 'On Defining the Phoneme', *Language Monographs*, **16**;reprinted in Joos (1957) pp. 55–80.

Welmers, W. (1973) *African Language Structures* (Berkeley and Los Angeles: University of California Press).

Whitney, W. D. (1889) *Sanskrit Grammar* (Cambridge, Mass.: Harvard University Press).

Wickens, G. M. (1980) *Arabic Grammar: A First Workbook* (Cambridge: CUP).

Wilbur, R. (1973) 'The Phonology of Reduplication', doctoral dissertation, MIT (distributed by IULC).

Williams, E. (1981a) 'On the Notions "Lexically Related" and "Head of a Word" ', *LI*, **12**, pp. 234–74.

Williams, E. (1981b) 'Argument Structure and Morphology', *Linguistic Review*, **1**, pp. 81–114.

Woodbury, A. (1977) 'Greenlandic Eskimo, Ergativity and Relational Grammar', in Cole and Sadock (1977).

Wright, W. (1967) *A Grammar of the Arabic Language* (Cambridge: CUP; 3rd edn). (First edition, vol. 1 (1859); vol. 2 (1862).

Yip, M. (1982) 'Reduplication and CV-skeleta in Chinese Secret Languages', *LI*, **13**, pp. 637–61.

Yip, M. (1988) 'The Obligatory Contour Principle and Phonological Rules: A Loss of Identity', *LI*, **19**, pp. 65–100.

Younis, M. H. (1975) 'A Brief Description of Syrian Colloquial Arabic', MA dissertation, Lancaster University.

Zimmer, K. E. (1964) *Affixal Negation in English and Other Languages: An Investigation of Restricted Productivity* (Supplement to *Word* 20, Monograph no. 5, New York).

Zwicky, A. (1977) *On Clitics* (IULC).

Zwicky, A. (1985) 'Clitics and Particles', *Lg*, **61**, pp. 283–305.

Zwicky, A. and Pullum, G. (1983) 'Cliticisation vs. Inflection: English *n't*', *Lg*, **59**, pp. 502–13.

Zwicky, A. and Pullum, G. (1987) 'Plain Morphology and Expressive Morphology', *BLS*, **13**, pp. 330–40.

Index of Languages

346

Subject Index

348

Author Index

353